PROCESS
ALGEBRA FOR
PARALLEL AND
DISTRIBUTED
PROCESSING

Chapman & Hall/CRC
Computational Science Series

SERIES EDITOR

Horst Simon

Associate Laboratory Director, Computing Sciences
Lawrence Berkeley National Laboratory
Berkeley, California, U.S.A.

AIMS AND SCOPE

This series aims to capture new developments and applications in the field of computational science through the publication of a broad range of textbooks, reference works, and handbooks. Books in this series will provide introductory as well as advanced material on mathematical, statistical, and computational methods and techniques, and will present researchers with the latest theories and experimentation. The scope of the series includes, but is not limited to, titles in the areas of scientific computing, parallel and distributed computing, high performance computing, grid computing, cluster computing, heterogeneous computing, quantum computing, and their applications in scientific disciplines such as astrophysics, aeronautics, biology, chemistry, climate modeling, combustion, cosmology, earthquake prediction, imaging, materials, neuroscience, oil exploration, and weather forecasting.

PUBLISHED TITLES

PETASCALE COMPUTING: Algorithms and Applications
Edited by David A. Bader

PROCESS ALGEBRA FOR PARALLEL AND DISTRIBUTED PROCESSING
Edited by Michael Alexander and William Gardner

PROCESS ALGEBRA FOR PARALLEL AND DISTRIBUTED PROCESSING

EDITED BY

MICHAEL ALEXANDER
WILLIAM GARDNER

CRC Press
Taylor & Francis Group
Boca Raton   London   New York

CRC Press is an imprint of the
Taylor & Francis Group, an **informa** business

A CHAPMAN & HALL BOOK

Cover Image Credit: Intel Teraflops Research Chip Wafer from Intel, used with permission.

Chapman & Hall/CRC
Taylor & Francis Group
6000 Broken Sound Parkway NW, Suite 300
Boca Raton, FL 33487-2742

© 2009 by Taylor & Francis Group, LLC
Chapman & Hall/CRC is an imprint of Taylor & Francis Group, an Informa business

No claim to original U.S. Government works
Printed in the United States of America on acid-free paper
10 9 8 7 6 5 4 3 2 1

International Standard Book Number-13: 978-1-4200-6486-5 (Hardcover)

Library of Congress Cataloging-in-Publication Data

Process algebra for parallel and distrubuted processing / editors, Michael
  Alexander and William Gardner.
      p. cm. -- (Chapman & Hall/CRC computational science series)
    Includes bibliographical references and index.
    ISBN 978-1-4200-6486-5 (alk. paper)
      1. Parallel processing (Electronic computers) 2. Electronic data
processing--Distributed processing. 3. Formal methods (Computer science) I.
Alexander, Michael, 1970 Sept. 25- II. Gardner, William, 1952-

    QA76.58.P7664 2009
    004.01'51--dc22   2008029295

Visit the Taylor & Francis Web site at
http://www.taylorandfrancis.com

and the CRC Press Web site at
http://www.crcpress.com

Contents

Foreword

This book brings together the state of the art in research on applications of process algebras to parallel and distributed processing.

Process algebras constitute a successful field of computer science. This field has existed for some 30 years and stands nowadays for an extensive body of theory of which much has been deeply absorbed by the researchers in computer science. Moreover, the theoretical achievements of the field are to a great extent justified by applications. The applications, in turn, strongly influence how the field evolves; some of the field's success may be attributed to frequently addressing needs that arose in practice.

Meanwhile, an explosion of complex systems of interacting components has been going on since the emergence of parallel and distributed processing. The complexity of the systems in question arises to a great extent from the many ways in which their components can interact. In developing a complex system of interacting components, it is important to be able to describe the behavior of the system in a precise way at various levels of detail, and to analyze it on the basis of the descriptions. Process algebras were and are developed for that purpose. Roughly speaking, a process algebra provides a collection of operators, a collection of equational laws for these operators, and a mathematical model of these laws. The latter allows for the behavior of a system to be described as composed of the emergent behaviors of several interacting components and for the described behavior to be analyzed by mere algebraic calculations.

The advent of process algebras was marked by the introduction of CCS in the seminal monograph *A Calculus of Communicating Systems* by Milner, published as volume 92 of Springer's Lecture Notes in Computer Science in 1980; the elaboration of CSP in the influential paper "*A theory of communicating sequential processes*" by Brookes, Hoare, and Roscoe, published in the *Journal of the ACM* in 1984; and the presentation of ACP as a strict algebraic theory in the paper "*Process algebra for synchronous communication*" by Bergstra and Klop, published in *Information and Control* in 1984.

The very first applications of process algebras were mostly concerned with the description and analysis of communication protocols. Later on, the applications became more and more advanced. Often, they were concerned with the description and analysis of embedded systems, and more recently with Internet-based distributed systems.

The applications of process algebras led to a number of developments. The very first applications brought about the development of basic algebraic verification techniques, i.e., basic techniques to establish—on the basis of algebraic calculations—whether the actual behavior of a system is in agreement with its expected behavior. The construction of basic tools to facilitate description and analysis followed. Later applications led to extensions of existing process algebras and the development of more advanced algebraic verification techniques. Both make it easier to describe and analyze the behaviors of the systems often encountered in practice nowadays: systems that may change their communication topology dynamically, systems that must react within a certain amount of time under certain circumstances, systems that exhibit at certain stages behavior that is stochastic in nature, systems that in their behavior depend on continuously changing variables other than time, etc. Experience with the existing process algebras led also to the development of special process algebras for the definition of the semantics of programming languages that support parallel programming or the design of microprocessors that utilize parallelism to speed up instruction processing.

It is difficult to foresee future developments and applications, yet some tendencies are noticeable. One area that is gaining momentum is specialized process algebras that are tailored to a certain paradigm for parallel or distributed computing or even to a certain technology for parallel or distributed computing. This fits in with the tendency to apply process algebras to describe and analyze a prototype of a certain class of systems. Such applications can be useful in understanding certain aspects of the systems of an emerging class. However, the adequacy of the prototypes may be a major issue, because often drastic simplifications are needed to keep the description manageable. There is also a tendency to apply process algebras outside the realm of computing, which shows promise.

Many theoretical developments in the field of process algebras were collected by Bergstra, Ponse, and Smolka in the *Handbook of Process Algebra*, published in 2001. In 2005, a workshop was organized in Bertinoro, Italy, to celebrate the first 25 years of research in the field. Several special issues of the *Journal of Logic and Algebraic Programming* are devoted to this workshop. Despite the importance of applications of process algebras for the success of the field, both the handbook and these special issues concentrate strongly on the theoretical achievements. This shortcoming is compensated for in a splendid way by this book, which brings together the state of the art in research on applications of process algebras.

Kees Middelburg
Programming Research Group
University of Amsterdam, the Netherlands

Acknowledgments

The editors are very grateful to those listed below who served as peer reviewers for contributions to this book. Their diligent and conscientious efforts not only helped us to make the final selection of chapters but also provided numerous valuable suggestions to the authors, resulting in a high-quality collection.

Luca Aceto
School of Computer Science
Reykjavik University
Reykjavik, Iceland

Lorenzo Bettini
Dipartimento di Informatica
Universita' di Torino
Torino, Italy

Tommaso Bolognesi
CNR—Istituto di Scienza e Tecnologie
 dell'Informazione "A. Faedo"
Pisa, Italy

Gerhard Chroust
Systems Engineering and Automation
Institute of Systems Sciences
Johannes Kepler University of Linz
Linz, Austria

Philippe Clauss
Scientific Parallel Computing
 and Imaging
Université Louis Pasteur
Strasbourg, France

Pedro R. D'Argenio
Department of Computer Science
Facultad de Matemática,
 Astronomia y Fisica
Universidad Nacional de
 Córdoba—CONICET
Córdoba, Argentina

John Derrick
Department of Computer Science
University of Sheffield
Sheffield, South Yorkshire,
 United Kingdom

Gaétan Hains
Laboratoire d'Algorithmique,
 Complexité et Logique
University of Paris-Est
Créteil, France

and

SAP Labs France
Mougins, France

Michael G. Hinchey
Lero—The Irish Software Engineering
 Research Centre
University of Limerick
Limerick, Ireland

Thomas John
Department of Information Systems
Vienna University of Economics and
 Business Administration
Vienna, Austria

Kenneth B. Kent
Faculty of Computer Science
University of New Brunswick
Fredericton, New Brunswick, Canada

Felix Mödritscher
Institute for Information Systems
  and New Media
Vienna University of Economics and
  Business Administration
Vienna, Austria

Raymond Nickson
School of Mathematics, Statistics,
  and Computer Science
Victoria University of Wellington
Wellington, New Zealand

Frederic Peschanski
Laboratoire d'Informatique de Paris 6
UPMC Paris Universitas
Paris, France

Luigi Romano
Department of Technology
University of Naples Parthenope
Naples, Italy

Gwen Salaün
Department of Computer Science
University of Málaga
Málaga, Spain

Emil Sekerinski
Department of Computing
  and Software
McMaster University
Hamilton, Ontario, Canada

Kenneth J. Turner
Department of Computing Science
  and Mathematics
University of Stirling
Scotland, United Kingdom

Huibiao Zhu
Software Engineering Institute
East China Normal University
Shanghai, China

Introduction

Parallel processing is a rich and rapidly growing field of interest in computer science. Its spectrum ranges from small-scale, fine-grained multithreaded parallelism on single- and multicore processors to coarse-grained parallel execution on large, geographically dispersed distributed systems. With the rapid commoditization of multicore computers, developers are increasingly eager to exploit those multiple cores. Yet, this requires them to adopt parallel execution models that in turn need to explicitly address the underlying concurrency issues.

At the other end of the spectrum are applications that are spread out over a network of computers. They may be tightly coupled, such as for multiple computers cooperating on a single application in a high-performance cluster. Or, they may be loosely organized, as for mobile agents or service-oriented architectures (SOA). Regardless of the scale, a common requirement in parallel execution models is for carrying out interprocess communication and synchronization. Programmers know—or soon learn—that when this requirement is handled carelessly, a host of unpleasant failure modes tend to manifest: deadlocks, unrepeatable errors from race conditions, along with corruption of shared data.

One root cause is that the usual tools that programmers know best—mutexes, semaphores, condition variables, monitors, and message-passing protocols—are difficult to use correctly. Ad hoc design is common, and too often, success seems like a matter of good luck. In reality, many programmers new to parallel programming have an insufficient theoretical basis for what they are trying to do. Meanwhile, designers have learned that *modeling* is extremely helpful for all phases of system development—from understanding the requirements to expressing the specifications, and for systematically, even automatically, deriving an implementation "correct by construction" from the model. Yet, most popular modeling tools do not seem to give sufficient guidance for programming in concurrent environments.

Help is coming forth from computer scientists, who are fond of developing rigorous and logical ways of thinking and reasoning about computing. The purpose of this book is to show how one formal method of reasoning—that of *process algebra*— has become a powerful tool for solving design and implementation challenges of concurrent systems. Its power stems from providing a sound theoretical basis for concurrency, along with a formal notation that—in contrast to popular modeling techniques—is unambiguous. The price for obtaining the benefits is the necessity of coping with mathematical notation, symbols, and equations. Admittedly, this prospect may appear daunting to conventionally trained developers, who likely feel more comfortable drawing UML diagrams than puzzling out obscure equations. Fortunately, practitioners of process algebraic techniques are building domain-specific languages and application-specific tools that can be utilized by developers who then need to know less about the underlying theoretical elements. In such tools, the process algebra may be largely kept under the hood.

This book is not intended as a tutorial in process algebras, but there is space here for a brief orientation: Just as with the "algebra" we all learned in grade school, its ingredients include symbols standing for constant values and for variables, and operators that act on the symbols. While elementary algebra is concerned with manipulating numbers, a process algebra—or the synonymous term *process calculus*—is concerned with the creation, life, and death of processes that carry out computations. The emphasis is on interactions of processes among each other and with their environment. Symbols are used to stand for individual actions or events that a process may engage in; the processes themselves; channels, an abstraction used for interprocess communication; and data transmitted over the channels. Operators specify a sequential ordering of events; a choice between several events; if/then decisions; looping or recursive invocation of processes; composition of processes so that they execute concurrently, either synchronizing cooperatively on specified events, or running independently; and syntax for parameterizing and renaming so that process definitions can be reused in a modular fashion. Practitioners can use one of the classical process algebras mentioned in the Foreword—CCS, CSP, and ACP—or a more recent one such as the pi-calculus. Established process algebras have the advantage of the availability of automated tools that can be used for exploring a specification's state space and proving properties such as absence of deadlock (see Section 9.1 for a good explanation of these provable properties). Alternatively, practitioners can extend any of the base algebras by adding new symbols and operators, or even invent a new process algebra from scratch. A key advantage of formal, mathematical notations with rigorous semantics is that one can prove that particular conditions such as deadlock states do or do not exist, not just hope for the best, or discover them at runtime.

While the notion of process algebras goes back over two decades, what is new is the rapid proliferation of parallel computing environments that need their help in transforming sequential programming models to new ones better suited to parallelism. The urgent necessity for reliably exploiting concurrent computing resources has caused researchers to press classical process algebras into practical service, along with their invention of new ones. This book is intended as a showcase for recent applications of process algebras by current researchers from diverse parts of the international computer science community. While these contributions may appear largely theoretical due to the quantities of symbols and formulas, they are, in reality, process algebras applied to specific problems. The formalism is needed to establish the soundness of the theoretical basis, and to prove that the resulting tools are properly derived.

This book will be of special interest to students of process algebras, to practitioners who are applying process algebras, and to developers who are looking for fresh approaches to software engineering in the face of concurrency. The chapters are worth studying from two perspectives: first, those who identify with the problem domains (e.g., middleware systems or multicore programming) may ask, "Are the authors doing something that I could use, or can I adapt their approach?" and second, those who are interested in process algebras as a tool can ask, "How did the authors use a process algebra in their solution? What role did it play? How did they formally define it, and what did they prove in order to give it a sound basis?" A common pattern is to first create a process algebra whose symbols and operators are tailored to the problem domain (e.g., mobile agents). Next, a virtual (or abstract) machine (VM) is defined that executes specifications in the process algebra. Because both are formally defined, it is possible to prove that the VM is correct with respect to the algebra. The last step is to implement the VM, and this may be done by transferring it into a language that is semantically close, e.g., a functional programming language. Since the target language already possesses a compiler and/or runtime framework, the job of implementation is done, at least for prototyping and demonstration purposes.

The applied nature of the contributions is emphasized by organizing them into three target areas:

Part I Parallel Programming, Part II Distributed Systems, and Part III Embedded Systems.

We will now introduce each section and the chapters within it.

Part I Parallel Programming

The specific problem in view here is how to parallelize an algorithm, so as to take advantage of, say, multiple processor cores. In an ideal world, parallelization would be accomplished automatically, perhaps by compilers that are able to detect implicit parallelism in source code and generate instructions for concurrent threads on their own. Bearing in mind that modern processors already do this, in effect, with instruction streams—selecting independent instructions for out-of-order execution by multiple logic units—it may seem surprising that compilers have largely failed to match this at the source code level. But processors are detecting and exploiting implicit fine-grained parallelism. In contrast, fine-grained parallelism in software algorithms is often not worth exploiting, since there are significant overheads in spawning, managing, and collecting results from concurrent threads, plus potential bottlenecks for access to shared data. Furthermore, automatically identifying higher-level, coarse-grained parallelism implicit in source code, while desirable from an efficiency perspective, has proven to be a very challenging problem.

Therefore, in the real world, parallelization is still done on a best effort basis by hand, and then it becomes a question of ensuring that a parallel version is truly equivalent to the serial version. Our contributors have developed formal methodologies for achieving precisely that result.

In Chapter 1, Anand and Kahl address programming the Cell BE (Broadband Engine) processor from Sony, Toshiba, and IBM. Its heterogeneous multicore architecture features eight Synergistic Processor Units (SPUs) with their own local storage on a token ring under the control of a general-purpose Power Processor Element (PPE) core. The SPUs are intended to act as coprocessors for the PowerPC, being loaded on-the-fly with instructions and data, and coordinating via signals. The authors'

Coconut tool set provides the means to take an algorithm written in a domain-specific language (DSL) embedded in Haskell, and parallelize it for the Cell BE. The key to their approach is utilizing a graph-based internal representation of the program's data and control flows, which are then targeted to a VM that deals strictly with concurrency issues, e.g., data transfers and interprocessor signaling. The programmer manipulates the graph to create a high-performance schedule on the eight SPUs, with the authors' tool being used to verify that the scheduled version is correct, i.e., independent of a parallel execution order. The role of process algebra in this approach is to define the VM language, and then carry out correctness verification. While currently targeted to the Cell BE, their approach can potentially be ported to other multicore platforms.

In Chapter 2, Loidl et al. take an approach similar to that in Chapter 1 in that they also develop a runtime environment, Glasgow parallel Haskell (GpH), that can be ported to different parallel platforms, and they also focus on a functional programming language. Rather than attempting to automatically extract parallelism from GpH source code, they allow programmers to insert "par" and "seq" constructs into a program to give "semi-explicit" direction while carrying out successive steps of refinement to a parallel version. Their runtime environment, GUM, has been ported to a number of different parallel platforms.

These two chapters describe tools that are currently being used to program parallel systems. In comparison, Chapter 3 is more theoretical. Degerlund and Sere present an approach to taking algorithms described using another formal model called *action systems*, and developing an equivalent parallel version useful for scientific computing. An action system is specified using a process algebra called *refinement calculus*. The steps of refinement are used to introduce parallelism into the action system, with execution on a parallel target platform in view. The formal semantics of the refinement calculus ensure that the transformations are correct.

Part II Distributed Systems

Process algebras find their natural application in terms of formally modeling and verifying the behavior of distributed systems. Distributed systems are quite diverse, and this section also has the largest number of chapters.

We start with the work of Groote et al. in Chapter 4, who have developed a process algebra, mCRL2, which is specifically targeted at distributed applications. Its provision for local communication scope (i.e., restricted to a hierarchy of processes), as opposed to purely global scope, makes it useful for describing component-based architectures. mCRL2 also accommodates true concurrency "multiactions" distinct from interprocess communication, and supports the specification of abstract data types and action times. The authors have built tools capable of analyzing properties and simulating applications specified using mCRL2. This chapter has examples of visualizing the state space of a specification by means of a generated graphic.

One category of distributed systems that is currently gaining attention is the SOA that enables a business process to be automated by invoking software components located across a network. However, business processes and services described in words are subject to misinterpretations, leading to errors in integrating SOAs. A key area for formalization is turning prose descriptions into unambiguous specifications. Chapters 5 and 6 make contributions in this area.

Chapter 5 shows Nestmann and Puhlmann using an existing process algebra, the pi-calculus, to formally specify business processes. Their approach allows business processes to be captured in the form of Business Process Modeling Notation (BPMN) process graphs—whose nodes specify interactions such as parallel split, synchronizing merge, exclusive choice, and others—and then converted to pi-calculus agents enhanced by the authors' "trio" construct. Interactions of business processes and services can be formally modeled, and the models analyzed for various soundness properties. The authors have created a tool to automate the property analysis.

In Chapter 6, Rosa uses an ISO-standard process algebra, LOTOS, to formalize the construction of middleware systems. Each architectural component is specified as a single LOTOS process, which defines the component's structure—the ports available to connect to other components—and behavior in process-algebraic terms. The use of a formal notation as an architecture description language, in contrast to ambiguous prose descriptions, aids both would-be service providers and service integrators, and makes it possible to prove temporal properties of an architecture. The author further employs this technique to create a library of abstract message-oriented components for use in defining middleware. A third demonstration formalizes middleware for wireless sensor networks.

Another manifestation of distributed systems is based on the notion of mobile agents moving around a network. The purpose is to send software to the data rather than pulling the data down to a computation node. As argued in Chapter 8, mobility helps to minimize the impact of two problems common to distributed systems: network latency and network failure. Systems based on mobile agents will also benefit from formal descriptions, especially when it comes to ensuring security. Chapters 7 and 8 deal with mobility. As in Chapters 1 and 2, the authors of Chapters 7 and 8 take the approach of formally defining a VM.

In Chapter 7, similar to Chapters 5 and 6, Paulino targets SOA and middleware. His VM, service-oriented mobility abstract machine, is based on an extension of the pi-calculus. By programming for the VM, programmers can be abstracted from the details of the network while still utilizing mobility. The author's strategy for deployment is to execute the machine on network nodes running an existing framework called DiTyCO.

In Chapter 8, Phillips presents his Channel Ambient (CA) calculus for specifying mobile applications. The textual notation also has a helpful graphical counterpart. Based on the abstract notion of an "ambient"—which may stand for a machine, a mobile agent, or a software module—the CA calculus provides operations whereby ambients may interact, and migrate in and out of each other, via channels. Security properties can be verified for a given specification. The execution target is called the Channel Ambient Machine (CAM), and its correctness with respect to the CA calculus is proven by the author. His implementation strategy is to map the CAM to a functional programming language, OCaml, that can then be executed on network nodes.

Part III   Embedded Systems

While an exact definition of an embedded system is debatable, it is generally described as a product that contains within it some combination of software running on a general- or special-purpose processor, plus associated custom digital logic. The latter is commonly used to accelerate specific portions of calculations, which would otherwise require a more expensive processor in order to meet timing constraints. The term "embedded" refers to the fact that the end user is not necessarily aware that the product contains a computer, and, in any case, cannot utilize the computer for some other purpose. Embedded systems have special design constraints because, unlike software for desktop or server computers, the products have a significant recurring cost, such as parts, assembly, packaging, labor, etc., in addition to the nonrecurring engineering cost that goes into developing and interfacing the software and hardware components. Furthermore, they may have to meet rigorous requirements of power consumption (battery life), size, and weight. Cell phones and digital cameras are common examples. Robotic devices, such as autonomous vacuum cleaners, are embedded systems. Some are safety-critical, such as an antilock brake controller. The aim of choosing the best combination of hardware and software is to meet all the performance requirements at the lowest manufacturing cost, i.e., offering a sufficient set of features at a competitive price.

Parallelism in embedded systems comes in several forms: embedded devices are often designed in terms of concurrent threads, some monitoring sensor inputs, others computing outputs, or actuating control outputs. Some have multiple processors, which may be heterogeneous, such as a 16-bit microcontroller plus a DSP, and some have hardware/software concurrency. They are often designed and marketed in a family of related products with more/fewer features, or adding features over time, or that utilize differing HW technology more favorable to different production quantities. Formal methods are very attractive for ensuring reliability, especially for safety-critical products, and those in hard-to-access locations.

In Chapter 9, McEwan leads off with a contribution targeted at formally deriving a hardware implementation, although his technique is also applicable to software. As with Chapter 3, which combines the state-based formalism of action systems with the event-based refinement calculus, McEwan combines state-based Z with the event-based process algebra CSP, in a formal methodology called Circus. Z is used to provide a formal model for the data. Refinement toward an implementation proceeds by applying laws that safely inject parallelism into a sequential specification. The target language, Handel-C, used to synthesize digital logic, is close to CSP. Circus is flexible enough to allow engineering choices to guide the refinement, while still ensuring that the resulting implementation is correct.

Typical parallel systems leave the scheduling of multiple processes to the operating system under the assumption of adequate CPU time and memory. Accordingly, many process algebras used to specify parallel systems do not have visibility into process scheduling, yet real-time embedded systems must guarantee responses within certain time constraints, and do so in the context of limited resources (chiefly CPU time and memory). Formal notations to specify resource requirements such as timing constraints are therefore of great utility in the embedded domain.

In Chapter 10, Mousavi et al. address this problem by developing a pair of process algebras, together called PARS: one to specify processes, and another to specify scheduler behavior. The two, combined, produce a scheduled system.

The final chapter, Chapter 11, considers embedded systems through the lens of product lines, in which reuse of concurrent artifacts across different hardware platforms is emphasized. Yovine's solution is to use an algebraic language, FXML, to specify concurrent behavior, associated control and data dependencies, and timing constraints. FXML specifications are then processed by a software synthesis tool, Jahuel, to yield an implementation customized for a given platform. Platforms may differ in the means by which concurrency is supported and interprocess synchronization and communication are carried out. FXML can also be used in a design automation toolset as an intermediate specification with formal semantics. Moreover, one can define a translation of a nonformal language into FXML, thus giving the language a formal semantics and opening up the possibility of verification.

We would be remiss not to acknowledge a "dark cloud" in the picture: the challenge of scaling up some of these techniques for industrial-sized applications. Such specifications may have so many states that automated verification tools cannot reach them all in reasonable time. This drawback may not be evident from these chapters themselves, since the authors were forced to use small examples, both for the sake of clarity and due to space limitations. Yet some claim to have applied their techniques to larger-scale problems. For more details, consult the references to the authors' own related work, and they will be pleased to answer queries.

The process of collecting these chapters has highlighted for us, as educators, that colleges and universities have further to go in training undergraduates to be comfortable with concurrency and skilled in reliable methods of parallel programming. More exposure to formal methods is also needed, so that these approaches do not appear so foreign. As it is, in most software development curriculums, particularly in North America, formal methods are more "honored in the breach" than by systematic instruction. At the same time, researchers will do well to embed their process algebraic techniques into tools suitable for users without special training, in order to widen their prospects for adoption.

To conclude, we offer this book as an early collection of research fruits in what will undoubtedly become a burgeoning growth industry in the coming years, and to which the editors are eager to contribute their own research.

Editors

Michael Alexander holds degrees in electrical engineering (Technologisches Gewerbemuseum [TGM]), business administration (University of Southern California), and economics (University of Vienna). He is a supervisor in the Vienna, Austria KPMG IT Advisory practice. His experience includes teaching and research at Wirtschaftsuniversität Wien; as a lecturer and product management at IBM, Siemens, Nortel Networks, and Alcatel; as a product line manager for ADSL and Optical Access Networks. He has authored a textbook on networks and network security [1] published by Hüthig/Verlagsgruppe Süddeutsche, and has edited a special issue on mathematical methods in network management of the *Wiley International Journal of Network Management*. For the last four years, Dr. Alexander has served as the program committee chair for the Workshop on Virtualization in High-Performance Cluster and Grid Computing (VHPC). His current research interests include computer networking, formal methods, network management, concurrency, payment systems, machine learning methods in operations research, and e-learning.

William Gardner received his bachelor's degree in computer science from the Massachusetts Institute of Technology. Subsequently, he pursued his career in software engineering at Litton Systems (Canada) Ltd., Toronto, working primarily on embedded systems for defense customers. Returning to graduate school later in life, he did his doctoral research in the VLSI Design and Test Group in the Department of Computer Science at the University of Victoria, British Columbia, Canada. His interests are chiefly in design automation starting from formal specifications, particularly targeting embedded systems via hardware/software codesign. His software synthesis tool CSP++ [2] translates formal CSP specifications into executable C++. Its design flow features "selective formalism," where a system's control backbone is specified and verified in CSP, and then the principal functionality is provided by plugging C++ user-coded functions into the backbone, to be actuated by CSP events and channel communications. He was formerly on the faculty of Trinity Western University, British Columbia, Canada, and moved in 2002 to the University of Guelph, Ontario, Canada, where he is currently an associate professor in the Department of Computing and Information Science. With his graduate students, he has been participating in the R2D2C project at the NASA Goddard Space Flight Center, which involves automatic code generation from scenarios via under-the-hood formal specifications [3]. Dr. Gardner teaches courses on software development, operating systems, embedded systems, and hardware/software codesign.

References

[1] M. Alexander, *Netzwerke und Netzwerksicherheit—Das Lehrbuch*, Hüthig Telekommunikation, Heidelberg, 2006.

[2] W.B. Gardner, Converging CSP specifications and C++ programming via selective formalism, *ACM Transactions on Embedded Computing Systems* (TECS), Special Issue on Models and Methodologies for Co-Design of Embedded Systems, 4(2), 302–330, May 2005.

[3] J. Carter and W.B. Gardner, Converting scenarios to CSP traces with Mise en Scene for requirements-based programming, *Innovations in Systems and Software Engineering*, 4(1), April 2008.

Contributors

Abdallah Al Zain
School of Mathematical
 and Computer Sciences
Heriot-Watt University
Edinburgh, United Kingdom

Christopher Kumar Anand
Department of Computing and Software
McMaster University
Hamilton, Ontario, Canada

Ismail Assayad
VERIMAG
Grenoble, France

Clem Baker-Finch
Department of Computer Science
Australian National University
Canberra, Australia

Twan Basten
Department of Electrical Engineering
Eindhoven University
 of Technology
Eindhoven, the Netherlands

Ananda Basu
VERIMAG
Grenoble, France

Michel Chaudron
Leiden Institute of Advanced
 Computer Science
Leiden University
Eindhoven, the Netherlands

Francois-Xavier Defaut
VERIMAG
Grenoble, France

Fredrik Degerlund
Turku Centre for Computer
 Science
Åbo Akademi University
Turku, Finland

and

Department of Information
 Technologies
Åbo Akademi University
Turku, Finland

Jan Friso Groote
Department of Mathematics
 and Computer Science
Eindhoven University
 of Technology
Eindhoven, the Netherlands

Kevin Hammond
School of Computer Science
University of St Andrews
St Andrews, Scotland
United Kingdom

Wolfram Kahl
Department of Computing and Software
McMaster University
Hamilton, Ontario, Canada

Hans-Wolfgang Loidl
Institut für Informatic
Ludwig-Maximilians-Universität
Munich, Germany

Aad Mathijssen
Department of Mathematics
 and Computer Science
Eindhoven University
 of Technology
Eindhoven, the Netherlands

Alistair A. McEwan
Department of Engineering
University of Leicester
Leicester, United Kingdom

MohammadReza Mousavi
Department of Mathematics
 and Computer Science
Eindhoven University
 of Technology
Eindhoven, the Netherlands

Uwe Nestmann
School of Electrical Engineering
 and Computer Science
Berlin Institute of Technology
Berlin, Germany

Hervé Paulino
Computer Science Department
New University of Lisbon
Lisbon, Portugal

Andrew Phillips
Microsoft Research
Cambridge, United Kingdom

Frank Puhlmann
Hasso-Plattner-Institut
Potsdam, Germany

Michel A. Reniers
Department of Mathematics
 and Computer Science
Eindhoven University
 of Technology
Eindhoven, the Netherlands

Nelson Souto Rosa
Centre of Informatics
Federal University of Pernambuco
Recife, Brazil

Kaisa Sere
Department of Information
 Technologies
Åbo Akademi University
Turku, Finland

Phil Trinder
School of Mathematical
 and Computer Sciences
Heriot-Watt University
Edinburgh, Scotland, United Kingdom

Yaroslav S. Usenko
Department of Mathematics
 and Computer Science
Eindhoven University
 of Technology
Eindhoven, the Netherlands

Muck van Weerdenburg
Department of Mathematics
 and Computer Science
Eindhoven University
 of Technology
Eindhoven, the Netherlands

Sergio Yovine
VERIMAG
Grenoble, France

Marcelo Zanconi
VERIMAG
Grenoble, France

Part I

Parallel Programming

Chapter 1

Synthesizing and Verifying Multicore Parallelism in Categories of Nested Code Graphs

Christopher Kumar Anand and Wolfram Kahl

Contents

1.1   Introduction

Current trends in high-performance processor technology imply that the programming interface is becoming more complicated. The pronounced shift toward pervasive Single Instruction Multiple Data (SIMD) parallelism, and the introduction of heterogeneous computing elements, including hierarchically distributed memory (encompassing conventional distributed memory, streaming memory, software-directed caches, and private memories), will require changes in software development tools.

Among the first generation of such processors, the Cell Broadband Engine (BE) is the most widely deployed. Using its resources efficiently requires explicit synchronisation of both code and data, and packaging of both to fit in small memory spaces and constrained communication infrastructures.

The COde CONstructing User Tool (Coconut) project has the aim of capturing the entire design and development process for safety-critical, high-performance scientific software, e.g., medical imaging, in a sequence of formal specifications. The hardware-specific parts of Coconut currently target a single Cell BE. See Section 1.1, for background on the Cell BE and Coconut.

This chapter describes the support Coconut provides for multicore parallelism, including the way that concurrent control flow is represented in nested hypergraphs, the way multicore issues are abstracted into a virtual machine, and the support Coconut provides for developers of parallelization patterns. This support includes a linear-time algorithm to verify the soundness of a parallelization. For the developer, verification eliminates the need for time-consuming parallel debugging, and the analogy between the virtual machine and a superscalar CPU provides a familiar way of thinking about scheduling.

The key to being able to provide an efficient verification tool is the abstraction provided by modeling multicore communication as a simple process algebra. Simplicity is justified because parallelism is not an inherent part of algorithm specifications in the domain of scientific computation, it is only added to implementations to meet performance targets, so the process algebra only needs to model the patterns of communication which support this. Adding additional features to support general control flow would significantly complicate verification, and could be used by developers to implement program logic, making programs much harder to understand, and thereby reducing programmer productivity.

The Coconut tool set uses as intermediate representations of programs and program fragments, a variant of term hypergraphs we call "code graphs." Term graphs have a long history as an efficient implementation of symbolic expressions. A certain kind of term graph also provides a natural model of a resource-conscious kind of categories (gs-monoidal categories, [8]); this provides a useful justification of many kinds of algebraic data-flow graph transformations. To combine control flow and data flow, we nest the simple data-flow code graphs inside an outer-level code graph with control-flow semantics, similar to the two different semantics of flownomials [19]. Software pipelining, i.e., instruction reordering to make better use of instruction pipelines and multiple-issue queues, is an algebraic transformation involving both levels of such nested code graphs.

This approach naturally extends to modeling (restricted) concurrency: We introduce a third, outer layer for explicit concurrency, with a semantics more on the data-flow side (this is related with the implicit concurrency in pure functional programs), but again with some control-flow characteristics.

Algebraic transformations crossing the outer two levels in such three-level nested hypergraphs are used to map potential concurrency to explicit concurrency, and to hardware parallelism. On the level of multicore processors, this will involve the automatic introduction of appropriate synchronization instructions or, in the opposite direction, the elimination of such concurrency primitives for the purpose of verifying against a concurrency-independent specification.

This chapter presents the control flow at the middle multicore level. In conventional programs, this level of control flow is usually encoded in library calls (e.g., MPI [33,38]) in the program text. Our approach is to make this disjoint in a separate level of the graph. This allows for a simple verification of correctness for deeply pipelined data-flow graphs through a network of cores (processors). As a result, whether multicore parallelism is directed by user code, user-code generators, or automatic parallelization, the programmer/compiler writer can focus on tuning for performance, and check correctness in a computation which grows linearly with program size, and quadratically in the complexity of the hardware.

The distinguishing feature of code graphs at this level is that they are presented as ordered lists of instructions. Proofs of sound parallelization work by going through the list in order, maintaining a limited set of state information (status of signaling and data transfers, contents of buffers). Any ambiguity (from a race condition) or deadlock can be flagged in the presented program.

1.1.1 Related Work

This work is the first step in bringing together disconnected research efforts in (1) verification of concurrent programs, (2) scheduling for distributed processors, and (3) decomposition of large numerical problems.

Practical parallel computation is a special subset of the set of concurrent programs. We have defined a precise subset of the vaguely defined set "parallel computation." Verification of correctness in this subset should be simpler than in the larger set, but many important ideas come from efforts to verify general concurrent programs. We make fundamental use of partial orders in analyzing concurrency within our programs. Godefroid and several collaborators have demonstrated that partial order methods can be much more efficient than other methods for general concurrent problems [16,20]. In the language of Ref. [16], we construct a partial order in such a way as to show that there are no backtracking points, i.e., no points where a different execution sequence would have led to a different result. Our implementation is much simpler, with tight complexity and memory bounds, because it is tailored to the restricted set of programs we consider. The trade-off is that the programs we are able to consider are orders of magnitude larger than typical concurrency benchmarks.

We target the Cell BE processor, using matrix multiplication as our first benchmark, as do [5]. Our approach differs in that we present the programmer with an unconventional interface via domain-specific languages (DSLs) embedded in Haskell. DSLs [14] can be a useful tool for making special-purpose software development accessible to domain experts; Ref. [13] and, in particular, Hudak [25,26] showed that embedding DSLs in Haskell has significant advantages both for implementability and for usability.

There is a lot of literature related to loop decomposition, and specifically the decomposition of large operations in linear algebra. The generation of consumers used in our scheduling algorithm implicitly encodes a loop tiling (see Refs. [21,22]). Research in this direction is currently considering much more complex computational examples than we are able to at this time, and will provide a rich set of benchmarks for the future. That said, some of the work on tiling will have to be reconsidered in light of the shift from computational to main-memory bottlenecks. We hope that our tool will accelerate the development of patterns of communication involving many point-to-point pipelined transfers.

1.1.2   Cell BE Architecture and the Virtual Machine

The Cell BE architecture exposes both computational and data-flow resources to user-level programming. In Figure 1.1, one sees the main computational resources. The synergistic processing units (SPUs) do not have direct access to the multilayer token ring, but must go through the attached memory flow controller (MFC), which operates most efficiently when executing reorderable lists of direct memory access (DMA) instructions. The PowerPC Unit (PPU) is capable of general computation, but has much lower performance than the SPUs. The initial implementations of the Cell BE include interfaces to extend the ring architecture to multichip rings or switched networks [7].

In implementing the approach to concurrent control flow presented in this chapter, the PowerPC (PPC) core manages run-time start up (address resolution, code loading, etc.).

Our virtual machine model was designed with the Cell BE in mind, but could easily be adapted to other architectures. If an architecture does not support signals,

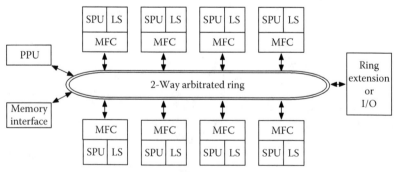

FIGURE 1.1:   Bus diagram for Cell BE.

they could be emulated using data transfers. Just as object code optimized for one generation of a processor family may run poorly on the next generation, the performance of a generated program will depend on it being optimized for the right implementation of the virtual machine. If signals are not supported, some patterns of communication may become more expensive relative to other patterns.

The current Cell BE is a tightly coupled single-chip device. For this reason, fault tolerance is not designed into the virtual machine, and efficiently handling partial equipment failure would require extensions to the virtual machine model.

1.1.3　Background on Coconut

Coconut is architected with both long- and short-term correctness requirements. In the long-term, we believe that safety-critical applications like medical imaging should come with meaningful correctness guarantees. Meaningful, to the patient, means that the image which comes out is proven to be the result of applying algorithms with known specifications and understandable physical inputs. Algorithms can be incorrect, but they should be specified by experts in the field. No low-level optimizations should be taken without floating up their consequences to the algorithm level for approval. Specifications at this level must include concrete anchors in the physical universe, including not just units, but oriented coordinate systems, etc. High level User Specification of Constraints (HUSC) [40] is a prototype including special implementations of type inferencing and alternatives suited to these applications.

Current practice starts with an informal specification of the algorithm, possibly supplemented with a partial implementation in a mathematical scripting language. This specification is translated into different types of low-level specifications and finally implemented in an imperative language, possibly including assembly language.

The Coconut workflow involves no unverifiable human translation of specifications or manual editing of assembly code. Code optimization experts contribute by formally specifying patterns of efficient implementation as code transformations or code generators. At this time, all transformations and leaf functions are specified in Haskell. For the lowest-level patterns, which are implemented without control flow (although they may contain control flow synthesized using logical and permutation instructions) we have defined a DSL embedded in Haskell, including a number of support tools such as simulation and symbolic execution. Such tools are important not just for verifying of correctness, but for pinpointing errors in code under development.

Much of the technical difficulty in optimizing code (whether by human or by compiler) is the recognition of patterns of control flow expressed using conventional patterns. The Coconut strategy for avoiding this is to provide more expressive control constructions at the language level. One of these is the MultiLoop [3]. Novel control structures require novel instruction scheduling. Explicitly Staged Software Pipelining [39] is a novel approach to scheduling loops which supports the MultiLoop.

This work defines a language for concurrent control flow, and the principal tool for both debugging patterns and verifying correctness.

Coconut is being implemented from the bottom up to insure that intermediate representations do not interfere with the identification of optimizable patterns of execution. One advantage of this approach is that we are able to produce high-performance C-callable single-core functions, and using the infrastructure defined here, we will be able to provide libraries which encapsulate optimal parallel execution. At this point, Coconut will be able to generate the most difficult aspects to verify in our target applications. For many applications, verification of this level of computation is enough. There is no plan to allow Coconut-generated code to call functions written in other languages, since this would make verification much more difficult.

This strategy allows us to earn early dividends from the considerable investment required to build verifiably correct software at this level of performance.

Coconut is implemented in Haskell, except for type inferencing in HUSC. Haskell is used for developer-productivity reasons, mostly strong and flexible typing. The current DSL embedding takes advantage of Haskell language features, and the existence of support for literate code, and these features are appreciated by novice users, but another pure functional language could have been used.

1.1.4   Overview

We explain in Section 1.2 how directed hypergraph syntax naturally integrates data flow and concurrent control flow, and summarize how we use these code graphs in nested arrangements for different aspects of the generation of high-performance code.

We then turn our attention to implementation aspects of multicore parallelism (Section 1.3), contrasting it with other kinds of parallelism, to provide the background for the decision (Section 1.4) to define a virtual machine for encapsulating certain low-level aspects of targeting multicore architectures. The details of this virtual machine are then presented in Section 1.5, and in Section 1.6 we show how we produce, partially guided by graph transformations, a multicore program scheduled to hide communication latencies as far as possible. To independently verify and certify the concurrency aspects of the result, we introduce, in Section 1.7, an efficient algorithm that checks satisfaction of the relevant safety and liveness properties in a single pass over the fuzzy multicore schedule.

1.2   Nested Code Graphs

Graphs are frequently used as internal presentations for code generation and analysis purposes. We use code graphs as uniform graph syntax. One use of our code graphs is as data-flow graphs generalizing term graphs, and another use is as control-flow graphs generalizing finite-state machines and Petri nets. To express concurrent processes and parallel execution, we use the same code-graph syntax at different levels of nesting to combine data-flow aspects, control-flow aspects, and concurrency aspects in nested code graphs.

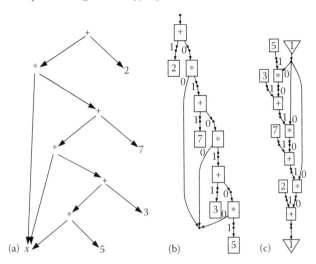

FIGURE 1.2:   (a) Horner term graph, (b) jungle, and (c) code graph.

Term graphs, the abstract version of data-flow graphs, are usually represented by graphs whose nodes are labeled with function symbols and edges connect function calls with their arguments [37]. We use the dual approach of directed hypergraphs where nodes are labeled with type information (if applicable), function names are hyperedge labels, and each hyperedge has a sequence of input tentacles and a sequence of output tentacles (each incident with a node).

For example, Figure 1.2 shows first a conventional term graph (as presented by HOPS [27]) for calculating the polynomial $5x^3 + 3x^2 + 7x + 2$ using Horner's rule, i.e., as $x * (x * (x * 5 + 3) + 7) + 2$, and second a conventional jungle with nodes drawn as black discs, and edges drawn in boxes with a result tentacle from the result node to the edge and numbered* argument tentacles from the edge to the argument nodes. Variables are represented by nodes that are not the result of any edge.

In contrast, the tentacles of our code graphs follow the direction of the data flow, and we draw the data flow downward, as shown in the right-most drawing of Figure 1.2. In our code graphs, we also explicitly flag input nodes (corresponding to jungle variables) by attaching a numbered triangle with an arrow to the node, and output (result) nodes by attaching a numbered triangle with an arrow from the node.

Such data-flow code graphs are the central data structure of the SPU scheduler of Coconut. The assembly-level data-flow code graph for the same polynomial is shown to the left in Figure 1.3, with edges containing in this case either the SPU SIMD opcode fma (fused floating-point multiply-add), or SIMD constants replicating (splatting) the same 32-bit floating-point number four times into 128-bit SPU registers. In this drawing, we show node numbers (normally for debugging purposes),

---

* Since all these code-graph drawings have been automatically generated via dot [29], these numbers are necessary to indicate argument sequence.

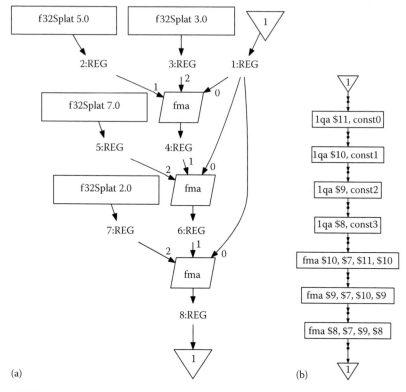

(a)   (b)

FIGURE 1.3:   (a) Horner assembly-level data-flow and (b) control-flow code graphs.

and indicate the node type "REG" standing for 128-bit SPU register values. Since the SPU has a unified register file, there are no further type distinctions at this level, although a refinement makes sense for the purpose of analyzing and verifying the data flow of individual register components, see Ref. [4].

To the right in Figure 1.3, we show the assembly instructions that the left graph is compiled to, presented as a control-flow code graph. Assembly instructions are opcodes together with argument and result registers, so the data flow can be reconstructed from this.* For this chapter, it is central that code graphs can be used for both data-flow and control-flow graphs.

In the light of the above, a hypergraph in the sense of this paper is therefore directed, with sequences of input (argument) and output (result) tentacles at each edge:

---

* The constants are transformed into data segments for the assembler, and referenced by the labels "const0", ..., "const3". This is necessary since there are no SPU instructions with immediate arguments of 32 bits.

DEFINITION 1.1

A hypergraph $H = (\mathcal{N}, \mathcal{E}, \text{src}, \text{trg}, \text{nLab}, \text{eLab})$ *over a node label set* NLab *and an edge label set* ELab *consists of*

- *A set* $\mathcal{N}$ *of* nodes *and a set* $\mathcal{E}$ *of* hyperedges *(or* edges*)*

- *Two functions* src, trg $: \mathcal{E} \rightarrow \mathcal{N}^*$ *assigning each hyperedge the sequence of its* source nodes *and* target nodes *respectively*

- *Two functions* nLab $: \mathcal{N} \rightarrow$ NLab *and* eLab $: \mathcal{E} \rightarrow$ ELab *assigning labels to nodes and hyperedges*

Hypergraph homomorphisms are pairs consisting of a total node mapping and a total edge mapping, preserving src, trg, nLab, and eLab, and induce a category.

If a hypergraph in some sense represents a program, it has an input/output interface:

DEFINITION 1.2

A code graph $G = (H, \text{In}, \text{Out})$ *over a node label set* NLab *and an edge label set* ELab *consists of*

- *Hypergraph* $H = (\mathcal{N}, \mathcal{E}, \text{src}, \text{trg}, \text{nLab}, \text{eLab})$ *over* NLab *and* ELab

- *Two node sequences* In, Out $: \mathcal{N}^*$ *containing the* input nodes *and* output nodes *of the code graph*

A code-graph homomorphism is a hypergraph homomorphism that additionally preserves In and Out.

Typing aspects are dealt with in the usual way:

DEFINITION 1.3

A hypergraph signatur*e is a hypergraph where the labeling functions* nLab *and* eLab *are the identities on the respective label sets.*

A typed hypergraph *is a hypergraph together with a homomorphism into a hypergraph signature.*

The resulting categories of (typed) hypergraphs have pushouts, and can therefore be used as basis for the double-pushout approach to graph transformation [11], where rules are induced by spans in the category of code graphs [4].

Code graphs also give rise to a gs-monoidal category with sequences of node labels as objects and code graphs as morphisms; Ref. [8] uses this to establish functorial semantics for code graphs, and Refs. [9,28] show examples where this semantics uses locally ordered categories, such that code-graph homomorphisms $F : G_1 \rightarrow G_2$ document ordering relations $[\![G_2]\!] \sqsubseteq [\![G_1]\!]$ on the semantics.

1.2.1 Data-Flow Graphs and Parallelism

Using code graphs as term graphs follows the jungle view of term graphs, initially put forward by Hoffmann and Plump [24,34], and equipped with functorial semantics by Corradini and Gadducci [8]. In such code graphs, nodes are labeled with types and in a certain sense represent values of the respective types; hyperedges are labeled with operations taking input values as arguments and producing results as outputs.

Data-flow semantics of hypergraphs implies the following:

- If an edge has multiple input tentacles, the computation associated with it has multiple argument positions, one for each input tentacle.

  This is the only kind of branching present in term trees. This kind of branching is also the source of potential parallelism, particularly in functional programming, since it represents a split of the computation of the outputs of this edge into potentially independent computation of its inputs.

- If a node is attached to multiple input tentacles, the value represented by this node is used by several consumers.

  This sharing marks the transition from term trees to term graphs, and is frequently only a tool for efficiency (shared effort) without semantic relevance, for example, in most implementations of functional programming. (Sharing does have semantic relevance for example in functional-logic programming [1] and in multialgebras [9,10].)

- If an edge has multiple output tentacles, the computation associated with it produces multiple results, one for each output tentacle.

  This presents a relatively unproblematic generalisation of term trees or term graphs.

- Possibility that a node is attached to multiple output tentacles was used in Ref. [28] to represent joins in data flow, i.e., different equivalent ways to produce a value.

Cycles in term graphs are frequently used to encode recursions; this normally uses a least fixed-point semantics and relies on the presence of nonstrict constructs, in particular conditionals, or, in lazy functional programming, data-type constructors.

In the central assembly-level components of Coconut, assembly op-codes (which have strict semantics) serve as edge labels, with inputs corresponding to argument registers and outputs corresponding to result registers. We use acyclic data-flow code graphs with these op-codes as edge labels to represent loop bodies of scientific computation applications, and use graph transformation and analysis to exploit instruction-level parallelism (ILP) and achieve optimal schedules [2,4].

1.2.2 Nonconcurrent Control-Flow Graphs

The archetypal control-flow graphs are finite-state machines; nodes are labeled with state types and represent states, and edges labeled with actions, sometimes

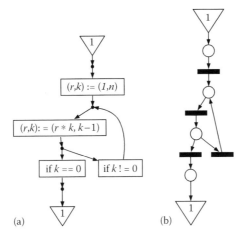

FIGURE 1.4:   (a) Factorial control-flow graph and (b) underlying Petri net.

perceived as active (operations), sometimes as passive (recognition of input symbols), and sometimes as both (for example, in Mealy machines).

For conventional edges, finite-state machines serve as the standard control-flow model, and represent nonconcurrent control flow.

Cycles in control-flow graphs express iteration, and correspond to the asterate operator of Kleene algebras.

In Figure 1.4, we show the control-flow graph of an imperative implementation of the factorial function presented as a code graph. Even though from a flow-chart point of view, it may seem strange to have two separate edges for the conditional, this is in fact the way such programs are handled in relational semantics (see Refs. [17,36]), and fits well with the concurrent semantics below—indeed, abstracting all edges to transitions and considering the nodes to be places produces the place-transition Petri net (drawn to the right in Figure 1.4) representing the control flow of this program.

At the lowest (output) level in Coconut, assembly instructions (except conditional branches), i.e., op-codes together with register references, serve as control-flow edge labels, with the state before as single input, and the state after as single output.

The conversion of op-code-labeled hyperedges to instruction-labeled control-flow edges happens as part of scheduling of the data-flow graphs; since scheduling essentially means adding of control dependencies, and scheduling of a data-flow graph together with a register allocation enforces the data-flow dependencies via the stronger control-flow dependencies.

At a higher level, we represent computational kernels as sequential control-flow graphs where most of the edges are labeled with larger data-flow graphs, and implement software pipelining essentially as graph transformation on these control-flow graphs [3].

1.2.3   Control-Flow Graphs and Concurrency

The standard model for concurrent control flow are Petri nets, which are, in more theoretical work (e.g., [18]), formalized as directed hypergraphs, with places represented by hypergraph nodes and transitions by hyperedges (but with multisets of input and output places for each transition; Sassone introduced the concept of prenet which has sequences instead of multisets [6,35]).

Putting everything together, control-flow semantics of directed hypergraphs therefore implies the following:

- If a node is attached to multiple input tentacles, it is normally understood that control propagates along any applicable outgoing edge.

  This kind of branching implements decisions by making different edges applicable depending on the circumstances. In finite-state automata recognizing regular languages, those edges whose labels match the next available input are applicable; more generally, conditions depending on the current state can be used, as for example "$k == 0$" in Figure 1.4. If several outgoing edges of the same node are applicable, then this kind of branching introduces nondeterminism.

  If this is the only kind of branching, the resulting graphs can be used as decision trees, or as tree structures for branching-time temporal logics [15].

- If a node is attached to multiple output tentacles, the corresponding different operations pass control to the same state.

  This corresponds to moving from tree structures for branching-time temporal logics to graph structures, and similarly from infinite regular languages to finite automata; in imperative programming, this is typically induced by jumps or higher-level control structures such as loops.

- If an edge has multiple output tentacles, we understand this as forking the current thread of control into multiple threads of control.

  This corresponds to the fact that Petri net transitions with multiple outgoing edges place tokens into each successor place upon firing.

- If an edge has multiple input tentacles, we understand this as introducing synchronization of concurrent threads.

With this background, it becomes clear that, in the nonconcurrent case, conditionally branching assembly instructions should not be represented by branching hyperedges, but instead as two conventional edges of which in each state only one is applicable, as can be seen in the example shown in Figure 1.4.

1.2.4   Concurrent Interpretation of Data-Flow Graphs

From the theoretical literature [6,18,35] one notices that Petri net computations, which are finite acyclic versions of Petri nets equipped with a start and end interface, give rise to essentially the same kinds of categories as term graphs, namely variants of gs-monoidal categories [8].

The description in the previous section was accordingly tuned to emphasize the compatibility between the concurrent control-flow view of a hypergraph with the data-flow view, as long as no cycles are present.

Indeed, Sassone's prenets [6], designed to reflect the "individual-token philosophy" to composition of Petri net computation, can be understood as making individual threads of control explicit.

The two views combine easily into data-flow graphs with concurrency aspects, where the distribution aspects are most easily made explicit by adding host identifiers to nodes and hyperedges. For an example, see Figure 1.7.

Combining this representation of concurrency aspects of distribution with the nested hypergraph representation of sequential computational kernels, we obtain a nesting of hypergraphs with at least three levels. Implementing iteration of concurrent behavior will even require wrapping the concurrent control-flow code graph at the third level into a cyclic sequential control-flow graph, which produces a four-level nesting.

1.3   Efficient Multicore Parallelism

Different levels of parallelism raise different issues with respect to efficiency and correctness. It is useful to put multicore parallelism into a context categorized in order to contrast current issues with parallelism issues in the past.

Distributed parallelism: It involves execution on multiple hosts, also called nodes, each running an operating system.

Long and variable latencies, and network and node failures during computation may be significant issues in this context. The message-passing library MPI [33,38] is most commonly used for distributed scientific processing, typically implementing a Single Program Multiple Data (SPMD) model. The most feature-rich tools for distributed parallelism are marketed as grid computing.

The SPMD approach allows programmers to write a single program which runs on multiple nodes. Each instance therefore contains code for all instances and knowledge of the data distribution across nodes. This model is not suited to heterogeneous multicore architectures with small private memories.

ILP: CPUs need multiple execution units for different types of computation to reduce design complexity. ILP takes advantage of this hardware design imperative by dispatching instructions to multiple units in parallel. In the vast majority of cases, the CPU has circuitry to introduce this parallelism into a sequential stream of instructions, and may make other modifications (e.g., register renaming) to improve throughput. Using this mechanism, sequential programmers and basic compilers get some degree of parallel execution for free, while advanced compilers can produce significant levels of parallelism (e.g., four-way on recent Power architectures). Very Long Instruction Word architectures expose this level of parallelism explicitly to the compiler.

SIMD: It properly refers to architectures with hardware support for large-vector arithmetic operations. The Cray-1 was a successful computer using this model.

Vector-in-register parallelism (VIRP): In current architectures, SIMD most often refers to short-vector-in-a-register architectures designed to accelerate the manipulation of digital media. Typical instructions treat a 128-bit instruction as four 32-bit integers and perform operations on them in parallel. Although some early implementations could accurately be described as SIMD, current instruction sets have added rich horizontal instructions which operate across elements in a single register values, for example, by shuffling bytes within a single register value [7].

Multicore architectures: They are evolving rapidly, nevertheless, we can assume the following characteristics: computational resources are organized into units similar to conventional CPUs, with asynchronous execution in the multiple cores, and weakly coherent access to common memory resources. The programming model for these resources differs greatly: private memory, user-directed caching of global memory, or hardware-directed caching of global memory. The models range from highest potential performance to simplest programming model.

ILP has been so successful that almost no programmer considers it when writing code. Large-vector SIMD is too restricted to apply to a wide enough range of applications to make it viable. Distributed parallelism always requires significant programmer intervention, and a whole industry has grown up around providing tools to write and debug distributed parallel programs. We have previously shown that taking advantage of VIRP requires the development of new patterns of efficient execution, and that an approach marrying a DSL programmer interface to a system based on graph transformation can produce highly efficient, highly readable, and high-assurance code [2].

Our goal is to duplicate the success of ILP and VIRP at the multicore level, avoiding as many of the problems associated with distributed parallelism as possible. Current SPMD models for distributed parallelism raise the cost of developing software above what most applications can support. Run-time requirements for library and operating-system support do not scale well to large numbers of cores, and exceed the capabilities of some lightweight cores.

Why is ILP universally exploited while SPMD is considered an arcane art? Accepting that it makes sense to compare such different levels of parallelism, we believe that the answer lies in the way the two are understood (both formally and informally). SPMD is implemented as a library, while ILP is implemented as a language (machine language). SPMD can be used to express any programming construct and implemented at any scale, and most support libraries try to abstract away the scale and architecture of the underlying hardware. The level of parallelism supported is unbounded. Any conceivable program can be implemented. Unfortunately, most conceivable programs are not correct, and it is difficult in practice to tell which ones they are.

ILP, on the other hand, is defined in terms of a language (machine or assembly language). Both resources and the amount of parallelism are bounded. All resources whose efficient use determines performance are accessible to the programmer through instruction (re)scheduling. For the most common architectures, the programming model is strictly sequential within the CPU. Some architectures allow external observers to see limited nonsequential execution (by reordering external memory accesses for performance reasons).

Obviously, presenting a sequential model to the programmer makes ILP easier to use. Note, however, that programmers (and more commonly compiler writers) aiming for peak efficiency must be aware of nonsequential (out-of-order) program execution. The key point is that correctness can be determined using the sequential programming model or the data-flow-graph-based model, even if significant changes in scheduling are required to efficiently use resources. Furthermore, the limited resources in support of parallel execution make it possible to characterize the set of schedules and design effective heuristics for searching within it.

1.4 Motivation for Multicore Virtual-Machine Model

Our aim is to encapsulate the important features of multicore parallelism in a virtual machine (language with semantics) with the hope that techniques developed for ILP will apply to parallelism at this level. This is our attempt to find a balance, exposing enough details to enable high-performance computation while hiding enough details to make it easy for the programmer to understand program semantics.

The most important feature we need to capture is data locality. The hardware required to present the programmer with a flat memory model across multiple execution threads is very expensive, and introduces performance-reducing nondeterminacy. Processors designed for signal processing have for many years provided user-directed cache modes, including directly addressable caches, in which the programmer can fix part of the cache to memory mapping in order to guarantee fast access to important data structures. Because such optimizations are nonportable and require support from the operating system, they have not been used in multiuser servers or desktop processors.

The Cell SPUs [23] have no cache, but rather have private local storage. Communication with the rest of the processor must be performed explicitly using DMA transfers. This can be seen as integrating on-chip the architecture of distribution that has existed for decades on the circuit boards of workstations, where microprocessor and graphics processor units execute independently, and communicate via DMAs.

Data locality poses a challenge for synchronization for both cached and noncached memory architectures. This is because access to memory which appears to be local may require distant access. In a cached architecture, this occurs when two physical cores access the same physical memory causing the bus hardware to transfer cache lines between the two processors, introducing delays. To avoid this, our language does not contain access to nonlocal memory, even as part of interprocess signaling, which implies that we do not use mutexes.

Our virtual machine must contain multiple threads of execution corresponding to individual cores. Each thread has a small list of local resources: data buffer to store blocks of data for computation and signal registers which can be set by remote processes. These are the registers of our virtual CPU. The former usually contain values visible at the high level, like registers containing the results of arithmetic. Signal registers are only visible to the parallelizer, just as registers for floating point exceptions and PowerPC condition registers are only visible to the compiler or low-level assembly language coder.

The execution of procedures implementing register assignments based on pure,[*] local functions is the arithmetic execution unit of our virtual CPU, and asynchronous communication is the data movement execution unit.[†]

We can synthesize branching control flow inside the pure functions just as we did for VIRP, and we can move it to a higher level in the Coconut code graph. Restricting computation to pure functions, and control flow to thread synchronization, significantly simplifies the process of reasoning about parallelization for the programmer or compiler and the verifier.

By separating control and data flow, this allows a single-threaded control process which may be distant from some or all of the computation engines to originate the complex data-flow patterns necessary in scaling up multicore computations, while separating algorithm-intrinsic control flow from control flow required by parallel computation. In this way, the algorithm designer can solve one problem at a time: writing a correct algorithm and writing an efficient parallelization strategy. Unlike ILP and like SPMD, to tune performance, the programmer must explicitly insert parallelization, although, in Coconut, this is done in program generators, not programs, and can be simplified by using higher-level patterns. Like ILP and unlike SPMD, verification ensures that the program as a whole is a pure function, i.e., independent of parallel execution order.

1.5   Virtual-Machine Language

A sequence of computations is a list of pairs (c, i) of execution core identifiers and instructions. In the present description, the instructions on each core execute in order, although just as in the ILP case, it would be easy to introduce limited out-of-order execution managed by the run-time system or dedicated hardware.

---

[*] Pure functions are functions without side effects, which here, in particular, includes no invocations of communication primitives, and implies referential transparency. The latter means that results depend only on the explicitly provided arguments, and no other circumstances, in particular, not on execution order of subcomputations in cases where this order is not fully determined.

[†] Although a single communication primitive would be sufficient, we separate data flow and signals because machine architectures can (and should) optimize one for throughput and the other for latency. On the Cell BE, signals are implemented as both signals and messages, and data flow as DMA transfers.

Data movement instructions, however, are assumed to run asynchronously from other computations, so the exposed instructions only initiate data movement/signaling or wait for it to complete. DMA engines to handle DMA are almost always implemented as autonomous processors executing transfer lists. We embed the data movement with the purely local computations on a core because this is analogous to the interleaving of load/store instructions and register-based computation on a Reduced Instruction Set Computer (RISC). Unlike load/store instructions in a RISC CPU, we have to separate the request for transfer from the acknowledgment of completion—in RISC processors, issuing the instruction using the target register of a load instruction will stall the execution pipeline until the data have been transferred, while DMA engines do not provide equivalent synchronization mechanisms.

We choose to expose this for several reasons. Signaling may compete with data transfers on a common bus, and this allows the instruction scheduler to take this into account. Data transfers themselves introduce ordering, so signaling may not always be necessary. The separation makes it easier to maximize latency hiding for two-way communication within given resource constraints. Each transmission executes asynchronously with a variable latency, so it is useful to separate them enough to hide the expected latency.

While computation within a core is in order, computation on different cores is a priori unordered, as are all data transfers. For this to work, the computational threads on the send and receive sides must refrain from writing to the source and target areas, respectively, and the receive side must also refrain from reading from the target area, until completion of the transfer has been signaled.

To take advantage of the higher efficiency of large transfers, signaling of DMA completions is handled via additional data—a tag—contained in the transfer. This is the method we use on the Cell BE; it is possible to do this since the sequence of writes to DMA target addresses can be ordered using fences.* Detection of a tag (which must be different than the tag associated with the previous buffer contents) indicates to the receiver that the transfer has completed. We will assume that the sender also has a mechanism for detecting completion,[†] and that the architecture contains a signaling mechanism, with signal status stored locally (in local private memory, or a special hardware register) on each core.[‡]

---

* DMA engines usually support lists of blocks to transfer. Transfer out of order may be more efficient, so order is usually ignored, but different constructions (bit flags on the DMA list instructions) may be available to enforce ordering within a particular list of blocks, and within all lists of blocks queued up for transfer. By putting the tags in a separate atomic unit for transfers (aligned 16 bytes), and putting this line in a separate DMA block, separated by a fence, following the list of blocks needed for the data transfer, the tag will not appear to the receiver until after all data have been transferred.

[†] On the Cell BE processor, DMA engines are paired with computational cores, and detecting DMA completion is low cost and does not involve bus traffic. This is the preferred setup from a hardware point of view, because each private local memory needs an interface to the on-chip network fabric, and this requires that most of the functionality of a DMA engine is implemented in any case.

[‡] On the Cell BE, we use the SPU signal registers.

Cores are referenced by *CoreId*. To each core, we associate the following state:

State	Reference	Function	Values
Signal registers	*SignalId*	Enumeration of a small, predetermined number of Boolean values local to each core, but writable from any core	Can be set or unset
Buffers	*LocalAddress*	Store blocks of data	Readable, or writable by set of *CoreId*s, or exclusively for local access
Data tags	*DataTag*	Tags associated with data buffers	Contain integers

Two examples are helpful at this point. In Figure 1.5, we introduce a graphical representation of the virtual-machine language. Following universal modelling language (UML) convention, communication is asymmetric, so it is shown with half arrowheads. There is both synchronization and data transmission, so thin and thick lines are used to represent the respective low and high bandwidth requirements. Signals are labeled by their *SignalId*, and data are labeled by a symbol representing the data being transferred, subscripted by the data tag.

In a simple transfer, the receiving core sends a preassigned signal to indicate that the receiving buffer is available for writing. It is up to the scheduler to ensure that no local or remote reads or writes to this buffer occur. In the case of local reads and writes, this is easy to verify visually, by checking the inputs and outputs of all pure

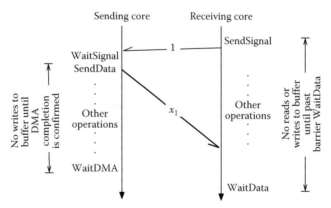

FIGURE 1.5:　Simple data transfer requires one ready signal before data are transferred.

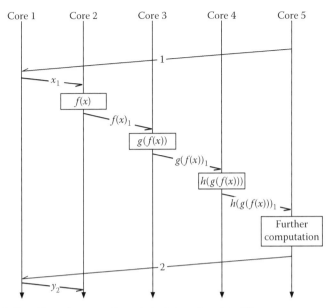

FIGURE 1.6: Complex transfer, minimizing signal traffic.

functions between the SendSignal and WaitData. Ensuring that no DMA activity sources or targets this buffer during this time is more difficult, and requires the establishment of a partial order on the instructions executing on different cores. Similarly, the scheduler must ensure that the source buffer is not written to between the Wait-Signal and WaitDMA instructions, although reading is allowed, and is effective in pipelining instructions.

In the more complex example shown in Figure 1.6, a single signal is sent from core 5 to core 1 to indicate that the destinations for a whole chain of transfers are available. This amortizes the cost of the signaling over multiple transfers, at the expense of potentially tying up buffers for longer periods of time. Using such a single signal may be easier for the scheduler to software pipeline than a sequence of simple transfers, because there are fewer latencies to worry about. For algorithms requiring numerous transfers of small amounts of data, the reduction in signaling bandwidth may also be significant. In this example, we introduce conventional mathematical notation for pure functions, f, g, and h. Since the data transfers in one cycle have distinct target cores, a single data tag can be used for all transfers. To achieve good performance, the signals and data transfers would be software pipelined within each node, so that, for example, in a simple mapping of $h \circ g \circ f$ over an array $\{x_i\}$, with minimal pipelining, the evaluation of $f(x_{i+2})$ would overlap the evaluation of $g(f(x_{i+1}))$ and the calculation of $h(g(f(x_i)))$.

1.5.1 Virtual-Machine Instruction List

We present the syntax of the virtual-machine instructions, together with high-level description of their intended effects. We also summarize how we implement these for the current Cell BE architecture, how they could be implemented on other architectures, and their performance implications.

There are no nested instructions. Each instruction takes a sequence of operands from the finite types *CoreId*, *SignalId*, *DataTag*, *DMATag*, *LocalAddress*, *GlobalAddress*. The sizes of these types are determined by the specific hardware/software architecture. Different architectures have different addressing schemes, but we assume in general that each core uses a *LocalAddress* to refer to the memory it can access with low latency, and a *GlobalAddress* to refer to other memory. On the Cell BE, local memories can be mapped into the originating process's memory space, but on other architectures, global addresses may require different representations depending on its location.

SendSignal *CoreId SignalId*

Send the specified signal to the specified core. If the signal is already set, it remains set. The receiver must unset it (with a WaitSignal) in order to be able to detect the next signal. Any core can send any signal to any other core. It is up to the programmer/generator to assign signals just as registers are assigned so that different uses do not overlap.

One use of this command is as a "clear to send" notice so that a remote SPU is able to initiate a DMA data transfer. This is illustrated in Figure 1.5. When used as a "clear to send" notice, the buffer to which the requested data will be written should have already been marked as free with the FreeBuffer command, unless the programmer/compiler has arranged that the new data tag is different from the current data tag.

WaitSignal *SignalId*

Pause execution on this core until the specified signal has been received.

Efficiency of the schedule depends on this instruction only occurring in the execution list for a core after the signal has already been received by the underlying hardware. In this case, the latency of the transfer is completely hidden.

SendData *LocalAddress DataLength GlobalAddress DataTag DMATag*

Initiate a DMA transfer to send a data block from the memory of this core to the private memory of a second core, using its address in the global address space.

In the current implementation, the *DataTag* is written to a reserved memory word preceding the buffer, and the data transmission is adjusted to include it in the transmission.

The *DMATag* is an identifier used to identify one or more DMA commands, so that the local thread of execution can query the DMA engine about completion. The *DMATag* must be used on the same core with a WaitDMA instruction to verify the end of the transfer before local writes are allowed. On the Cell BE, it is possible to interrupt on DMA completion, which can be used to reduce the latency between initiation of a send and the next use of the buffer.

This introduces interrupt overhead into the run-time system, and significantly complicates the verification of the run-time system.

WaitDMA *DMATag*

Wait for completion of the DMA transfer with the given tag. Only the initiating core can check this tag. Another core, the controller, or off-node I/O must infer that this transfer is complete from other synchronizing instructions.

WaitData *LocalAddress DataTag*

Wait for the data block located at the given local address to have fully received an inbound transfer identified by the given data tag.

In the current Cell BE implementation, this is done by repeatedly reading the data tag from the local private memory until the expected tag is read. This is convenient because it is possible to put fences in DMAs, ensuring that the data tag is transferred after the block data is transferred.

More efficient hardware would support sending a signal after a DMA transfer is complete. To implement this on the current Cell BE, one would have to wait for the local signal for DMA completion and then send a signal, or use DMA list elements separated from the data write by a fence to write to memory-mapped IO registers controlling signaling. The first approach would introduce either a lot of extra latency or, if interrupts are used, extra overhead on the sending core for the interrupt handler, and significantly more-complicated verification of the run-time system for the cores. The second approach may be more efficient than the tag scheme, and we plan to benchmark it.

FreeBuffer *LocalAddress*

Indicate that the buffer at the given local address is free to be written to.

This has to occur before an incoming data-transfer request is initiated when the previous value of the data tag is not known, in order to be able to tell when the data-block transfer has completed. For simplicity of generation or verification, each WaitData may be paired with a FreeBuffer.

In the current Cell BE implementation, this instruction causes a byte to be written to a reserved memory location preceding each buffer. If light weight signals are supported for signaling DMA completion to remote locations, this instruction might be required to reset a hardware register.

LoadMemory *LocalAddress DataLength GlobalAddress DMATag*

Initiate a DMA transfer of the specified data from main node memory to the core's private memory. The source buffer can be written to, and the target used after checking the *DMATag*.

StoreMemory *LocalAddress DataLength GlobalAddress DMATag*

Initiate a DMA transfer of the specified data from the core private memory to node main memory The source buffer can be written to, and the target used after checking the *DMATag*.

LoadComputation *Computation GlobalAddress DMATag*

> Load new computational kernel from location in main memory specified in the *Computation* structure.

> This duplicates the functionality of **LoadMemory** with a different type. Separating them makes verification a lot easier, and takes advantage of the encapsulation of *Computation*.

> In the Cell BE implementation, **LoadComputation** can be converted to **LoadMemory**, because the SPU cores do not distinguish between text and data.

RunComputation *Computation LocalAddress* LocalAddress* Parameter**

> Run the specified computational kernel, using a list of addresses for input and output buffers, and a list of parameters. The signature of the computation includes information on the sizes of the buffers and which input and output buffers may coincide (for in-place computations). This information is needed to evaluate the instruction stream into a data-flow graph.

> The computation is a compiled pure function, adhering to a defined binary interface. The binary interface allows a list of input data buffers, a list of output data buffers, and a list of parameters to be passed to the function. In the Cell BE implementation, the pure functions are generated from source code in an existing DSL.

1.6   Scheduling Algorithm

The advantage of the approach developed in this chapter is that users can develop their own schedulers for multicore parallelism, taking advantage of the most efficient low-level asymmetric messaging, without having to cope with distributed debugging. In this section we outline the algorithm for the scheduler we are using to test this framework.

This scheduling problem is NP-complete, and finding good approximation algorithms is a current research problem, especially for the (harder) case of multiprocessor scheduling [31]. Devising a good heuristic algorithm depends on understanding the properties of a good solution. Just as in the single-pipelined-processor ILP problem, the most important feature of a good schedule is the separation of loads and instructions consuming the results. We use a greedy algorithm, which seeks to move the signal/data transfers as far from the computations as possible.

We divide the process into three stages, guided by the code-graph transformation sequence shown in Figure 1.7. These graphs present a multicore implementation of matrix multiplication, specialized, for the sake of readability, for multiplying a matrix A tiled into 1×1 blocks with a matrix B tiled into 1×2 blocks, splitting the two resulting block multiplications, labeled "$*$", over two cores 0 and 1. Allocation to a core i is denoted by prefixing edge and node labels with "i:".

These edges labeled with "$*$" should be considered as nested edges, labeled with a control-flow graph representing the loop structure of a software-pipelined submatrix

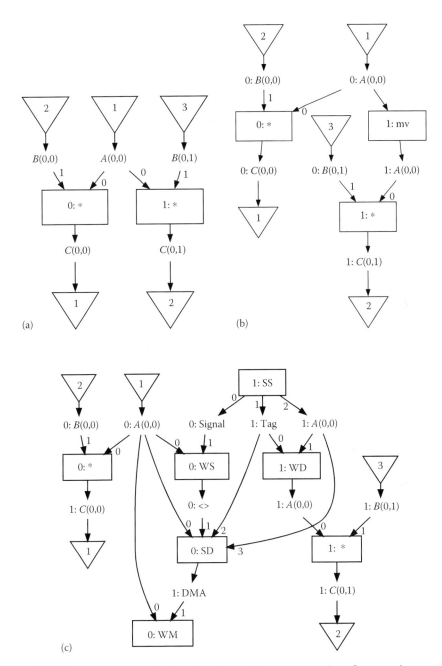

FIGURE 1.7: (a) Pure, (b) distributed, and (c) concurrent data-flow graphs.

multiplication program; the loop body in that control-flow graph is labeled with a pure data-flow graph assembled from several stages of an unrolled submatrix multiplication loop body [3].

The three stages, described in more detail in Sections 1.6.1 through 1.6.3, essentially produce a schedule for the final, concurrent data-flow graph by approximating it first with a schedule for the data-flow graph and then with one for the intermediate, distributed data-flow graph.

Payload scheduling (see Section 1.6.1) enumerate the data blocks with sources, in the order in which they are used in computations (as inputs or outputs of pure functions, or as sources for transfers to main node memory), and enumerate the computations in the order they will be executed (this can be done once for each core or in a unified time-line.)

We consider this as scheduling a pure data-flow graph, which is shown for a trivial 2-core example in Figure 1.7 (left); this drawing shows that the two multiplications are already assigned to different cores.

Payload buffer allocation and communication generation (see Section 1.6.2) generate ordered lists of instructions for each core from the initial lists, converting pure functions into RunComputations, and inserting data transfers and signaling as required to get data blocks into local buffers before pure computations.

We consider this as refining the previous schedule for the corresponding distributed data-flow graph mainly by also allocating the nodes to cores, and simultaneously performing resource allocation (corresponding to register allocation in simple data-flow graphs). Nodes adjacent to computations that are all on the same core are directly allocated to the same core as those computations; the other nodes are split into node sequences connected by explicit "mv" edges representing data transfer between cores (we consider "mv" to be executed by the transfer target, but do not consider it as a computation). This implies that an input node allocated to core i has its data directly transferred from memory to core i, and analogously for output nodes.

The result of this is shown (for the same simple example) as the middle code graph in Figure 1.7.

Communication scheduling (see Section 1.6.3) reorder the instruction lists to better hide latency of signals.

Since the "mv" edges of the previous graphs are implemented using sequences of instructions with widely varying latencies and limited dependencies, these instructions must be scheduled separately, which amounts to producing a schedule for the concurrent data-flow graph shown (again for the same example) as the right-most drawing of Figure 1.7. For that graph, note that the WaitDMA instruction "WM" does not produce a (data) output, but still needs to be scheduled since it is strict like all other instructions here. It also does not have a control output since nothing directly depends on its completion; its only effect is to keep the buffer allocation for the buffer containing $A(0, 0)$ on core 0 alive until after its completion: the outgoing control edge added to WM by scheduling it will, together with a correct allocation, ensure that there are no writes to that buffer before completion of that WM.

We now expand on each of these three steps:

1.6.1 Payload Scheduling

This first stage is separated out because the key to efficiently utilising resources in high-efficiency multicore architectures is the maximization of data reuse within the cores, and hiding of latency.

Access bandwidth to main node memory (and to distant memory on other nodes) is bounded by hardware interfaces which cannot grow at the rate computational performance is growing. The first stage orders computations so that input and intermediate data can be maximally reused. In the future, on multicore processors with more processors and topologically nontrivial on-chip networks, this is the point at which operations can be reordered to distribute intercore communication sufficiently over fabric bottlenecks. On the Cell BE, main node memory traffic is the bottleneck for all of our target computations. We will refer to the list of buffer users as consumers.

1.6.2 Payload Buffer Allocation + Communication Generation

The second stage maximizes latency hiding by filling buffers with data blocks required for future computations as soon as buffers become available. This is done by iterating the following six stages:

1. Look for unoccupied buffers on each core and where found assign consumers, generating instructions to send and receive signals and initiate DMA transfers, except for output blocks, which require a buffer reservation, but no action.

2. Age all pending transfers by the estimated time to initiate transfers.

3. Look at the next instructions on each core, and if all of the required blocks have been assigned to local buffers, and inputs are available, generate the RunComputation or StoreData instruction.

4. Adjust use counts on data blocks following computation.

5. Age all pending transfers by the estimated computation time for executed computations.

6. Find the oldest pending data transfer; add a WaitData instruction to the instruction stream for its core, and mark the data block in the target buffer as available.

1.6.3 Communication Scheduling

The resulting schedule makes maximal use of the available buffers to hide the latency of data transfers. It does not hide the latency of signals. If signals are implemented as high-priority, low-bandwidth side-band communication, it would not be necessary to worry about this. If signals share a single data bus, signal latency may be as long as data latency. On the Cell BE, the situation is somewhere in between. Signals are transmitted on the same bus as data, but the bus is implemented as multiple token rings running in different directions. So shorter signals running in the opposite direction to data will tend to have significantly shorter latency. Nevertheless, it is important to be able to hide this latency as well. We do this in a final stage, using what is akin to peephole optimization, by locating the matching WaitSignal and SendData pair and moving the WaitSignal forward across a number of instructions representing the smallest amount of computation above the minimum of a fixed expected latency, and a fraction of the total computation time between the WaitSignal and SendData in the input instruction stream.

1.7   Concurrency Verification

Recall that verification serves two customers: the developers and users (possibly represented by certification authorities and government regulators). The user wants to know whether safety-critical software is correct. The developer also needs to understand the source of defects, so that they can be fixed, or used to develop a new design. Although errors in concurrent parallelism are inherently nonlocal, the algorithm in this section identifies the type of error and one of the instructions involved in creating an unsound program.

We envisage two verification steps, which make sense in the context of multicore parallelism: Reduction of the parallel program to a pure data-flow graph, and matching the data-flow graph to a specification. For the types of problems we are immediately interested in, like structured linear algebra, it is easy to generate single- and multilevel pure data-flow graphs and convert them to an appropriate normal form. This makes the second step easy, and the first step the more interesting.

The first step requires the verification that the introduction of multicore parallelism does not modify the semantics of a program. Even where normal forms for pure data-flow graphs are not available, this means that testing can be as efficient as it is for single-threaded programs. This is similar to the graph-transformation strategy we are applying successfully to the development of code transformations for VIRP [4].

In practice, incorrect programs are as interesting to the programmer whose job it is to fix them as are correct programs. On failure, our verifier will identify a program location and type of error. This is much easier to do with an ordered input program than with a set of communicating processes.

This brings us back to the superscalar, out-of-order model. Although compilers know that instructions are executed out of order, they have to be presented in a linear order, and the semantics must match the sequential program semantics. This is the case for verified programs presented in our DSL. The difference from the superscalar CPU case is that some programs cannot be verified, and therefore do not have the same semantics as the equivalent sequential programs.

We will use the following terminology:

- A program is a list of pairs (*CoreId, Instruction*).

  We assume that the instructions originate in this order and are transmitted to cores in this order. Since each core buffers its instructions and executes them without additional synchronization, the original program order is not necessarily the same as the resulting execution order.

  However, due to limited buffer sizes there is a limit to the distance within the list between any two instructions whose execution overlaps in time. Therefore, considering a single list is more restrictive than a separate list of instructions for each core.

  The situation here corresponds to the dispatch queues and instruction-fetch buffers on superscalar CPUs, which result in a maximum number of instructions in-flight.*

  We do not use the maximum in-flight property, but do use the presentation order, and we will add restrictions to the presentation order below.

- A program is order independent if, given the same input (in main memory input locations), all possible execution orders terminate and produce the same output (in main memory output locations), although intermediate values computed and temporarily stored in private or global memory may differ.

- A program is locally sequential if every $(c_1, \mathsf{SendSignal}\ c_2\ i)$ is followed by a corresponding $(c_2, \mathsf{WaitSignal}\ i)$, and every $(c_1, \mathsf{SendData}\ b\ c_2\ t\ m)$ is followed by corresponding $(c_2, \mathsf{WaitData}\ t)$ and $(c_1, \mathsf{WaitDMA}\ m)$ instructions.

  Note that it is easy to construct order-independent programs without this property, simply by ordering the instructions by core.

- A program is safely sequential if it is locally sequential and order independent.

The programmer wants to know whether the programs are order independent. Checking this is very expensive. The programmer understands sequential programs, and with some difficulty can understand parallel programs close to sequential programs. Weakly sequential programs are our attempt to formalize this concept, and we may restrict it to improve programmer understanding in the future as we gain experience writing and generating a wider range of programs. The key to usability, however, is the fact that we can efficiently identify the safely sequential programs within the class of locally sequential programs.

To summarize these definitions using relational language, instructions on a single core are totally ordered; instructions on different cores have a partial order derived

---

* The instructions in-flight at a given time are all of the instructions in the process of being decoded by the CPU, dispatched to execution units, executing (pipelined or nonpipelined), or being completed (having their results recorded in register files, store queues).

from the use of synchronizing primitives; the partial order is the intersection of all possible execution orders; the presentation order (order in the list) is a total order, and we will give conditions that imply that the partial order is a subrelation thereof, and in particular, that the presentation order is a valid execution order.

One reason we can identify the safe subset efficiently is that we can identify what the intended result should be by using the pairing of send and receive instructions, and from there we can verify that these are the only results by keeping track of a bounded set of state information (with the size of the information being $\mathcal{O}(\text{cores}^2)$).

1.7.1 Motivating Example

Before explaining the efficient method of verification, it is important to understand that locally sequential programs are not necessarily sound.

For simplicity, the following locally sequential program uses signals, but the same issues arise with data transmission:

Index	Core 1	Core 2	Core 3
1		Long computation	
2	SendSignal $s \to c_2$		
3		WaitSignal s	
4		computation	
5			SendSignal $s \to c_2$
6		WaitSignal s	

Remember that each core executes independently of the other cores, except where explicit wait instructions block execution until some kind of communication (signal, change in data tag, DMA) is confirmed to have completed. Therefore, in this case the most likely instruction completion order has core 3 executing the SendSignal as soon as it is queued, allowing the signal to be sent before core 2 has received the core 1's signal and cleared the signal hardware:

Index	Core 1	Core 2	Core 3
2	SendSignal $s \to c_2$		
5			SendSignal $s \to c_2$
	Second signal overlaps the first, only one registered		
1		Long computation	
3		WaitSignal s	
4		computation	
	No signal is sent, so the next WaitSignal *blocks*		
6		WaitSignal s	

To be precise, completion of the SendSignal means that the signal has been initiated by the sender, and reception may be delayed, so the signal from core 3 could even arrive before the signal from core 1. In either case, neither signal will arrive after the first WaitSignal, so the second WaitSignal will wait forever, and this program execution will not terminate.

The problem is caused because there are no signals or data transmissions enforcing completion of instruction 5 to follow completion of instruction 3.

This example, when considered as part of a longer program, also demonstrates a possible safety violation with the valid completion order:

Index	Core 1	Core 2	Core 3
1		Long computation	
5			SendSignal $s \rightarrow c_2$
3		WaitSignal s	
4		Computation **using** **wrong assumptions**	
2	SendSignal $s \rightarrow c_2$		
6		WaitSignal s	

In this case, the computation 4 might rely on assumptions that are only available once the SendSignal 2 completes—for example, it might initiate a DMA to core 1, which could arrive before some earlier instructions on core 1 are still writing to the DMA's target area, thus invalidating the DMA transfer.

1.7.2 Strictly Forward Inspection of Partial Order

We have seen that locally sequential programs are not always sound. The advantage that locally sequential programs have over general programs is that order-inducing instructions must occur before instructions whose well-definedness depends on them. Using simple inference to maintain a list of known ordering relations between instructions on different cores, we can inspect the program in order (even though it can execute out of order).

Referring to Figure 1.8, we see that after instruction 3, we know that instructions on core 2 will complete after instruction 2 on core 1 (and any previous instructions). We know nothing about the relative execution of instructions other than this until instruction 6. At this point we know that instructions after and including 6 on core 1 must execute after any instruction before instruction 5 on core 2. After instruction 8, we have the relation that this and further instructions on core 3 execute after instructions up to instruction 7 on core 1, but since instruction 7 on core 1 executes after instruction 5 on core 2, we also know that instruction 8 on core 3 executes after instruction 5 on core 2.

Contrast this with the example in Section 1.7.1, where we never have any relations between instructions on core 3 and other cores. Keeping track of these relations is the key to efficient verification of correctness.

There is one other class of (bad) example to keep in mind: the situation where, in presentation order, two cores send the same signal to a third core without a wait

Good Program

Index	Core 1	Core 2	Core 3
1		Long computation	
2	SendSignal $s \to c_2$		
3		WaitSignal s	
4		computation	
5		SendSignal $s \to c_1$	
6	WaitSignal s		
7	SendSignal $s \to c_3$		
8			WaitSignal s
9			SendSignal $s \to c_2$
10		WaitSignal s	

FIGURE 1.8:   Weakly sequential program.

on the third core to separate them. This case is visually apparent, because the sends overlap in program order, but we have to safeguard against it nonetheless.

Clearly, the same considerations apply to data transfers. With data transfers, we must additionally check the values of the data tags, because if the existing tag matches the tag being transferred in, the WaitData instruction will never block execution, even if the data are not available.

1.7.3 State

Recall that a locally sequential program is an sequence of instructions (communication primitives) with a totally ordered presentation. The order does not determine the execution order, but it does define the analysis order for the verification algorithm. Without this order, there would be no linear-time verification algorithm. Given the order, it makes sense to talk about the state maintained by the verification algorithm. This state encodes only the information about the possible execution states required to verify the soundness of the program in this way.

For brevity, we will now write I instead of *Instruction* for the set of instructions, C instead of *CoreId*, S instead of *SignalId*, and L instead of *LocalAddress*. Let D be the set of DMA Tags. Let T be the set of data tags.

In the following, we assume a fixed program P. Depending on this program P, let $N = \{0, \ldots, \text{length}(P) - 1\} \subseteq \mathbb{N}$ be the set of indices into P considered as a list, so that we can now consider the program as a total function $P : N \to C \times I$ instead. N is considered to be ordered with the standard ordering of the natural numbers.

For each core $c : C$, define $N_c = \{n : N | P(n)_1 = c\}$ to be the corresponding restriction of the index set to instructions for c.

We are now interested in the partial strict-ordering $\prec$ (depending of course on P) on N where $n_1 \prec n_2$ means the instruction (of P) at n_1 necessarily completes before the instruction at n_2. The reflexive closure of this is frequently known as "happens-before relation" [16], or Mazurkiewicz's trace [20,32].

For a fixed core, $c : C$, the instructions in the program assigned to c are completed in order, so we demand that, for each core $c : C$, the restriction of $<$ to N_c is included in $\prec$:

$$\forall c : C . \forall n_1, n_2 : N_c . n_1 < n_2 \Rightarrow n_1 \prec n_2$$

There is no a priori order between instructions assigned to distinct cores, and this may lead to unsound programs like the example in Section 1.7.1. Ordering is determined by transition dependencies, which in our case means blocking communication.

It would be sufficient to calculate $\prec$ completely, but only a small portion of this partial order is required at a time. Since instructions on a single core are totally ordered, it is easy to calculate the entire relation $\prec$ if, for each instruction index n and distant core, the latest instruction known to complete before n, and the earliest instruction known to complete after n are stored.

We will see that it is sufficient to store only the latest instruction known to complete before the current instruction, and convenient to repeat information for all cores, in the form of a "Φollows" map indexed by $n \in N$:

$$\Phi_n : C \times C \to N$$
$$\Phi_n(c_1, c_2) = \max_{\leq}\{m : N_{c_1} \mid \forall n' : N_{c_2} . n \leq n' \Rightarrow m \prec n'\}.$$

Note, in particular, that if $n \notin N_{c_2}$ this definition does not say that is the latest instruction on c_1 such that all future instructions on c_2 follow m. This property could be used instead, but is more expensive to compute since it requires information about future instructions. Figure 1.9 shows that the instruction at index $\Phi_n(c_1, c_2)$ is always a send instruction on core c_1.

For the verification method, we need to define a function

$$\widetilde{\bigvee} : (C \times N \times F) \times (C \times F) \to F,$$

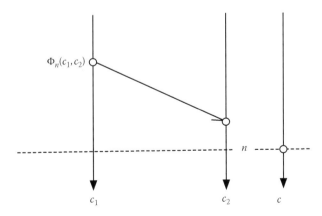

FIGURE 1.9: For convenience, $\Phi_n(c_1, c_2)$ is defined for all n, even if $n \notin N_{c_2}$, to be the index of the last instruction on c_1 known to complete before an instruction on c_2 up to and including index n.

where $F = C \times C \to N$ models a slice of Φ. $\widetilde{\bigvee}$ combines the slices of the partial order at a corresponding pair of send/wait instructions that induce an order on the completion of the wait and instructions on other cores. This potentially strengthens what we know about $\Phi_{n_2}(c, c_{\text{Wait}})$ for all c, i.e., about the last instructions known to complete before the wait and future instructions on the destination core.

$$\widetilde{\bigvee}((c_1, n_1, \Phi_{n_1}), (c_2, \Phi_{n_2}))(c', c'')$$

$$= \begin{cases} \max\{\Phi_{n_2}(c', c''), n_1\} & c' = c_1, c'' = c_2 \\ \max\{\Phi_{n_2}(c', c''), \Phi_{n_1}(c', c_1)\} & c'' = c_2 \\ \Phi_{n_2}(c', c'') & \text{otherwise.} \end{cases}$$

Our method of verification is a construction of the following functions inductively in the instruction index, n.

$$\sigma_n : C \times S \to \begin{pmatrix} C \times N \times F \\ \oplus \\ N \end{pmatrix}$$

$$\Delta_n : C \times L \to \begin{pmatrix} T \times C \times N \times F \\ \oplus \\ T \times N \\ \oplus \\ T \times 2^D \\ \oplus \\ T \times D \end{pmatrix}$$

$$\Phi_n : C \times C \to N$$

where
  $\oplus$ means disjoint union
  2^D is the power set of D

To calculate X_{n+1}, where $X_n = (\sigma_n, \Delta_n, \Phi_n)$, it is only necessary to have access to X_n, because we make copies of slices of Φ which are needed in future steps.

The disjoint unions can be understood in the following way: σ, which records the last instruction to send the given signal to the given core, has two cases for signal status: the signal has already been received by a WaitSignal or the receiver is still pending. Case I: a signal is expected from a core $c : C$, at instruction number $n : N$, with $\Phi' : F$ being the known partial order at n. Case II: the signal has been received at instruction $n : N$. To simplify the cases, we assume that all signals have last been received at instruction -1 (before the program starts).

For Δ, these are the following cases:

1. An incoming transfer with tag in T, from core in C started at instruction in N with known partial order in F is expected.

2. The data tag is in T, and the data were last used at an instruction in N.

3. The data buffer is the source of one or more DMA transfers whose list of (completion) tags which have not yet been consumed by a WaitDMA are in 2^D, and the data tag is in T.

4. An incoming DMA transfer with tag in D was started, and the data tag is in T. Such a transfer can only have come from main memory, and been initiated by a LoadMemory.

Initially, σ_0 maps every argument to -1 (indicating that the last signal was received before the beginning of the program), and Δ_0 maps to the pair ($\langle$unused data tag$\rangle$, -1). Φ_0 maps every argument to -1 as well, indicating that nothing is known about ordering.

The maps σ_{n+1}, Δ_{n+1}, and Φ_{n+1} are constructed from the maps at n and the instruction at n executing on core c using the universal rule $\Phi_{n+1}(c, c) = (n)$, and specific rules for the following cases:

SendSignal $c'\, s'$

We need to verify that there is no current use of this signal in any program order. Dividing into two cases according to the form of $\sigma_n(c', s')$, the first case excludes programs in which the signal is in use in the presentation order. This is not sufficient because the presentation order is not the only possible execution order, so we need to use the follows relation to check this.

If $\sigma_n(c', s') = (c'', n', \Phi')$, this is an error, because this signal s' may be sent by both cores (from c at the present instruction n and from c'' at n') at the same time.

If $\Phi_n(c', c) \geq \sigma_n(c', s')$, then we know that the last instruction to consume the signal s' on c' at $\sigma_n(c', s')$ completes before $\Phi_n(c', c)$ on c' which (by definition) occurs before the current instruction on c. (This includes the case that this signal has never been used, in which case $\sigma_n(c', s') = -1$.) Let $\Phi_{n+1} := \Phi_n$, $\Delta_{n+1} := \Delta_n$, and $\sigma_{n+1}(c', s') := (c, n, \Phi_n)$, $\sigma_{n+1}(c'', s'') := \sigma_n(c'', s'')$ for other values of (c'', s'').

If the inequality is not satisfied, the program is not sound, because we do not have $n' \prec n$, meaning that this signal could arrive before the previous WaitSignal s' on c', which is the case of the bad example program.

WaitSignal s'

If $\sigma_n(c, s') = (c'', n', \Phi')$, then we know that the signal s' will come from core c'', and we can use the fact that the wait completes after the send, to update the follows map. Let $\Phi_{n+1} := \tilde{\vee}((c'', n', \Phi'), (c, \Phi_n))$, $\Delta_{n+1} := \Delta_n$, $\sigma_{n+1}(c, s') := (n)$ otherwise $\sigma_{n+1}(c''', s''') := \sigma_n(c''', s''')$.

If $\sigma_n(c, s') = (n')$, this is an error, or the program is not locally sequential, as required for this analysis.

SendData $l' \lambda c'' l'' t' d$

> At the source: We need to ensure that the source data are not about to be rewritten and that the use of t' and d does not conflict with another use. If d appears in $\Delta_n(c, \tilde{l})$ for some $\tilde{l}$, then the DMA tag is already in use, and the completion of this DMA would be indistinguishable from the completion of another DMA already originated from this core, so this is an error.

> If $\Delta_n(c, l') = (t'', n'')$, then no DMAs (SPU to SPU transfers or transfers from main memory) are pending for this buffer, so record the send by letting $\Delta_{n+1}(c, l') := (t'', \{d\})$ (with Δ_{n+1} equal to Δ_n at all other values).

> If $\Delta_n(c, l') = (\tilde{t}, d')$, then on the sending core c an incoming DMA is in progress from main memory with data tag $\tilde{t}$ and DMA tag d'. Therefore it is not safe to start an outgoing DMA as this data could be overwritten while it is being sent, so this is an error.

> If $\Delta_n(c, l') = (\tilde{t}, D')$ for some subset $D' \subset D$, then there are already outgoing DMAs in progress, which is valid if we do not try to reuse DMA tags; or if $\tilde{t} \neq t''$, then two different tags are being used for outgoing DMAs, which is not supported (because it would result in an unknown data tag being sent for the previously initiated transfer). Otherwise, record its use by $\Delta_{n+1}(c, l') := (t, D' \cup \{d\})$, and copy the other maps from n to $n + 1$.

> Otherwise, this buffer is the target of a DMA from another core, which is an error.

> At the destination: Between cores, execution order does not have to follow presentation order. So we need to check that all use of the previous contents of the target buffer must have completed before the current instruction.

> If $\Delta_n(c'', l'') = (t'', n'')$, $t'' \neq t'$, and $n'' \leq \Phi_n(c'', c)$, then the destination buffer has no pending DMAs after instruction n'', the new tag does not match the old tag (so it will be detectable by the WaitData), and the current instruction is known to complete ($\Phi_n(c'', c)$) after the last instruction (n'') to have used the target buffer, so it is safe, and we set $\Delta_n(c'', l'') := (t', c, n, \Phi_n)$. Otherwise, report the error that multiple incompatible DMAs with source and target (c'', l'') were detected, or the data tag was illegally reused.

WaitData $l' t'$

> If $\Delta_n(c, l') = (t', c'', n', \Phi')$, then we are waiting for this data, and have stored the follows map at the send instruction, which we can combine with the map at the current instruction: $\Phi_{n+1} := \widetilde{\vee}((c'', n', \Phi'), (c, \Phi_n))$. Let $\sigma_{n+1} := \sigma_n$, and $\Delta_{n+1} := \Delta_n$ except $\Delta_{n+1}(c, l') := (t', n)$, which indicates that this buffer is safe to use after the current instruction.

> Otherwise, no incoming DMA from another core preceded this instruction, so the program is not locally sequential, or this DMA overlaps DMAs initiated locally, which is unsafe.

LoadMemory l' λ g' d

Check that d does not appear in the image of $\Delta_n(\{c\} \times L)$, otherwise the uses of this tag overlap, which is an error.

If $\Delta_n(c, l') = (t, n')$, then the buffer has no pending IO, and we can initiate an incoming DMA by setting $\Delta_{n+1} := \Delta_n$ except $\Delta_{n+1}(c, l') := (t, d)$, $\sigma_{n+1} := \sigma_n$ and $\Phi_{n+1} := \Phi_n$.

Otherwise, the target buffer is the source or target for transfer, which may not have completed, and this will produce an indeterminate result.

StoreMemory l' λ g' d

Check that d does not appear in the image of $\Delta_n(\{c\} \times L)$, otherwise the uses of this tag overlap, which is an error.

If $\Delta_n(c, l') = (t, n')$, then no other IO is pending, and the store is safe. Set $\Delta_{n+1} := \Delta_n$ except $\Delta_{n+1}(c, l') := (t, \{d\})$ $\sigma_{n+1} := \sigma_n$ and $\Phi_{n+1} := \Phi_n$.

If $\Delta_n(c, l') = (t, D')$, other outgoing DMAs are pending, which is allowed. So set $\Delta_{n+1} := \Delta_n$ except $\Delta_{n+1}(c, l') := (t, D' \cup \{d\})$ (to record the new DMA tag we are waiting for), $\sigma_{n+1} := \sigma_n$ and $\Phi_{n+1} := \Phi_n$.

Otherwise, the source buffer is the target for a pending transfer, and this will produce an indeterminate result.

WaitDMA d

The DMA tag could be associated with incoming or outgoing IO. In either case $\exists l$ such that $\Delta_n(c, l) = (t, d)$ (incoming), or $\Delta_n(c, l) = (t, D')$ and $d \in D'$ (outgoing), and l is unique. Otherwise, there is an error, because no DMAs are pending using this tag. This either indicates an unsound program or a nonlocally sequential presentation.

If $\Delta_n(c, l') = (t', \{d\})$ then $\Delta_{n+1}(c, l') := (t', n)$.

If $\Delta_n(c, l') = (t, D')$ then $\Delta_{n+1}(c, l') := (t, D' \setminus \{d\})$

If $\Delta_n(c, l') = (t', d)$ then $\Delta_{n+1}(c, l') := (t', n)$.

For other values $\Delta_{n+1} := \Delta_n$. Also, $\sigma_{n+1} := \sigma_n$ and $\Phi_{n+1} := \Phi_n$.

RunComputation x $(l', l'', ...)$ $(\tilde{l}', \tilde{l}'', ...)$ $(p', p'', ...)$

Inputs: If for every $l \in \{l', l'', ...\}$, $\Delta_n(c, l) = (t, n')$ or (t, D') then $\Delta_{n+1}(c, l) := (t, n)$ or (t, D') (respectively), otherwise, the buffer has pending incoming IO, which makes this computation unsound.

Outputs: If for every $l \in \{l', l'', ...\}$, $\Delta_n(c, l) = (t, n')$ then $\Delta_{n+1}(c, l) := (t, n)$, otherwise, there is pending IO and this modification makes the program unsound.

Other instructions have no effect, so $\Phi_{n+1} := \Phi_n$, $\Delta_{n+1} := \Delta_n$, and $\sigma_{n+1} := \sigma_n$.

THEOREM 1.1

A locally sequential program is order independent iff the inductive verification described in this section terminates without error.

For the proof of this theorem, we need the following lemmas:

LEMMA 1.1
  Each $d \in D$ occurs at most once in $\Delta_n(\{c\} \times L)$.

PROOF   This property is preserved because the verifier starts with no elements of D in the images of any of the maps, and all rules check for previous use of a tag before constructing the $n + 1$ maps from the n maps and signal an error if multiple occurrences of some d would result.   ⃞

LEMMA 1.2
  In a locally sequential program which passes the verification procedure, the only signal a $(n, c, \textsf{WaitSignal } s)$ can trap is the one in $\sigma_n(c, s)$.

PROOF   Assume that there is another signal which can be trapped. Since the program is locally sequential, the "matched" SendSignal c s must precede the WaitSignal s in presentation order. There are two cases to check: the other send precedes the wait, or follows the wait.

  1. If it precedes the wait, we have two different SendSignal c s instructions sending the same signal to the same core, $c \in C$, preceding the WaitSignal, with unrelated instructions potentially interleaved. However, this situation would not pass the verification, because at the second send, $(n, \textsf{SendSignal } c\ s)$, $\sigma_n(c, s) = (c', n', \Phi')$ where n', c' are the index and location of the first send, which would cause the verifier to return an error.

  2. If the second SendSignal instruction follows the wait in presentation order, but both sends can execute before the wait, then at the second send, the verifier will find $\Phi_n(c', c) < n'$, where (n, c) are the coordinates of the second send, and (n', c') of the intermediate wait. This would have caused the verifier to signal an error, precisely because it could not guarantee the safety of the second send.   ⃞

LEMMA 1.3
  In a locally sequential program which passes the verification procedure, the only data a $(n, c, \textsf{WaitData } l\ t)$ can observe arriving is the one in $\Delta_n(c, l)$.

PROOF   Analogous to the proof of Lemma 1.2.   ⃞

LEMMA 1.4
  The presentation order of a locally sequential program is a valid execution order up to n if verification produces no error up to instruction n.

PROOF Assume the contrary, and let m be the first instruction which is not executable after instruction $m - 1$ and all previous instructions have completed. The only instructions which may not be executable are the wait instructions (for a signal, for a data tag, or for a DMA completion tag). Since m is the first nonexecutable instruction, all previous instructions have completed, which includes the corresponding send instruction (since the program's presentation is locally sequential). So m is also executable, which is the contradiction. ☐

LEMMA 1.5

For all $c', c \in C$, $n \in N_c$, $n' \in N_{c'}$, if $n > \Phi_{n'}(c, c')$, then there exists an execution order ϵ in which $n' <_\epsilon n$.

PROOF Construct an execution order ϵ as follows. Let $U_{c''} = \{m'' \in N_{c''} : m'' \leq \Phi_{n'}(c'', c')\}$, and $U = \bigcup_{c'' \in C} U_{c''}$. A modification of the proof of Lemma 1.4 shows that the presentation order is a valid execution order for the subset U. The only new point is that the instructions not in U are not required to make the instructions of U executable, because the follows maps capture the supremum of the necessary instructions for the subset U. Since n is not in the execution order up to this point, $n > n'$ in this order, as required. ☐

1.7.4 Proof of Theorem

To prove that a locally sequential program which is order independent will pass verification we consider the converse situation, a locally sequential in which an error is flagged. For each case of such an error, one must construct a possible execution order which results in a deadlock. We do this for the hardest case, the errors raised during the processing of a SendSignal. The other errors can be treated in the same way, or they obviously indicate a failure to be locally sequential (waits preceding sends).

$(n, c, \text{SendSignal } c' s)$

1. Error caused by $\sigma_n(c', s) = (c'', n', \Phi')$ where we have a signal s which could be sent by both c and c'' to c'. Presentation order is a valid execution order up to n if it produces no error up to instruction n by Lemma 1.4. Therefore both signals could arrive at c' before the next WaitSignal, in which case both would be consumed. This would leave a second WaitSignal without a paired SendSignal according to the 1-1 pairing of SendSignal and WaitSignal required in a locally sequential program, and a deadlock would result. (Note that even if the program is malformed in other ways, as long as the sends and waits are paired, each send can make at most one wait executable, so the pairing is the only required property to insure a deadlock, unless some other problem manifests itself first.)

2. Error caused by $\sigma_n(c', s') > \Phi_n(c', c)$. This tells us that the last instruction to consume signal s' on c' at $\sigma_n(c', s')$ does not necessarily complete before $\Phi_n(c', c)$ on c'. By Lemma 1.5, there is an execution order in which the signal sent at n completes before the wait at $\sigma_n(c', s')$, in which case the wait would capture two signals, and there would be an unpaired wait later in the program to cause a deadlock, as above.

To prove the converse, let P be a locally sequential program passing verification. We must first show that the program has no deadlocks, and always produces the same result.

Assume that there is an execution order with a blocking instruction. Let n be the first instruction in presentation order which blocks. In analogy with the proof of Lemma 1.4, this instruction must be a wait instruction, which by Lemma 1.2 and Lemma 1.3 is paired with a unique send instruction which precedes it in the presentation order. Since that instruction did not deadlock (because n is the first such instruction), it is executable, making n executable, implying that a verified program can have no deadlocks.

To show that the (nondeadlocking) result is order independent, we can symbolically execute it, checking at each step that the results do not depend on execution order. To do this, modify the image of Δ_n to include an expression type, E, with different meanings for each case in the disjoint union:

$T \times C \times N \times F \times E$—an incoming transfer is expected, and when it arrives, E will be the contents of the buffer;

$T \times N \times E$—the buffer contains the expression E, which was last used at instruction $n \in N$;

$T \times 2^D \times E$—the buffer contains the expression E, and is the source for outgoing transfers;

$T \times D \times E$—the buffer is the target of an incoming transfer from memory, and when it completes the buffer will contain E.

New expressions are formed by a RunComputation at instruction n from the expressions in the input buffers in Δ_n and the new expressions are stored in Δ_{n+1}. The verification mechanism ensures that the source and destination buffers in Δ are in the definite state $T \times N \times E$. Sourcing buffers in other (indefinite) states would have been flagged as an error. Input expressions are formed by memory loads, and an additional map from memory store locations to expression values represents the output of the program. Memory outside the on-core buffers is considered to be static input, or write-once output data. This concludes the proof of the theorem.   ☐

For the second step in verification, one must compare the expressions generated in this way with specifications for the program. For computations of limited size, this can be done by using a nodes in a directed acyclic graph (DAG) with sharing as the type E. The result is a term forest, which can easily be compared to a specification in the same form. This procedure would be considerably more complicated for programs whose corresponding DAGs would not fit in program memory, but it is the nature of this approach that computation, and specifications are nested, so that the size of DAGs at each level are manageable. So far, this has always been the case.

1.8 Conclusion

We have shown that the control flow required to efficiently utilize multiple lightweight cores can be abstracted out of the program into a simple process algebra, which we use to label the middle layer of Coconut nested code graphs.

We have drawn an analogy between this language and a RISC instruction set used in a superscalar processor. This analogy leads naturally to the idea that the multicore level of the graph be presented in order, for out-of-order execution. With this presentation, it was possible to verify the correctness of this level of parallelism in an efficient manner.

Above all, Coconut is designed to make compiler optimizations transparent to the users who are responsible for performance. Out-of-order execution in superscalar cores is well understood and widely exploited by both the optimizing compiler community, and high-performance code tuners. The main difference between out-of-order execution within a CPU and our multicore model is that we do not see a hardware solution to the problem of maintaining in-order semantics. This turns the verification algorithm of this chapter from a requirement for exceptional safety-critical applications into an everyday tool required by anyone optimizing multicore applications under this model. Users working on parallelization strategies within Coconut will come to view verification as a useful build/debug tool. As evidenced by the extensive involvement of undergraduates in developing Coconut programs, even users who do not understand the theoretical underpinnings can derive significant benefit from tools which integrate verification mechanisms. At the SIMD level, verification was provided by symbolic evaluation and simulation. At the multicore level, users will be indirectly using process algebras to check the soundness of multicore schedules.

Although we have developed this approach to multicore parallelism within the framework of Coconut, it could quite easily be adapted to other languages, replacing existing remote-procedure-call and synchronization libraries. The main challenge would be expressing patterns of concurrent parallelization. All functional programming languages have constructions which lend themselves to parallelization, as do environments like Matlab, Octave, Maple, and Mathematica. Other languages have been designed to support specific patterns of parallelism, notably MapReduce [12,30]. The next challenge for Coconut is to find combinations of patterns and target architectures for which more complex data flows can produce higher levels of performance.

In the future, we will extend this language to handle nested parallelism (including multiple nodes each containing multiple cores). We will also address multiphase computations, including the changing use of memory, loading and unloading of computations, and the overlapping of fill/drain between phases.

Acknowledgments

We thank Kevin E. Browne and Shiqi Cao for significant help with the proof in Section 1.7. We are also grateful to the editors and the anonymous reviewers for numerous constructive comments. We especially thank Robert Enenkel of IBM Canada, Inc., for support and advice over several years.

Cell Broadband Engine is a trademark of Sony Computer Entertainment, Inc., in the United States, other countries, or both. IBM and PowerPC are registered trademarks of International Business Machines Corporation. Matlab is a trademark of The MathWorks, Inc. Maple is a trademark of Waterloo Maple Inc. Mathematica is a trademark of Wolfram Research, Inc.

References

[1] E. Albert, M. Hanus, F. Huch, J. Oliver, and G. Vidal. Operational semantics for declarative multi-paradigm languages. *Journal of Symbolic Computation*, 40(1):795–829, 2005.

[2] C. K. Anand and W. Kahl. *A Domain-Specific Language for the Generation of Optimized SIMD-Parallel Assembly Code*. SQRL Report 43, McMaster University, Hamilton, Ontario, Canada, May 2007. Available from http://sqrl. mcmaster.ca/sqrl_reports.html.

[3] C. K. Anand and W. Kahl. Multiloop: Efficient software pipelining for modern hardware. In *CASCON '07: Proceedings of the 2007 Conference of the Center for Advanced Studies on Collaborative Research*, ACM, New York, 2007, pp. 260–263.

[4] C. K. Anand and W. Kahl. Code graph transformations for verifiable generation of SIMD-parallel assembly code. In A. Schurr, M. Nagl, and A. Zundorf, (Eds.), *Applications of Graph Transformations with Industrial Relevance, AGTIVE 2007, Lecture Notes in Computer Science*, 2008. (to appear).

[5] P. Bellens, J. M. Perez, R. M. Badia, and J. Labarta. CellSs: A programming model for the Cell BE architecture. *IEEE*, Tampa, FL, November 2006.

[6] R. Bruni, J. Meseguer, U. Montanari, and V. Sassone. Functorial models for petri nets. *Information and Computation*, 170(2):207–236, 2001.

[7] I. B. M. Corporation, S. C. E. Incorporated, and T. Corporation. *Cell Broadband Engine Programming Handbook*. IBM Systems and Technology Group, Hopewell Junction, New York, 1.0 edition.

[8] A. Corradini and F. Gadducci. An algebraic presentation of term graphs, via gs-monoidal categories. *Applied Categorical Structures*, 7(4):299–331, 1999.

[9] A. Corradini, F. Gadducci, and W. Kahl. Term graph syntax for multi-algebras. Technical Report TR-00-04, Dipartimento di Informatica, Università di Pisa, Pisa, Italy, 2000.

[10] A. Corradini, F. Gadducci, W. Kahl, and B. König. Inequational deduction as term graph rewriting. *Electronic Notes in Computer Science*, 72(1):31–44, 2007.

[11] A. Corradini, U. Montanari, F. Rossi, H. Ehrig, R. Heckel, and M. Löwe. Algebraic approaches to graph transformation, part I: Basic concepts and double pushout approach. In G. Rozenberg, (Ed.), *Handbook of Graph Grammars and Computing by Graph Transformation, Vol. 1: Foundations*, Chapter 3. World Scientific, Singapore, 1997, pp. 163–245.

[12] J. Dean and S. Ghemawat. MapReduce: Simplified data processing on large clusters. *Communications of the ACM*, 51(1):107–113, 2008.

[13] A. van Deursen and P. Klint. Little languages: Little maintenance? *Journal of Software Maintenance*, 10:75–92, 1998.

[14] A. van Deursen, P. Klint, and J. Visser. Domain-specific languages: An annotated bibliography. *ACM SIGPLAN Notices*, 35(6):26–36, 2000.

[15] E. A. Emerson. Temporal and modal logic. In J. van Leeuwen, (Ed.), *Handbook of Theoretical Computer Science, Volume B: Formal Models and Semantics*. Elsevier Science Publishers, Amsterdam, the Netherlands, 1990, pp. 995–1072.

[16] C. Flanagan and P. Godefroid. Dynamic partial-order reduction for model checking software. In *Principles of Programming Languages, (POPL) 2005*. ACM, Long Beach, CA, January 2005, pp. 110–121.

[17] M. Frappier, A. Mili, and J. Desharnais. A relational calculus for program construction by parts. *Science of Computer Programming*, 26(3):237–254, May 1996.

[18] F. Gadducci and U. Montanari. Axioms for contextual net processes. In K. Larsen et al., (Eds.), *ICALP '98*, Aalborg, Denmark, *Lecture Notes in Computer Science*, 1443:296–308, 1998.

[19] Gheorghe Ştefănescu. *Network Algebra*. Springer, London, April 2000.

[20] P. Godefroid. *Partial-Order Methods for the Verification of Concurrent Systems: An Approach to the State-Explosion Problem*, Vol. 1032. Springer-Verlag, New York, 1996.

[21] G. Goumas, M. Athanasaki, and N. Koziris. Automatic code generation for executing tiled nested loops onto parallel architectures. In *SAC '02: Proceedings of the 2002 ACM Symposium on Applied Computing*, ACM, New York, 2002, pp. 876–881.

[22] G. Goumas, N. Drosinos, M. Athanasaki, and N. Koziris. Automatic parallel code generation for tiled nested loops. In *SAC '04: Proceedings of the 2004 ACM Symposium on Applied Computing*, ACM, New York, 2004, pp. 1412–1419.

[23] M. Gschwind, H. P. Hofstee, B. Flachs, M. Hopkins, Y. Watanabe, and T. Yamazaki. Synergistic processing in cell's multicore architecture. *IEEE Micro*, 26(2):10–24, 2006.

[24] B. Hoffmann and D. Plump. Jungle evaluation for efficient term rewriting. In J. Gabrowski, P. Lescanne, and W. Wechler, (Eds.), *Algebraic and Logic Programming, ALP '88, Mathematical Research*, 49. Akademie-Verlag, Gaussig, GDR (now Germany) 1988, pp. 191–203.

[25] P. Hudak. Building domain-specific embedded languages. *ACM Computing Survey*, 28(4es):196, 1996.

[26] P. Hudak. Modular domain specific languages and tools. In P. Devanbu and J. Poulin, (Eds.), *Proceedings of the Fifth International Conference on Software Reuse*. IEEE Computer Society Press, Victoria, British Columbia, Canada, 1998, pp. 134–142.

[27] W. Kahl. The term graph programming system HOPS. In R. Berghammer and Y. Lakhnech, (Eds.), *Tool Support for System Specification, Development and Verification*. Wien, 1999. Springer-Verlag, Vienna (Wien), Austria, pp. 136–149. ISBN: 3-211-83282-3.

[28] W. Kahl, C. K. Anand, and J. Carette. Control-flow semantics for assembly-level data-flow graphs. In W. McCaull et al., (Eds.), *8th International Seminar on Relational Methods in Computer Science*. St. Catherines, Ontario, Canada, *Lecture Notes in Computer Science*, 3929:147–160, 2006.

[29] E. Koutsofios and S. C. North. Drawing graphs with dot. Technical Report, AT&T Bell Laboratories, Murray Hill, NJ, 1993.

[30] R. Lämmel. Google's MapReduce programming model—revisited. *Science of Computer Programming*, 70(1):1–30, 2008.

[31] S. Leonardi and D. Raz. Approximating total flow time on parallel machines. *Journal of Computer and System Sciences*, 73(6):875–891, 2007.

[32] A. Mazurkiewicz. Trace theory. In *Petri Nets: Applications and Relationships to Other Models of Concurrency, Advances in Petri Nets 1986, Part II Lecture Notes in Computer Science*, 255:279–324, 1986.

[33] P. S. Pacheco. *Parallel Programming with MPI*. Morgan Kaufmann, San Francisco, CA, 1997.

[34] D. Plump. Term graph rewriting. In H. Ehrig, G. Engels, H.-J. Kreowski, and G. Rozenberg, (Eds.), *Handbook of Graph Grammars and Computing by Graph Transformation, Vol. 2: Applications, Languages and Tools*, Chapter 1. World Scientific, Singapore, 1999, pp. 3–61.

[35] V. Sassone. On the algebraic structure of petri nets. *Bulletin of the European Association for Theoretical Computer Science*, 72:133–148, 2000.

[36] G. Schmidt and T. Ströhlein. *Relations and Graphs, Discrete Mathematics for Computer Scientists. EATCS-Monographs on Theoretical Computer Science.* Springer, New York, 1993.

[37] M. Sleep, M. Plasmeijer, and M. van Eekelen, (Eds.). *Term Graph Rewriting: Theory and Practice.* Wiley, New York, 1993.

[38] M. Snir, S. W. Otto, S. Huss-Lederman, D. Walker, and J. Dongarra. *MPI— The Complete Reference. Scientific and Engineering Computation Series.* MIT Press, Cambridge, MA, 1996.

[39] W. Thaller. Explicitly staged software pipelining. Master's thesis, Department of Computing and Software, McMaster University, Hamilton, ON, 2006. Available at http://sqrl.mcmaster.ca/~anand/papers/ThallerMScExSSP.pdf.

[40] G. Uszkay. HUSC language and type system. Master's thesis, Department of Computing and Software, McMaster University, Hamilton, Ontario, Canada, 2006.

Chapter 2

Semi-Explicit Parallel Programming in a Purely Functional Style: GpH

Hans-Wolfgang Loidl, Phil Trinder, Kevin Hammond, Abdallah Al Zain, and Clem Baker-Finch

Contents

Declarative programming languages can play an important role in the process of designing and implementing parallel systems. They bridge the gap between a high-level specification, with proven properties of the overall system, and the execution of the system on real hardware. Efficiently exploiting parallelism on a wide range of architectures is a challenging task and should, in our view, be handled by a sophisticated runtime environment. Based on this design philosophy we have developed and formalized Glasgow parallel Haskell (GpH), and implemented it as a conservative extension of the Glasgow Haskell Compiler (GHC).

The high-level nature of declarative languages eases the task of mapping an algebraic specification down to executable code. In fact, the operational components of the specification can already be considered an implementation, with the associated properties acting as assertions in the program. Based on a formal model of the declarative language, the validity of these properties can be established by manual proof, which works on a level of detail similar to the specification language itself. Many operational aspects, usually complicating a proof of an implementation, do not come into the picture at this level. Most importantly, unnecessary sequentialization of the code is avoided.

However, the goal of implicit parallelism has proven an elusive one. Often the automatically generated parallelism is too fine-grained to be efficient. In other cases, the data-dependencies between expressions prohibit the generation of a sufficient amount of parallelism. Thus, we employ an approach of semi-explicit parallelism, where only potential parallelism has to be annotated in a program, and all aspects of coordination are delegated to the runtime environment. A corresponding formal model, in the form of a structured operational semantics, handling pools of both realized and potential parallelism, is used to establish the correctness of programs employing semi-explicit parallelism. The runtime environment itself is capable of synchronizing parallelism, using automatic blocking on data under evaluation, and by simulating virtual shared memory across networks of machines. Being embedded into the optimized runtime environment for a sequential language, we achieve efficient execution of a high-level language, close to the original specification language, while minimizing the programmer effort in parallelizing the code and being scalable to large-scale applications that can be executed on heterogeneous networks and computational grids.

This chapter summarizes research performed over more than a decade, covering language design [28], semantics [6], and implementation [21,22,29]. In particular this chapter elaborates on the semi-explicit programming model (Section 2.1), reflects on the mapping from specification to executable code (Section 2.2), presents a structural operational semantics (Section 2.3) for reasoning about these parallel programs (Section 2.4), discusses the main characteristics of the graph-reduction-based implementation (Section 2.5), underlines the usability of the system by assessing the performance of existing applications (Section 2.6), and concludes by recapitulating the role of our language and methodology as a tool for specification, transformation, and efficient parallel execution (Section 2.7).

2.1   Introduction

One of the key problems of parallel programming is to identify work that may be suitable for parallel execution. Because of their side-effect-free character, it is relatively easy to identify independent expressions in purely functional languages, such as Haskell [24], and to then construct independent threads to evaluate these expressions. However, the costs of creating and synchronizing even ultralightweight threads can be extremely high relative to their runtime, and even projected multi-core architectures such as Intel's 80-core research testbed [13] will be unable to extract effective performance from more than a few hundred simultaneously running threads.

Many approaches have therefore been proposed to introduce parallelism for functional languages [16], ranging from purely implicit parallelism to explicit, manual thread creation, communication, and load management. For the GpH variant of Haskell that is described here [29], and which targets a multithreaded parallel implementation, we have chosen a level of abstraction that hides most of the coordination aspects of the parallel execution, but that still enables the programmer to influence key aspects of thread creation. In this way it is possible to guide the granularity of GpH threads to avoid, for example, the creation of excessively fine-grained threads. The resulting semi-explicit parallel model of computation is elaborated below.

2.1.1   Semi-Explicit Parallelism

Semi-explicit parallel languages [16] form an important class of notations, between explicit parallel languages where all coordination, communication, and control is made explicit (e.g., C extended with the MPI communications library or Concurrent ML [26]), and purely implicit notations where no parallelism control at all is provided (e.g., Sisal [27], Id [23], or NESL [9]). While aiming to provide high levels of abstraction and automatically managing many aspects of parallelism, as with purely implicit approaches, semi-explicit approaches require the programmer to include some directives to specify important aspects of the parallel coordination. Examples of such languages include Skil [10], Concurrent Clean [25], MultiLisp [14], and our own GpH notation.

While semi-explicit approaches were often historically based around the use of annotations, more recent approaches, such as Caliban [18], provide compositional language constructs. This supports a more powerful, flexible, and often programmable, parallel programming methodology where (parallel) coordination is separated from (sequential) computation. Complete parallel programs are then orchestrated from lower-level components, which may themselves be broken down into parallel or sequential subcomputations, as required to implement the program.

GpH [29] is a modest extension of Haskell with parallel and sequential composition as its two basic coordination primitives (Figure 2.1). Denotationally, both the `par` and `seq` constructs are projections onto the second operand, that is, the value of the first operand is not returned as the result of the construct, though it will usually be shared as part of some subcomputation in the second operand. Operationally,

`par :: ` $a \to b \to b$	—Parallel composition
`seq :: ` $a \to b \to b$	—Sequential composition

FIGURE 2.1:   Types of the basic coordination constructs in GpH.

`seq` indicates that the first operand should be evaluated before the second operand, and `par` indicates that the first operand may be evaluated in parallel with the evaluation of the second. The latter operation is termed "sparking." Unlike most parallel notations, the act of sparking an expression in GpH does not immediately force thread creation; rather the runtime environment determines which sparks are chosen to become parallel threads based on load and other information. It follows that programmers simply need to expose expressions in the program that they believe can usefully be evaluated in parallel. The runtime environment manages the details of parallel execution including thread creation, communication, workload balancing, etc., as described in detail in Section 2.5. Parallel implementations of GpH are publicly available from http://www.macs.hw.ac.uk/~dsg/gph/.

These two basic operations are then used to build higher-level constructs that help simplify parallel programming. Our early experience of implementing nontrivial programs in GpH showed that unstructured use of `par` and `seq` operators could lead to unnecessarily obscure programs. This problem can be overcome by using evaluation strategies [28]: nonstrict, polymorphic, higher-order functions that influence both the degree of evaluation of and the parallelism to be found in a GpH expression. Evaluation strategies provide a clean separation between coordination and computation. The driving philosophy behind evaluation strategies is that it should be possible to understand the computation specified by a function without considering its coordination.

Figure 2.2 shows the basic operations that we provide for managing evaluation strategies. Strategies are defined as functions of type `Strategy` that take a polymorphic argument and return a unit value. The `using` construct then applies an evaluation strategy to a Haskell expression. We define two strategies for handling reduction of Haskell expressions to normal forms (NFs). The basic evaluation strategy `rwhnf`

type *Strategy* $a = a \to ()$	— Type of evaluation strategy
`using :: ` $a \to$ *Strategy* $a \to a$	— Strategy application
`rwhnf :: ` *Strategy* a	— Reduction to weak head normal form
class `NFData` a **where**	— Class of reducible types
`rnf :: ` *Strategy* a	— Reduction to normal form

FIGURE 2.2:   Basic evaluation strategy operations in GpH.

specifies that the associated expression is to weak head normal form (WHNF), in which no reduction is performed either under λ-expressions or within data structures. This corresponds to the default nonstrict evaluation order used by Haskell. The overloaded rnf strategy correspondingly specifies that the associated expression is to be reduced to full NF, representing a maximal evaluation degree. We provide instantiations of this strategy for all major standard Haskell types.

2.2   From Algebraic Specification to Executable Code

Functional programming languages are designed to provide a high level of abstraction, focusing on what should be computed without committing the machine as to how the computation should be organized. This design principle has made functional languages a popular choice as execution languages for algebraic specifications. The main language features that contribute to this high level of abstraction are higher-order functions, polymorphism, and advanced mechanisms for modularisation, such as ML's functors or Haskell's type classes. Tight links between algebraic specification languages and execution languages can be identified in Extended ML [17], which maps down to ML, or in SPECTRUM [11], which uses advanced features such as type classes as found in Haskell.

In our work, we exploit the proximity between algebraic specification languages and modern functional programming languages by attempting a direct mapping of the specification into Haskell code. In many cases, the algebraic specification can already be considered an executable specification without the need of further refinement steps. For more complex specifications, the process of mapping it to executable code may yield a number of proof obligations, stated in the axioms of the specification language. The proof obligations may be discharged using tools, including general theorem provers such as Isabelle or Coq, or specialized provers, such as Sparkle. Our model of software development does not tie the development to any particular tool, and the proof obligations may even be discharged manually.

Some advanced language features provided by Haskell facilitate the mapping of a specification into a program. In particular, the sophisticated typing mechanism in Haskell, including type classes and functional dependencies, provides a powerful tool. Recent developments into the direction of dependent types [5] offer even more power to encode value-dependent properties into the type of a function, and move the programming language even closer to a specification language. These latest research directions aside, our current practice in mapping specification to code is, however, one of manually refining the specification to bring it into a rule-based format, suitable for a Haskell-like language. The proof obligations imposed by the axioms in the specification are proven either manually or with the help of a specialized theorem prover.

With respect to the parallel execution of the resulting code, a functional language like GpH avoids the redundant sequentialization found in imperative languages. Nor do we mandate the specification of a parallelism structure in early stages of the

$$
\begin{aligned}
m \mid n &= \exists k \in \mathbb{N}.\ m \times k = n \\
m \perp n &= \neg\ \exists k \in \mathbb{N}.\ 1 < k \wedge k \mid m \wedge k \mid n \\
\varphi(i) &= \mid \{ m \in \mathbb{N} \mid m < i \wedge m \perp i \} \mid \\
sumEuler(n) &= \Sigma_{i=1}^{n}\ \varphi(i)
\end{aligned}
$$

FIGURE 2.3:   Specification of the Euler totient function φ.

refinement. As demonstrated by the running example used in this chapter, we generate rule-based code, which does not commit to a particular evaluation order. We then identify possible sources of parallelism and specify the parallel coordination using evaluation strategies [28]. Finally, we can exploit equational reasoning at this level to improve the parallel performance of the resulting program. It should be emphasized that this entire process is architecture-independent and Section 2.6 demonstrates good parallel performance of GpH applications on varying architectures from shared-memory machines to wide-area grid networks.

As a running example, we will use a simple algorithm that computes the sum of the values of the Euler totient function for integers in the range from 1 to n, for some given n. The Euler totient function of a given integer i is the number of integers less than i that are relatively prime to i, as shown in Figure 2.3. Figure 2.4 shows the sequential Haskell code that implements this example. It directly maps the specifications in Figure 2.3 to the functions `relprime`, `euler` and `sumEuler`. In the code for `relprime` the following property is exploited: $\forall\ m\ n.\ \gcd m\ n = 1 \implies m \perp n$. Each function is preceded by a type signature, for example, `mkList :: ` $Int \to [Int]$ specifies that `mkList` is a function that takes a fixed precision integer, of type Int, and returns a list of integers, of type $[Int]$. The notation `[1..(n−1)]` defines the

```
mkList  ::  Int → [Int]
mkList n = [1..(n-1)]

gcd  ::  Int → Int → Int
gcd x 0 = x
gcd x y = gcd y (rem x y)

relprime  ::  Int → Int → Bool
relprime x y = gcd x y == 1

euler  ::  Int → Int
euler n = length (filter (relprime n) (mkList n))

sumEuler  ::  Int → Int
sumEuler = sum . (map euler) . mkList
```

FIGURE 2.4:   Sequential Haskell definition of `sumEuler`.

```
sumEulerPar1 n = sum ((map euler (mkList n))
                      `using` parList rnf)

sumEulerPar2 :: Int → Int → Int
sumEulerPar2 c n =
sum ([sum (map euler x) | x ← splitAtN c (mkList n)]
                      `using` parList rnf)
```

FIGURE 2.5:   Two parallel versions of `sumEuler` in GpH.

list of integers between 1 and $n - 1$. In the definition of the greatest common divisor
(gcd) we use the remainder function on integers (rem). The higher-order function
filter selects all elements of a list that fulfil a given predicate, and map applies a
function to all elements of a list.

Figure 2.5 shows two simple parallel implementations of sumEuler that use
evaluation strategies. The first parallel version, sumEulerPar1, simply applies the
parList rnf strategy to the elements of the result list. The parList strategy is
a parameterized strategy for lists that applies its argument (another strategy) to each
element of the list in parallel. In this case, we will evaluate each element to full NF
using a different thread. The parList strategy can easily be defined using the GpH
par construct, as shown in Figure 2.6. The definition applies the argument strategy,
strat, to each element of the list x:xs in parallel, returning the unit value () once
the entire list has been traversed (: is list cons).

The second parallel version, sumEulerPar2, is more sophisticated. It first splits
the list elements into groups of size c using the splitAtN function. A list comprehension is used to bind each of the groups to the variable x in turn. Here, the list
comprehension syntax [e | v ← l] builds a list of expressions whose value
is e. The expression e depends on the value of v, which is drawn from each element
of the list l in turn. The inner sum of the Euler totients for each group is then calculated, before all these results are summed. The definition thus exploits associativity
of the underlying + operator. The reason for splitting the indexes in this way is so
that each group of c list elements may be evaluated in parallel by its own individual
thread, thereby increasing granularity. We will revisit this example in Section 2.4,
developing versions with improved parallel performance.

```
parList :: Strategy a → Strategy [a]
parList strat [] = ()
parList strat (x:xs) = strat x `par` (parList strat xs)
```

FIGURE 2.6:   Definition of `parList` in GpH.

2.3 Operational Semantics of GpH

In this section, we will present a parallel operational semantics for GpH, defined in terms of the call-by-need evaluation of a parallel extension to the λ-calculus, GpH-core. By making our semantics explicit in describing the way threads are managed and stored, we are able to reason accurately about the behaviour of GpH programs in terms of both coordination and computation.

2.3.1 Operational Semantics Overview

The GpH operational semantics is a two-level transition semantics. At the lower level, there are single-thread transitions for performing the ordinary evaluation of expressions through, e.g., β-reduction. All candidate single-thread steps are performed simultaneously and in lockstep. They are then combined into a parallel computation super-step using coordination relations defined at the upper level. We follow Launchbury's seminal work [19] by using a heap to allow sharing and by requiring all closures to be constructed by some let-binding. Where Launchbury uses a big-step natural semantics, in order to properly model coordination issues, we have chosen to use a small-step computational semantics.

GpH-core is a simple subset of GpH, comprising the untyped λ-calculus extended with numbers, recursive **let**s, sequential composition, **seq**, and parallel composition, **par**. Expressions are normalized so that all variables are distinct, and the second argument to application and the first argument to **par** must be variables.

$$x, y, z \in \text{Variable}$$
$$n \in \text{Number}$$
$$e \in \text{Expression}$$
$$e ::= n \mid x \mid e\,x \mid \lambda x.e \mid \textbf{let}\,\{x_i = e_i\}_{i=1}^{n}\,\textbf{in}\,e$$
$$\mid e_1\,\textbf{seq}\,e_2 \mid x\,\textbf{par}\,e$$

2.3.2 Heaps and Labeled Bindings

Following Ref. [19], we use a heap of bindings of expressions to variables. To deal with parallelism, each binding also carries a label to indicate its state. Thus, heaps are partial functions from variables to expression/thread-state pairs:

$$H, K \in \text{Heap} = \text{Variable} \multimap (\text{expression, state})$$
$$\alpha, \beta \in \text{State}$$
$$\alpha ::= \textit{Inactive} \mid \textit{Runnable} \mid \textit{Active} \mid \textit{Blocked}$$

We write individual bindings with the thread state appearing as an annotation on the binding arrow, thus

$$x \overset{\alpha}{\mapsto} e$$

A binding is *Active* (A) if it is currently being evaluated; it is *Blocked* (B) if it is waiting for another binding before it can continue its own evaluation; and it is *Runnable* (R) if it could be evaluated, but there are currently insufficient resources to evaluate it. All other bindings are *Inactive* (I). Bindings therefore correspond to heap closures and labeled bindings correspond to parallel threads. This is a rather simplified model of parallelism compared with the actual GUM implementation (Section 2.5): we assume idealized parallelism, with no communication costs; and unlike the actual implementation, threads are created instantly and may be migrated at no cost. However, it serves to provide limits on possible parallelism in the actual implementation.

The computational semantics is specified as a relation on heaps, $H \implies H'$. This is, in turn, defined in terms of a notion of single thread transitions (Section 2.3.3) and a scheduling relation (Section 2.3.4). The parallel operational semantics then builds on this to describe a reduction sequence from an initial global configuration to a final global configuration:

$$(H, main \overset{A}{\mapsto} e) \implies \ldots \implies (H', main \overset{I}{\mapsto} v)$$

where *main* identifies the root expression for the program. Values, v, are in weak head NF, that is

$$v ::= n \mid \lambda x.e$$

2.3.3   Single-Thread Transitions

The transition relation $\longrightarrow$ of Figure 2.7 describes the computational step taken by each active binding in the heap. The left-hand side in each rule represents a heap with the active binding distinguished by $H : z \overset{A}{\mapsto} e$. Multilabel bindings, such as $x \overset{RAB}{\mapsto} e$ in the *block₂* rule mean that the state is one of R, A, or B but not I. The right-hand sides of the rules are heaps comprising only those bindings changed or created by that computation step.

Let: The *let* rule populates the heap with new bindings. These bindings are inactive since under call-by-need they may not necessarily be evaluated.

Variables and blocking: In the *var* and *block$_i$* rules, z is a pointer to another closure (called x). If x has already been evaluated to WHNF (the *var* rule), then z simply receives that value. The notation $\hat{v}$ indicates that all bound variables in v are replaced with fresh variable names. If x is inactive and has not yet been evaluated (the *block₁* rule), then z blocks at this point and x joins the pool of runnable bindings. Finally, if x is not inactive (the *block₂* rule), then z blocks but x is unaffected.

Application: Evaluating $e\,x$ involves reducing e to a function abstraction using *app* and then substituting x for the bound variable y using *subst*.

Seq: The *seq* rule initially evaluates e_1 without considering e_2. When and if e_1 is reduced to WHNF, its value (but not any changes to the heap) is discarded by the *seq-elim* rule and evaluation then proceeds to e_2.

$$H : z \xmapsto{A} \textbf{let } \{x_i = e_i\}_{i=1}^n \textbf{ in } e \ \longrightarrow \ (\{x_i \xmapsto{I} e_i\}_{i=1}^n, z \xmapsto{A} e) \qquad (let)$$

$$(H, x \xmapsto{I} v) : z \xmapsto{A} x \ \longrightarrow \ (z \xmapsto{A} \hat{v}) \qquad (var)$$

$$(H, x \xmapsto{I} e) : z \xmapsto{A} x \ \longrightarrow \ (x \xmapsto{R} e, z \xmapsto{B} x) \qquad (block_1)$$

$$(H, x \xmapsto{RAB} e) : z \xmapsto{A} x \ \longrightarrow \ (z \xmapsto{B} x) \qquad (block_2)$$

$$H : z \xmapsto{A} (\lambda y.e)\, x \ \longrightarrow \ (z \xmapsto{A} e[x/y]) \qquad (subst)$$

$$\frac{H : z \xmapsto{A} e \ \longrightarrow \ (K, z \xmapsto{\alpha} e')}{H : z \xmapsto{A} e\, x \ \longrightarrow \ (K, z \xmapsto{\alpha} e'\, x)} \qquad (app)$$

$$H : z \xmapsto{A} v \ \textbf{seq}\ e \ \longrightarrow \ (z \xmapsto{A} e) \qquad (seq\text{-}elim)$$

$$\frac{H : z \xmapsto{A} e_1 \ \longrightarrow \ (K, z \xmapsto{\alpha} e_1')}{H : z \xmapsto{A} e_1 \ \textbf{seq}\ e_2 \ \longrightarrow \ (K, z \xmapsto{\alpha} e_1' \ \textbf{seq}\ e_2)} \qquad (seq)$$

$$(H, x \xmapsto{RAB} e_1) : z \xmapsto{A} x \ \textbf{par}\ e_2 \ \longrightarrow \ (z \xmapsto{A} e_2) \qquad (par\text{-}elim)$$

$$(H, x \xmapsto{I} e_1) : z \xmapsto{A} x \ \textbf{par}\ e_2 \ \longrightarrow \ (x \xmapsto{R} e_1, z \xmapsto{A} e_2) \qquad (par)$$

FIGURE 2.7:   Single thread transition rules.

Par: The *par* rule potentially introduces parallelism, i.e., suggests that an inactive binding could be made active by putting it into a *Runnable* state which may be promoted to *Active* later if sufficient resources are available. Nothing needs to be done if the binding is not inactive (*par-elim*).

2.3.4   Multithread Transitions and Scheduling

The changes required for all active bindings are combined by the *parallel* rule (Figure 2.8) to create a full heap-to-heap transition. This is the key point in the semantics where reductions are carried out in parallel. We write H^A to represent

$$\frac{H^A = \{\, x_i \xmapsto{A} e_i \,\}_{i=1}^n \qquad \{H : x_i \xmapsto{A} e_i \ \longrightarrow \ K_i\}_{i=1}^n}{H \xRightarrow{p} H[\bigcup_{i=1}^n K_i]} \qquad (parallel)$$

FIGURE 2.8:   Combining multiple thread transitions.

$$(H, x \overset{RA}{\mapsto} v, z \overset{B}{\mapsto} e^x) \overset{u}{\longrightarrow} (H, x \overset{RA}{\mapsto} v, z \overset{R}{\mapsto} e^x) \qquad (unblock)$$

$$(H, x \overset{RA}{\mapsto} v) \overset{d}{\longrightarrow} (H, x \overset{I}{\mapsto} v) \qquad (deactivate)$$

$$\frac{|H^A| < N}{(H, x \overset{R}{\mapsto} e) \overset{a}{\longrightarrow} (H, x \overset{A}{\mapsto} e)} \qquad (activate)$$

FIGURE 2.9:   Single-thread scheduling rules.

all the active bindings in H, i.e., $H^A = \{x \overset{A}{\mapsto} e \in H\}$. Hence in Figure 2.8 there are precisely n active bindings in H. The notation $H[K]$ updates heap H with all new or changed bindings given by K. A more precise definition, together with a proof that conflicts do not arise between bindings can be found in Ref. [6].

The scheduling actions for individual threads are defined in Figure 2.9 as follows: (i) any binding that is immediately blocked on a completing thread is made runnable (*unblock*); (ii) any active or runnable binding that is in WHNF is made inactive (*deactivate*); and (iii) as many runnable bindings as resources will allow are made active (*activate*). In the *unblock* rule, the notation e^x represents an expression that is immediately blocked on x, i.e., one of the three forms:

$$e^x ::= x \mid x\ y \mid x\ \mathbf{seq}\ e'$$

Note that in the *activate* rule, N is a parameter to the semantics, indicating the total number of processors. This ensures that no more than N bindings are activated in any step. These rules do not, however, specify which bindings are activated: bindings are chosen nondeterministically during the activation phase. Since, however, this choice is at the coordination level, it does not change the actual values that are computed.

The rules of Figure 2.10 extend the scheduling rules to multiple bindings. To achieve the maximum possible parallelism with respect to the available processors, it is necessary that all candidate threads are unblocked before deactivation and that deactivation takes place before activation. This sequence of actions is captured by the schedule relation of Figure 2.11.

$H \overset{\dagger}{\Longrightarrow} H'$ if:

i) $H \overset{\dagger}{\longrightarrow}^* H'$ and   ii) there is no H'' such that $H' \overset{\dagger}{\longrightarrow} H''$.

($\dagger$ is u, d or a.)

FIGURE 2.10:   Component scheduling relations.

$$\xRightarrow{s} \;=\; \xRightarrow{a} \circ \xRightarrow{d} \circ \xRightarrow{u} \qquad\qquad (schedule)$$

$$\Longrightarrow \;=\; \xRightarrow{s} \circ \xRightarrow{p} \qquad\qquad (compute)$$

FIGURE 2.11:   Overall scheduling and computation relations.

2.3.5   Computation Relation

Finally, our full semantic computation relation, *compute*, is defined as a parallel transition $\xRightarrow{p}$ followed by a scheduling of bindings $\xRightarrow{s}$ (Figure 2.11). This ordering ensures that the heaps that appear in a reduction sequence are always fully scheduled. Since our semantics is parameterised on the number of processors, we decorate the computation relation with the number of processors where necessary: $\underset{N}{\Longrightarrow}$, so $\underset{1}{\Longrightarrow}$ indicates the single-processor case.

2.3.6   Properties of the Operational Semantics

Abramsky's denotational semantics of lazy evaluation [1] models functions by a lifted function space, thus distinguishing between a term Ω (a nonterminating computation) and $\lambda x.\Omega$ to reflect the fact that reduction is to weak head NF rather than head NF. This is a widely used, simple, and abstract semantics. The properties and results developed in this section are expressed relative to this denotational semantics.

Launchbury [19] shows a number of results relating his natural semantics of lazy evaluation to Abramsky's denotational semantics. We borrow much of his notation and several of our proofs are inspired by him. Previously we showed that the 1-processor case of our semantics corresponds to Launchbury's.

There are three main properties that we expect of our semantics: soundness: the computation relation preserves the meanings of terms; adequacy: evaluations terminate if and only if their denotation is not $\bot$; determinacy: the same result is always obtained, irrespective of the number of processors and irrespective of which runnable threads are chosen for activation during the computation.

The denotational semantics of our language is given in Figure 2.12. The *Val* domain is assumed to contain a lifted version of its own function space. The lifting injection is *lift* and the corresponding projection is *drop*.

The semantic function:

$$[\![\ldots]\!] : Exp \to Env \to Val$$

naturally extends to operate on heaps, the operational counterpart of environments:

$$\{\!\{\ldots\}\!\} : Heap \to Env \to Env$$

The recursive nature of heaps is reflected by a recursively defined environment:

$$\{\!\{x_1 \mapsto e_1 \ldots x_n \mapsto e_n\}\!\}\rho = \mu\rho'.\rho[x_1 \mapsto [\![e_1]\!]_{\rho'} \ldots x_n \mapsto [\![e_n]\!]_{\rho'}]$$

$$\rho \in Env = Var \rightarrow Val$$

$$[\![\lambda x.e]\!]_\rho = lift\ \lambda\epsilon.[\![e]\!]_{\rho[x \mapsto \epsilon]}$$

$$[\![e\ x]\!]_\rho = drop([\![e]\!]_\rho)([\![x]\!]_\rho)$$

$$[\![x]\!]_\rho = \rho(x)$$

$$[\![\textbf{let}\ \{x_i = e_i\}_{i=1}^n\ \textbf{in}\ e]\!]_\rho = [\![e]\!]_{\{\!\{x_1 \mapsto e_1 \ldots x_n \mapsto e_n\}\!\}\rho}$$

$$[\![e_1\ \textbf{seq}\ e_2]\!]_\rho = \begin{cases} \bot & \text{if } [\![e_1]\!]_\rho = \bot \\ [\![e_2]\!]_\rho & \text{otherwise} \end{cases}$$

$$[\![x\ \textbf{par}\ e]\!]_\rho = [\![e]\!]_\rho$$

FIGURE 2.12:   Denotational semantics.

We also require an ordering on environments: if $\rho \leq \rho'$ then ρ' may bind more variables than ρ but they are otherwise equal. That is:

$$\forall x\ .\ \rho(x) \neq \bot \Rightarrow \rho(x) = \rho'(x)$$

The arid environment ρ_0 takes all variables to $\bot$.

2.3.6.1   Soundness

Our computational relation $H \Longrightarrow H'$ can be considered sound with respect to the denotational semantics in Figure 2.12 if the denotations of all the bindings in H are unchanged in H'. The $\leq$ ordering on environments neatly captures this notion.

PROPOSITION 2.1
  If $H \Longrightarrow H'$ then for all ρ, $\{\!\{H\}\!\}\rho \leq \{\!\{H'\}\!\}\rho$.

PROOF   Induction on the size of H and the structure of expressions.   ⧠

2.3.6.2   Adequacy

We wish to characterize the termination properties of our semantics and Propositions 2.2 and 2.3 show an agreement with the denotational definition. The proofs are modeled on the corresponding ones in Ref. [19].

PROPOSITION 2.2
  If $(H, z \overset{A}{\mapsto} e) \Longrightarrow^* (H', z \overset{I}{\mapsto} v)$ then $[\![e]\!]_{\{\!\{H\}\!\}\rho} \neq \bot$.

PROOF   For all values v, $[\![v]\!]_{\{\!\{H'\}\!\}\rho} \neq \bot$ so by Proposition 2.1 $[\![e]\!]_{\{\!\{H\}\!\}\rho} \neq \bot$.   ⧠

PROPOSITION 2.3

If $[\![e]\!]_{\{\!\{H\}\!\}\rho} \neq \bot$, there exists H', z, v such that $(H, z \overset{A}{\mapsto} e) \implies^* (H', z \overset{I}{\mapsto} v)$.

A proof of Proposition 2.3 is outlined in Ref. [6]. It is closely based on the corresponding proof in Ref. [19], working with a variant of the denotational semantics which is explicit about finite approximations.

2.3.6.3 Determinacy

We now turn to the question of obtaining the same result irrespective of the number of processors and irrespective of which runnable threads are chosen for activation during the computation. Clearly, since the results above hold for any number of processors it follows that *if* an evaluation with N processors gives *main* a value then, depending on which threads are activated, an evaluation with M processors *can* give the same result in the sense of Proposition 2.1.

However, a consequence of the definition of $\overset{a}{\implies}$ is that the main thread may be left runnable but never progress. It is possible that the main thread could be delayed or suspended indefinitely, if there is a constant supply of unneeded speculative threads being generated and scheduled in place of the main thread. This corresponds to the implementation of GpH, with the management of speculative evaluation the programmer's responsibility [29]. It is possible to define an alternative activation relation $\overset{a'}{\implies}$ that requires that a runnable thread on which *main* is blocked (in a transitive sense) will be activated in preference to other runnable threads. We can be sure that there will always be a free processor in this circumstance because the blocking action has made one available.

With this version of the activation relation, we can show that if any evaluation gives an answer for *main* then they all do, irrespective of the number of processors. For the 1-processor case, it is clear that the definition of $\overset{a'}{\implies}$ in Figure 2.13 ensures that there is always exactly one active binding and that the blocked bindings form a chain from *main* to that active binding.

$$H \overset{a'}{\implies} H' \quad \text{if:}$$

1. $H \overset{a}{\longrightarrow}^* H'$;

2. there is no H'' such that $H' \overset{a}{\longrightarrow} H''$ and

3. $req(main, H')$ is active in H'.

$$req(x, K) = \begin{cases} x, & \text{if } x \overset{RA}{\mapsto} e \in K \\ req(y, K), & \text{if } x \overset{B}{\mapsto} e^y \in K \end{cases}$$

FIGURE 2.13: Stronger activation relation.

The following proposition demonstrates that all the closures activated in the one processor case will also be activated in the multiprocessor case. Recall that $\underset{N}{\Longrightarrow}$ is the computation relation assuming a maximum of N processors.

PROPOSITION 2.4

Given $N \geq 1$ processors, suppose

$$(H, main \overset{A}{\mapsto} e) \underset{1}{\Longrightarrow} H_1 \underset{1}{\Longrightarrow} H_2 \dots \text{ and}$$

$$(H, main \overset{A}{\mapsto} e) \underset{N}{\Longrightarrow} K_1 \underset{N}{\Longrightarrow} K_2 \dots$$

If x is active in some H_i then there is a j such that x is active in K_j.

PROOF   Suppose z_k is active in some H_i. By $\overset{a'}{\Longrightarrow}$ there is a chain $main \overset{B}{\mapsto} e^{z_1}, z_1 \overset{B}{\mapsto} e^{z_2}, z_2 \overset{B}{\mapsto} e^{z_3}, \dots z_k \overset{A}{\mapsto} e$ in H_i.

By induction on the length k of this chain we can show that there must be some K_j where z_k is active in K_j.   ⬚

Finally we can bring all these results to bear to prove that evaluation is deterministic in the sense that we get the same answer every time, for any number of processors, assuming the $\overset{a'}{\Longrightarrow}$ activation relation.

COROLLARY 2.1

For any number of processors $N \geq 1$, if $(H, main \overset{A}{\mapsto} e) \underset{1}{\Longrightarrow}^* (H', main \overset{I}{\mapsto} v)$

and $(H, main \overset{A}{\mapsto} e) \underset{N}{\Longrightarrow} K_1 \underset{N}{\Longrightarrow} K_2 \dots$ then:

1. there is some $i \geq 1$ such that $K_i = (K_i', main \overset{I}{\mapsto} v')$;

2. $[\![v']\!]_{\{\!\{K_i'\}\!\}\rho_0} = [\![v]\!]_{\{\!\{H'\}\!\}\rho_0}$

PROOF

1. If there is no such K_i then *main* must remain active or blocked forever. In either case there must be some binding $z \overset{A}{\mapsto} e$ that remains active and does not terminate. In that case the denotation of e in the context of the corresponding heap must be $\bot$ by Proposition 2.3. But by Proposition 2.4 at some stage in the 1-processor evaluation z will be active and *main* will be (transitively) blocked on z. By Proposition 2.2 e will not reach a WHNF so *main* will remain blocked. (Unless $main = z$ in which case the result follows immediately.)

2. $\{\!\{H, main \overset{A}{\mapsto} e\}\!\}\rho_0 \leq \{\!\{H', main \overset{I}{\mapsto} v\}\!\}\rho_0$ by Proposition 2.1, so in particular $[\![v]\!]_{\{\!\{H'\}\!\}\rho_0} = [\![e]\!]_{\{\!\{H\}\!\}\rho_0}$.
   Similarly, $[\![v']\!]_{\{\!\{K_i'\}\!\}\rho_0} = [\![e]\!]_{\{\!\{H\}\!\}\rho_0}$.   ⬚

2.4 Program Equivalences and Reasoning

We will now demonstrate how we can perform program transformations on parallel programs in order to improve performance. We will use the semantic properties of the GPH par and seq constructs, defined in Section 2.3, to derive versions of the program that expose improved parallel behaviour. Our overall goal is to reduce the total execution time on some specific parallel platform. In working toward this goal, we will increase the degree of parallelism that can be found in the program by judicious introduction of evaluation strategies. However, since unlike the ideal situation considered by our operational semantics, in the real-world the costs and overheads of parallel execution and communication mean that maximal parallelism does not automatically lead to a minimal runtime.

We continue with the sequential version of sumEuler below. Note that the three main worker functions (sum, map euler, and mkList) have been composed into a three-stage sequential pipeline using the function composition operator (.).

```
sumEuler :: Int → Int
sumEuler = sum . map euler . mkList
```

Despite the simplicity of this program, it is a good example because it exhibits a typical structure of symbolic applications: it uses fairly small auxiliary functions that are combined with function composition and higher-order functions. The overall program structure is a fold-of-map (where the sum function is a fold). Operationally, this structure suggests two possible ways of parallelisation: producer–consumer (or pipeline) parallelism, and data parallelism.

2.4.1 Pipeline Parallelism

The function compositions give rise to producer–consumer (or pipeline) parallelism, where the initial producer (mkList) runs in parallel with its consumer (map euler), and this runs in parallel with the final consumer of the list (sum). We can specify pipeline parallelism by using a parallel variant of the function composition operator (. ||), which computes both the producer and the consumer function in parallel. The sumEuler function can then be written as

```
sumEuler =    sum .|| map euler .|| mkList
```

where the pipeline operator is defined as

```
(f .|| g) x = let { x' = g x ; y = f x' } in x' `par` y
```

While pipeline parallelism can be easily expressed, in this case it is not very efficient, since the tight connection between producer and consumer leads to frequent synchronization between the generated threads, which can consequently result in poor performance.

This behavior can be improved by attaching a strategy to the . || combinator, ensuring that the producer generates whole blocks of list elements and thus reduces the amount of synchronization. In order to further increase parallelism, we can use a parallel evaluation strategy to not only specify evaluation degree but also parallelism on the result of the producer. Here we use parallel, strategic function application $ | |, which applies a strategy to the argument of a function:

```
sumEuler n =
        sum $|| (parList rnf) $ map euler $|| (parList rnf) $ mkList n
```

This is a mixture of pipeline and data parallelism, which we study in more detail in the following section. Notably, this pipeline construct makes it possible to specify parallelism on the top level, when combining sequential code: although mkList n is a sequential function, the application of parList rnf triggers the parallel evaluation of all list elements. For large-scale programming this approach of specifying parallelism when combining functions, rather than when defining them, reduces and localizes the amount of necessary code changes (see Ref. [22] for a more detailed discussion).

2.4.2   Data Parallelism

More promising in this example is to use data parallelism, where the same operation is applied to multiple components of a data structure in parallel. In this case, the euler function, which is mapped over the list returned by mkList, is a good candidate for data parallelism. We can define a parallel variant of the map operation that builds on the parList strategy, and which abstracts over the pattern of parallelism defined in sumEulerPar1, as follows:

```
parMap  :: Strategy b → (a → b) → [a] → [b]
parMap strat f xs = map f xs `using` parList strat
```

This clearly shows how higher level parallel constructs can be constructed in a layered manner, in order to maintain clear separation between coordination and computation: it is obvious from the definition that the value returned from a call to parMap is identical to a sequential map. We can use this equivalence to simply replace the map euler expression by parMap s euler, where s describes the evaluation degree on the elements of the result list. In this case, we choose rnf to increase the granularity of the generated tasks.

Examining the behavior of this data-parallel code we find a large number of very fine-grained threads, since every list element gives rise to a new thread. In order to improve performance further, we need to combine computations on neighboring list elements into one, coarse-grained thread, as in the definition of parSumEuler2. We call this process clustering of evaluation and demonstrate how it can be derived from this code, and how it improves parallel performance.

2.4.3   Strategic Clustering

Clustering is easily introduced into functions that return a list or other collection. For example, it is possible to construct a strategy that captures the splitting technique

```
parListChunk  ::  Int → Strategy a → Strategy [a]
parListChunk c strat [] = ()
parListChunk c strat xs = seqList strat (take c xs) `par`
                                parListChunk c strat (drop c xs)
```

FIGURE 2.14: A clustering strategy used in `sumEuler`.

used in the definition of `sumEulerPar2` above. The `parListChunk` strategy of Figure 2.14 introduces a number of parallel threads to operate on subsequences, or chunks, of an input list. Here, each thread evaluates a sublist of length `c`. The `seqList` strategy is a sequential analogue of `parList`, applying its argument strategy to every element of the list in sequence. Using a `parListChunk c s` xs strategy will therefore generate $\lfloor \frac{k}{c} \rfloor$ potential threads, where k is the length of the list xs. The expression map `f` xs can now be clustered as follows.

```
parMapChunk c strat f xs = map f xs `using` parListChunk c strat
```

As before, the advantage of a purely strategic approach is that the clustered coordination operations can be captured entirely by the evaluation strategy and isolated from the computation.

2.4.4 Cluster Class

As defined in Section 2.4.3, clustering works on the results of a function. However, many common functions do not return a list or other collection as their result. One common example is a `fold` function, that collapses a collection to a single value. In order to cluster such functions, we introduce a generic *Cluster* type class with functions to *cluster* the input data, *lift* the function to operate on the clustered data, and perhaps *decluster* the result. Although clustering changes the computation component of a parallel program, equivalence to the sequential program is maintained by introducing clustering systematically using semantics-preserving identities. We will discuss such transformations in the following sections.

To be more precise, we require that every collection type c that is to be clustered is an instance of the Haskell `MMonad` class, as shown by the subclass dependency for *Cluster* below. We use a formulation of monads based on the functions *munit*, *mjoin*, and *mmap* [32], which is more suitable for our purposes than the usual Kleisli category (see Section 10.4 of Ref. [7]) with *return* and *bind* operations that is used in many of the standard Haskell libraries.

```
class MMonad c where
    munit ::  a → c a
    mjoin ::  c (c a) → c a
    mmap  ::  (a → b) → (c a → c b)
```

We introduce a new Haskell class, *Cluster*, parametrized by the collection type c with four operations: *singleton* turns a collection into a 1 element collection of collections; the generalized variant *cluster n* maps a collection into a collection of subcollections each of size n; *decluster* flattens a collection of collections into a single collection; and *lift* takes a function on $c\ a$ and applies it to a collection of collections. For *singleton*, *decluster*, and *lift* we can use existing definitions in the *MMonad* class to provide default definitions.

```
class (MMonad c) => Cluster c where
   singleton ::  c a → c (c a)
   cluster    ::  Int → c a → c (c a)
   decluster ::  c (c a) → c a
   lift        ::  (c a → b) → (c (c a) → c b)

   singleton = munit
   decluster = mjoin
   lift = mmap
```

All instances of the monad class come with proof obligations for the monad identities (see Rules (I)–(III), (i)–(iv) of Ref. [32]). From these identities we obtain the following equations relating *lift*, *decluster*, and *singleton*.

$$decluster \circ singleton\ =\ id \tag{M I}$$
$$decluster \circ lift\ singleton\ =\ id \tag{M II}$$
$$decluster \circ decluster\ =\ decluster \circ lift\ decluster \tag{M III}$$

$$lift\ id\ =\ id \tag{M i}$$
$$lift\ (f \circ g) = (lift\ f) \circ (lift\ g) \tag{M ii}$$
$$lift\ f \circ singleton\ =\ singleton \circ f \tag{M iii}$$
$$lift\ f \circ decluster\ =\ decluster \circ lift\ (lift\ f) \tag{M iv}$$

We further require for each n that *cluster n* is a one-sided inverse of *decluster*.

$$decluster \circ\ cluster\ n = id \tag{C I}$$

We now examine the properties of functions that modify the structure of the base domain. We call a function *malg* :: $c\ a \to a$ an (Eilenberg–Moore) algebra for the monad c if the following two identities hold

$$malg \circ\ munit = id \tag{A I}$$
$$malg \circ\ mmap\ malg = malg \circ\ mjoin \tag{A II}$$

The identities for an algebra can be shown as the two commuting diagrams.

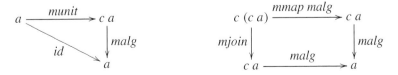

Given two algebras $\alpha : : c\ a \rightarrow a$ and $\beta : : c\ b \rightarrow b$, a *homomorphism* between them is a function $f : : a \rightarrow b$ such that $f \circ \alpha = \beta \circ mmap\ f$.

2.4.5 Transforming Clustered Programs

The categorical identities on collections are useful for transforming clustered programs. This section discusses the two main identities we use (lift1 and lift2), and shows that they follow from those of monads and algebras stated in Section 2.4.4, together with the identity for *cluster*.

We note that, for every a, the function $mjoin : : c\ (c\ a) \rightarrow c\ a$ is an algebra for the monad c (called the free algebra on a), and that, for every $f : : a \rightarrow b, mmap\ f$ is an algebra homomorphism between these free algebras.

Clustering can be introduced into a program by the following identity that holds for all algebra homomorphisms, $f : : c\ a \rightarrow c\ b$. Algorithmically the right-hand side splits the input into clusters, applies f to every cluster and flattens the result.

$$f = decluster \circ lift\ f \circ cluster\ n \qquad \text{(lift1)}$$

$$
\begin{array}{ccc}
c\ (c\ a) & \xrightarrow{\quad lift\ f \quad} & c\ (c\ b) \\
{\scriptstyle cluster\ n} \Big\uparrow & & \Big\downarrow {\scriptstyle decluster} \\
c\ a & \xrightarrow[\quad f \quad]{} & c\ b
\end{array}
$$

Recall that $mmap = lift$ and $mjoin = decluster$. Since f is an algebra homomorphism we know $decluster \circ lift\ f = f \circ decluster$, and (lift1) follows from (C I). The (A II) identity can be used as the following rewrite rule in order to apply the function *malg* to all clusters before combining the results using *malg* again.

$$f = f \circ lift\ f \circ cluster\ n \qquad \text{(lift2)}$$

Again, using $mmap = lift$ and $mjoin = decluster$, this identity has an easy proof. Since *malg* is an algebra for the monad, the identity $malg \circ lift\ malg = malg \circ decluster$ holds by (A II), and (lift2) follows again from (C I).

2.4.6 Strategic Clustering Version of sumEuler

We will now exploit this generic clustering mechanism to provide an improved parallel definition of sumEuler. We first summarize the performance analysis of the (unclustered) data parallel version (see also Section 2.6). In the map-phase of sumEuler it is easy to exploit data parallelism, by computing every euler function in parallel. However, an unclustered version such as this will generally yield a large number of very fine-grained threads, resulting in a speedup that may be close to one.

We have already seen how to improve the granularity of the algorithm by arranging for a whole list chunk to be computed by just one thread, using a parListChunk

```
sumEuler :: Int → Int → Int
sumEuler c n = sum (map euler (mkList n)
                    `using`
                    parListChunk c rnf )
```

FIGURE 2.15:   Strategic clustering version of `sumEuler`.

strategy. We now apply this strategy to the `sumEulerPar1` code in Figure 2.5 and arrive at a strategic clustering version of `sumEuler`, shown in Figure 2.15. This takes the cluster size `c` as an additional parameter and applies the `parListChunk` evaluation strategy to the inner expression. This then generates the list of result values that should be summed. Unfortunately, measurements in Ref. [21] show that the parallel performance of the resulting algorithm is still rather unsatisfactory on typical tightly connected networks. While sufficient parallelism is generated early on in the program, the sequential *fold* at the end of the computation becomes a serious bottleneck, giving typical speedups of only around a factor of 4 on a 16-processor network.

We can identify two reasons for this poor performance. First, the `sum` computation is sequential and this will therefore inevitably generate a sequential tail to the computation, even if it is partially overlapped by the parallel `map`. Second, both the argument to and the result of each parallel thread is a list, and these may require a relatively large amount of time to communicate where large chunks are concerned.

2.4.7   Generic Clustering Version of `sumEuler`

The derivation of a generic clustering version of `sumEuler` proceeds as follows. First we observe that a list is a monad with list-map as the *mmap* and list-append as the *mjoin* function. Since `sum` is defined as `fold (+) 0` in the Haskell prelude, it is an algebra over lists. Finally, `map euler` is an algebra homomorphism between free algebras.

$$
\begin{aligned}
sumEuler &= sum \circ map\ euler & \text{(unfold)} \\
&= sum \circ decluster \circ lift\ (map\ euler) \circ cluster\ z & \text{(lift1)} \\
&= sum \circ lift\ (sum) \circ lift\ (map\ euler) \circ cluster\ z & \text{(A II)} \\
&= sum \circ lift\ (sum \circ map\ euler) \circ cluster\ z & \text{(M ii)}
\end{aligned}
$$

The transformed `sumEuler` code in Figure 2.16 retains separation between the algorithmic and coordination code. In particular, the sequential code can be used as a `worker` function, and the clustering operations are wrapped around the (lifted) worker in order to achieve an efficient parallel version.

This version exhibits a much improved average parallelism on the same modern 16-processor network as before. The sequential tail in the computation has been vastly reduced, to about 15% of the total runtime compared with about 80% of the total runtime in the strategic clustering version. As a consequence of this improved

```
sumEuler :: Int → Int → Int
sumEuler z n = sum ((lift worker) (cluster z (mkList n))
                    `using` parList rnf)
               where worker = sum . map euler
```

FIGURE 2.16: Clustered version of `sumEuler`.

behavior we achieve a relative speedup of 14.3 in a 16-processor configuration. A detailed performance comparison can be found in Ref. [21]. On larger parallel machines it might be advantageous to use a different clustering scheme, which collects every zth element of the list into one block in order to achieve a better load balance. Notably, such a change would effect only the definition of `(de)cluster` but not the code of `sumEuler` itself.

In summary, we have seen how a (manual) program transformation process, controlled by basic monad and algebra laws, was able to significantly improve the performance of a data parallel algorithm. The developed clustering is generic and can be applied to arbitrary algebras and monads. The parallel performance obtained for the unclustered, and both strategic and generic clustered, versions of the program is reported in Section 2.6.

2.5 Implementation of GpH

GpH is implemented by the GUM runtime environment [29], a multithreaded parallel implementation that has been steadily evolved since the early 1990s, and which targets a variety of parallel and distributed architectures ranging from multicore systems through shared-memory machines and distributed clusters to wide-area computational grids [4,30].

2.5.1 Implementation Overview

GUM uses a virtual shared memory model of parallelism, implementing a parallel graph reduction mechanism to handle nonstrict parallel evaluation of GpH programs. In this model, the program constructs a graph structure during execution. Each node in the graph represents a possibly shared (sub)expression that may need to be evaluated. Nodes are evaluated if (and only if) they contribute to the result of the program, realizing lazy evaluation. Following evaluation, the graph node is updated with the value of the subexpression and thus sharing of results is preserved.

Parallelism is introduced by creating threads whose purpose is to evaluate nodes that have been marked using the `par` construct. This may cause other nodes to be evaluated if they are linked as subexpressions of the main node that is evaluated by the thread. Following evaluation, each graph node that has been evaluated by a thread is updated with the result of the evaluation. If one or more threads depend on this result and have therefore been blocked waiting for the node to be evaluated, they may now be notified of the value, and unblocked. We will elaborate on this below.

Since we use a virtual shared graph model, communication can be handled implicitly through accesses to globally shared nodes rather than through explicit system calls. When a thread needs the value of a globally shared node, and this is not available on the local processor, the processor which owns the master copy will be contacted. In the simplest case, when the master copy has already been evaluated, the value of the result is returned immediately, and recorded as a locally cached copy. Similarly, if the master copy has not yet been evaluated, it is returned in an unevaluated form to the requesting processor. The local node will now become the master copy, and the thread will continue by evaluating this node. If the master copy is currently under evaluation, the thread that requested the value becomes blocked until the result is produced, at which point it will be notified of the result.

An important feature of the GUM implementation is the use of a local heap in addition to a global heap. A local heap is used to hold cached copies of global values that will not change in future (the use of a purely functional language ensures there are no cache coherence issues—once produced, the value of a graph node is fixed and will never change), and to hold a graph that is not reachable globally. Since the majority of a graph that is produced falls into the latter category, this is a major advantage. The dual-level approach allows fast independent local collection, integrated with a slower, but much less frequently used global collection mechanism, currently based around distributed reference counting.

2.5.2   Thread Management

A *thread* is a virtual processor that executes a task to evaluate a given graph node. GUM threads are extremely lightweight compared to operating system constructs such as *pthreads*. They are implemented entirely within the language's runtime environment, and contain minimal state: a set of thread-specific registers plus a pointer to a dedicated stack object.

Threads are allocated to processing elements (PEs), which usually correspond to the cores or CPUs that are available on the target system. Each PE has a pool of runnable threads. At each scheduling step, the runtime scheduler selects one of these threads for execution. This thread then runs until either it completes, it blocks, or the system terminates as the result of an error condition (such as insufficient memory). This unfair scheduling approach has the advantage of tending to decrease both the space usage and the overall execution time [12], which is beneficial for parallel execution. However, it is not suitable for concurrency, or for handling speculative threads, since in both cases it is necessary to interleave thread execution.

In GPH, parallelism is introduced using `par` constructs in the source program. When the expression `e1 'par' e2` is evaluated, `e1` is *sparked* for possible future evaluation, and then `e2` is evaluated. Sparking involves recording the graph node associated with `e1` in the current PE's *spark pool*. At a future point, if there is insufficient workload, sparks may be selected from the spark pool and used to construct threads to actually evaluate the closure. Sparking a thunk (an unevaluated expression) is thus a cheap and lightweight operation compared with thread creation, usually involving adding only a pointer to the spark pool.

2.5.3 Spark Stealing

If a PE becomes idle, it will extract a previously saved spark from its local spark pool if it can, and use this to create a new thread. If the spark is no longer useful (e.g., because the node it refers to has already been evaluated by another thread), it will be discarded, and another one chosen for execution.

If the process fails and there are no useful local sparks, then the PE will attempt to find work from another PE. It does this by generating a FISH message that is passed at random from PE to PE until either some work is found, or it has visited a preset number of PEs. If no work is found, then the message is returned to the originating PE, and after a short, tunable delay, a new FISH message is generated.

If the PE that receives a FISH has a useful spark it sends a SCHEDULE message to the PE that originated the FISH, containing the corresponding graph node packaged with a tunable amount of nearby graph. The spark is added to the local spark pool, and an ACK message is sent to the PE that donated the spark.

2.5.4 Memory Management

Parallel graph reduction proceeds on a shared program/data graph, and a primary function of the runtime environment of a parallel functional language is to manage the virtual shared memory in which the graph resides.

In GUM, most sequential execution is exactly as in the standard sequential GHC implementation. This allows the GUM implementation to take advantage of all the sequential optimizations that have been built into GHC. Each PE has its own local heap memory that is used to allocate graph nodes and on which it performs local garbage collections, independently of other PEs. Local heap addresses are normal pointers into this local heap.

Global addresses (GAs) are used to refer to remote objects. A GA consists of a (PE identifier, local identifier) pair. Each PE then maintains a table of in-pointers mapping these local identifiers to specific local addresses. The advantage of this approach over simply using local addresses as part of a GA is that it allows the use of a copying garbage collector, such as the generational collector used in GHC. The garbage collector treats the table of in-pointers as additional roots to the garbage collection. The corresponding local address is then updated to reflect the new location of the object following garbage collection. An important part of this design is that it is easy to turn *any* local address into a GA (so that it can be exported to another PE), simply by creating a new in-pointer.

In order to allow in-pointers to be garbage collected, we use a weighted reference counting scheme [8], where GAs accumulate some weight, which is returned to the owner node if the remote copy of the node is garbage collected. If all weights are returned, the in-pointer becomes garbage, and the local node may be a candidate for recovery during garbage collection.

The only garbage not collected by this scheme consists of cycles that are spread across PEs. We plan ultimately to recover these cycles, too, by halting all PEs and performing a global collective garbage collection, but we have not yet found the need for this in practice, even in very long-lived applications.

2.6 Assessment of Parallel Performance

Taking `sumEuler` as our running example, Figure 2.17 compares the relative speedups of the unclustered version and both strategic and generic clustering versions developed in Section 2.4. The figure reports results on a 16-node Beowulf cluster of Linux RedHat 6.2 workstations with 533 MHz Celeron processors and 128 MB of DRAM connected through a 100 Mb/s fast Ethernet switch. It shows that the unclustered version produces hardly any speedup at all, due to the extreme fine granularity of the generated parallelism. A naive strategic version, which combines the execution of neighboring elements, produces only speedups up to 3.7, mainly owing to the sequential sum operation at the end. The improved generic clustering version avoids this sequential bottleneck at the end and shows a good speedup of 14.3 on 16 processors.

We have used our high-level program development approach on applications from different application areas, and Table 2.1 summarizes key aspects of some example programs. The first column shows the parallel paradigm. The second and third columns show the application domain and program name, respectively. The fourth and fifth columns present the total size of the program and size of the code specifying the parallel coordination, both measured in Source Lines of Code (SLOC). The last two columns report the maximal parallel performance and give a reference to detailed performance measurements. Parallel performance is measured as maximum relative speedup, i.e., speedup over the runtime of the *parallel* program executed on a single processor, on a given number of PEs.

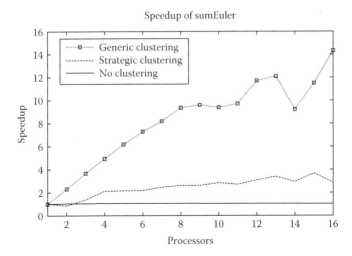

FIGURE 2.17: Speedups of different versions of `sumEuler`. (From Loidl, H-W., Trinder, P., and Butz, C., *Parallel Process. Lett.*, 11, 471, 2001. With permission.)

TABLE 2.1: Program characteristics and performance summary.

Parallel Paradigm	Applications		Size		Max Speed	
	Domain	Name	Code	Coord.	Speedup/PEs	Ref.
Data parallel	Symbolic computation	`linSolv`	121	22	11.9/16	[20]
		`smallGroup`	52	8	26.9/28	[2]
	Graphics	`raytracer`	80	10	8.9/16	[20]
	Numeric	`sumEuler`	31	5	9.1/10	[31]
		`matMult`	43	9	3.4/4	[20]
Divide and conquer	AI	`Queens`	21	5	7.9/16	[20]
Nested parallelism	Natural lang.-processing	`Lolita`	47k	13	1.4/2	[22]
		`Naira`	5k	6	2.5/5	[22]

A key aspect of GpH programming reflected by the fifth column of Table 2.1 is that the parallelization requires only small code changes, localized in only a few modules. This is in stark contrast to the pervasive changes required by a lower-level parallel programming models. The applications require a range of parallelism paradigms. The majority of the applications are symbolic in nature and exhibit irregular parallelism, i.e., both varying numbers and sizes of tasks. Hence the parallelism results should not be directly compared with more regular problems where near-optimal parallelism is achievable. The results show that, despite their very different computational structures, GpH delivers respectable parallel performance for all of the programs and on widely varying architectures, from modern multicores to very high latency computational Grids [4,22,30]. The performance results show that we can obtain good parallelism on local networks, such as Beowulf clusters, e.g., 26.9 on 28PEs [22]. More recent results exhibit almost linear speedup on an 8-core machine [15], and acceptable performance on a wide-area network composed of three local networks: we achieve speedups of 7 on 7 nodes, and up to 16 on 41 nodes for a challenging graphics application [3,4]. Thus, the parallel performance of GpH scales well even on heterogeneous high-latency architectures. We attribute the largely architecture-independent nature of the achieved speedup to the high-level of abstraction provided by our parallel programming model. Unlike most programming models it does not tie a concrete parallel implementation to the specifics of the underlying architecture. Rather, a parallel runtime environment is in charge of coordinating the parallelism by adjusting the coordination to dynamic properties such as the current workload.

2.7 Conclusion

On the one hand, the inherent complexity of designing parallel systems calls for a high-level approach of specifying its behavior. On the other hand, one of the main goals for using parallelism is to improve performance through the coordinated use of many processors on the same application. Thus, an efficient implementation of the specified parallelism is almost as crucial as the correctness of the implementation itself. To reconcile the tension between high-level design and low-level performance, we use GPH, a parallel extension of Haskell, as both specification and as programming language. As the former, it is a useful intermediate step from a high-level algebraic specification to tuned executable code. In this role, it exploits the formal foundations of the purely functional programming language Haskell, and enables the programmer to extensively use code transformations to modify the degree of parallelism in the code. As the latter, it profits from an efficient sequential implementation (based on the highly optimizing GHC), augmented with advanced concepts such as a virtual shared heap, yielding a system of efficient, largely architecture-independent parallelism.

We have demonstrated our high-level program development approach on applications from a range of application areas such as numerical analysis, symbolic computation, and natural language processing. These applications use a range of parallelism paradigms and deliver good parallel performance on a range of widely varying architectures, from modern multicores to very high latency computational Grids [4,22,30]. We have shown that GPH parallelization requires minimal, local refactoring rather than the pervasive changes required by lower-level approaches. We attribute the largely architecture-independent performance of GPH to its high-level parallel programming model. In Ref. [20] we have outlined how GPH's performance compares favorably with both conventional parallel technologies and with other parallel functional languages.

Our experience in realizing numerous parallel applications in various application areas underlines that efficient parallel code can be developed through a transformation-based programming methodology. In this process, a set of tools for controlling and examining the parallel behavior has proven indispensable. Our measurement results indicate good performance both on shared-memory machines and clusters of workstations. The emerging architectures of large-scale, computational grids on the one hand, and multicore, parallel machines on the other hand, are a hard stress-test for the underlying parallel runtime environment. Even though we already achieve acceptable performance on both of these new kinds of architectures, we are currently enhancing the features of the runtime environment to better deal with these new architectures, by adding support for dynamically adapting the chosen policies for work distribution and scheduling.

References

[1] S. Abramsky. The lazy lambda calculus. In *Research Topics in Functional Programming*, Reading, MA, Addison Wesley, 1990, pp. 65–117.

[2] A. Al Zain, K. Hammond, P. Trinder, S. Linton, H.-W. Loidl, and M. Costanti. SymGrid-Par: Designing a framework for executing computational algebra systems on computational grids. In *International Conference on Computational Science (2)*, Reading, MA. *Lecture Notes in Computer Science*, 4488:617–624, 2007.

[3] A. Al Zain, P. Trinder, H-W. Loidl, and G. Michaelson. Supporting high-level grid parallel programming: The design and implementation of grid-GUM2. In *UK e-Science All Hands Meeting*, Nottingham, U.K., EPSRC, September 10–13, 2007, pp. 182–189.

[4] A. Al Zain, P. Trinder, G. Michaelson, and H.-W. Loidl. Evaluating a high-level parallel language (GpH) for computational grids. *IEEE Transactions on Parallel and Distributed Systems*, 19(2):219–233, 2008.

[5] D. Aspinall and M. Hofmann. Dependent types. In *Advanced Topics in Types and Programming Languages*, Cambridge, MA, MIT Press, 2005, pp. 45–86.

[6] C. Baker-Finch, D. King, and P. Trinder. An operational semantics for parallel lazy evaluation. In *ICFP'00—International Conference on Functional Programming*, Montreal, Canada, ACM Press, September 2000, pp. 162–173.

[7] M. Barr and C. Wells. *Category Theory for Computing Science*, New York, Prentice-Hall, 1995. ISBN 0133238091.

[8] D.I. Bevan. Distributed garbage collection using reference counting. In *PARLE'87—Parallel Architectures and Languages Europe*, Eindhoven. *Lecture Notes in Computer Science*, 259:176–187, Eindhoven, June 12–16, 1987.

[9] G.E. Blelloch. Programming parallel algorithms. *Communications of the ACM*, 39(3):85–97, March 1996.

[10] G.H. Botorog and H. Kuchen. Skil: An imperative language with algorithmic skeletons for efficient distributed programming. In *HPDC'96— International Symposium on High Performance Distributed Computing*, Syracuse, New York, IEEE Computer Society Press, August 1996, pp. 243–252.

[11] M. Broy, C. Facchi, R. Grosu, R. Hettler, H. Hußmann, D. Nazareth, F. Regensburger, and K. Stølen. *The Requirement and Design Specification Language* SPECTRUM: *An Informal Introduction*. Technical Report TUM 19311/2, Institut für Informatik, TU München, 1993.

[12] F.W. Burton and V.J. Rayward Smith. Worst case scheduling for parallel functional programming. *Journal of Functional Programming*, 4(1): 65–75, January 1994.

[13] A.A. Chien. Parallelism drives computing. In *Manycore Computing Workshop*, Seattle, WA, June 2007. On-line proceedings.

[14] R. Halstead. Multilisp: A language for concurrent symbolic computation. *ACM Transactions on Programming Languages and Systems*, 7(4):106–117, October 1985.

[15] K. Hammond, A. Al Zain, G. Cooperman, D. Petcu, and P. Trinder. SymGrid: A framework for symbolic computation on the grid. In *EuroPar'07—European Conference on Parallel Processing*, Rennes, France. *Lecture Notes in Computer Science*, 4641:457–466, August 2007.

[16] K. Hammond and G. Michaelson, (Eds.) *Research Directions in Parallel Functional Programming.* Springer, 1999. ISBN 1-85233-092-9.

[17] S. Kahrs, D. Sannella, and A. Tarlecki. The definition of extended ML: A gentle introduction. *Theoretical Computer Science*, 173:445–484, 1997.

[18] P.H.J. Kelly. *Functional Programming for Loosely-Coupled Multiprocessors.* Research monographs in parallel and distributed computing, Cambridge, MA, MIT Press, 1989. ISBN 0262610574.

[19] J. Launchbury. A natural semantics for lazy evaluation. In *POPL'93— Principles of Programming Languages*, Charleston, SC, ACM Press, 1993, pp. 144–154.

[20] H.-W. Loidl, F. Rubio Diez, N. Scaife, K. Hammond, U. Klusik, R. Loogen, G. Michaelson, S. Horiguchi, R. Pena Mari, S. Priebe, A. Rebon Portillo, and P. Trinder. Comparing parallel functional languages: programming and performance. *Higher-order and Symbolic Computation*, 16(3):203–251, 2003.

[21] H.-W. Loidl, P. Trinder, and C. Butz. Tuning task granularity and data locality of data parallel GpH programs. *Parallel Processing Letters*, 11(4):471–486, December 2001.

[22] H.-W. Loidl, P. Trinder, K. Hammond, S.B. Junaidu, R.G. Morgan, and S.L. Peyton Jones. Engineering parallel symbolic programs in GpH. *Concurrency— Practice and Experience*, 11(12):701–752, 1999.

[23] R.S. Nikhil. The parallel programming language Id and its compilation for parallel machines. In *Workshop on Massive Parallelism: Hardware, Programming and Applications*, Amalfi, Italy, 1989. CSG Memo 313.

[24] S.L. Peyton Jones, editor. *Haskell 98 Language and Libraries: The Revised Report*. Cambridge University Press, 2003. ISBN 0521826144. Available at http://www.haskell.org/.

[25] M.J. Plasmeijr, M.C.J.D. van Eekelen, E. Nöcker, and J.E.W. Smesters. The concurrent clean system—Functional programming on the MacIntosh. In *International Conference of the Apple European University Consortium*, Paris, 1991, pp. 14–24.

[26] J.H. Reppy. Concurrent ML: Design, application and semantics. In *Functional Programming, Concurrency, Simulation and Automated Reasoning*, New York, Lecture Notes in Computer Science, 693:165–198, 1993.

[27] S. Skedzielewski. Sisal. In *Parallel Functional Languages and Compilers*, Frontier Series, ACM Press, New York, 1991, pp. 105–158.

[28] P. Trinder, K. Hammond, H.-W. Loidl, and S.L. Peyton Jones. Algorithm + strategy = parallelism. *Journal of Functional Programming*, 8(1):23–60, January 1998.

[29] P. Trinder, K. Hammond, J.S. Mattson Jr., A.S Partridge, and S.L. Peyton Jones. GUM: A portable parallel implementation of Haskell. In *PLDI'96—Programming Languages Design and Implementation*, Philadelphia, PA, May 1996, pp. 79–88.

[30] P. Trinder, H.-W. Loidl, E. Barry Jr., K. Hammond, U. Klusik, S.L. Peyton Jones, and A. Rebon Portillo. The multi-architecture performance of the parallel functional language GpH. In *Euro-Par 2000—Parallel Processing*, Munich. *Lecture Notes in Computer Science*, 1900:739–743, 2000.

[31] P. Trinder, H.-W. Loidl, and R. Pointon. Parallel and distributed Haskells. *Journal of Functional Programming*, 12(4, 5):469–510, July 2002. Special Issue on Haskell.

[32] P. Wadler. Comprehending Monads. *Mathematical Structures in Computer Science*, 2:461–493, 1992.

Chapter 3

Refinement of Parallel Algorithms

Fredrik Degerlund and Kaisa Sere

Contents

3.1 Introduction

In this chapter, we describe how to design correct parallel algorithms using the refinement calculus framework. This is a framework that supports the stepwise derivation of algorithms from their abstract high-level description down to an implementation on some platform using correctness-preserving program transformations. The abstract high-level description of the target parallel algorithm can be, for instance, a sequential algorithm or a mathematical formula that can be used as an initial specification of the problem. We take the view that the developed algorithms will be executed on grid- or cluster-type of networked platforms or processor farms where we have some central resource coordinating the computation, while the computation itself is divided into mutually independent tasks which can be executed on any node of the network. Tasks are typically computationally demanding. The methodology we present is, however, of a more general nature as will be discussed below, and indeed applies to a wide range of parallel and distributed systems.

The refinement calculus [6] generalizes the weakest precondition technique of Dijkstra [10] to prove the correctness of refinement steps during program construction. It is based on the weakest precondition predicate transformer semantics of the underlying programming language. All correctness reasoning is carried out within the calculus relying ultimately on the predicate transformers. However, within the calculus we can define transformation rules and base the derivation on these rules. The rules can be general purpose or targeted and domain specific. The correctness of the transformation rules is proven within the calculus. In this chapter, we will take this approach and define a small collection of transformation rules particularly suitable for derivations within the field of scientific computing.

Parallelism is often formalized in terms of processes. Each process consists of a more or less sequential program. The processes execute in parallel and communicate with each other by sending and receiving messages or through shared memory. Examples of such process-oriented process algebras are CSP [12] and CCS [16]. The reasoning about the behavior of an entire system is based on the behaviors of the single processes. We put forward an alternative approach and model the behavior of parallel and distributed programs in terms of the actions which the system carries out. The action systems formalism we use [4] is a state-based approach in contrast to the event-based process algebras. Moreover, an action system even though modeling parallel computation has an interpretation as a sequential program. The design of the logical behavior of the parallel system is carried out within the context of this purely sequential interpretation, hence simplifying considerably the tedious task of formally verifying and reasoning about parallel programs. The refinement calculus is used as the formal framework for action systems. Event B [15] and UNITY [8] are related formalisms to action systems. Event B has actually been heavily influenced by action systems and it relies on a very similar refinement basis whereas UNITY is based on a temporal logic.

We start in Section 3.2 by describing the refinement calculus for parallel programs and define a specification and programming language for this. Thereafter in Section 3.3, we describe a programming methodology together with a set of transformation rules for designing parallel algorithms. In Section 3.4, we illustrate the methodology with the derivation of a parallel algorithm for integer factorization. We end in Section 3.5 with some concluding remarks.

3.2   Refinement Calculus

In this section, we describe the foundations of the refinement calculus for parallel programs. We define the notion of correct refinement between statements and describe the method of refinement.

3.2.1 Specification Language

We use the classic guarded commands language of Dijkstra [10], with some extensions as our specification language. In this language, we have two syntactic categories, statements and actions.

Statements S are defined by

$S ::=$	$x := e$	((*multiple*) *assignment*)
	$\|$ $\{Q\}$	(*assert statement*)
	$\|$ $S_1; ...; S_n$	(*sequential composition*)
	$\|$ **if** $A_1[]...[]A_m$ **fi**	(*conditional composition*)
	$\|$ **do** $A_1[]...[]A_m$ **od**	(*iterative composition*)
	$\|$ $\|[$ **var** $x; S$ $]\|$	(*block with local variables*)

Here $A_1, \ldots, A_m$ are actions, x is a list of variables, e is a list of expressions, and Q is a predicate.

An *action* (or *guarded command*) A is of the form:

$$A ::= g \rightarrow S$$

where g is a boolean expression (the *guard* of A, denoted gA) and S is a statement (the *body* of A, denoted sA). We will say that an action is *enabled* in a certain state if its guard is true in that state.

The *assert statement* $\{Q\}$ acts as *skip* if the condition Q holds in the initial state. If the condition Q does not hold in the initial state, the effect is the same as *abort*. This statement can be used for specification purposes as it allows us to express conditions on the values of the state variables in a convenient way. The other statements have their usual meanings.

3.2.2 Weakest Precondition Semantics

The refinement relation between statements and between actions is defined with respect to the weakest precondition semantics of these. The *weakest precondition* $wp(S, R)$ for a statement S and any predicate R is a predicate that describes the set of all states such that execution of S begun in any of them is guaranteed to terminate in a state satisfying R. Hence, this is a total correctness semantics requiring statements to terminate. The weakest preconditions of a statement S are listed below [10]:

$wp(x := e, R)$	$=$	$R[e/x]$
$wp(\{Q\}, R)$	$=$	$Q \wedge R$
$wp(S_1; ...; S_n, R)$	$=$	$wp(S_1, wp(S_2; ...; S_n, R))$
$wp(\textbf{if } A_1[]...[]A_m \textbf{ fi}, R)$	$=$	$\left(\bigvee_{i=1}^{m} gA_i\right) \wedge \left(\bigwedge_{i=1}^{m} gA_i \Rightarrow wp(sA_i, R)\right)$
$wp(\textbf{do } A_1[]...[]A_m \textbf{ od}, R)$	$=$	$\exists k.k \geq 0.H_k(R)$
$wp(\|[\textbf{var } x; S]\|, R)$	$=$	$\forall x.wp(S, R)$

The conditions $H_k(R)$ above are given by

$$H_0(R) = R \wedge \neg(\exists j.1 \leq j \leq m.gA_j)$$

and for $k > 0$

$$H_k(R) = wp(\textbf{if } A_1[]...[]A_m \textbf{ fi}, H_{k-1}(R)) \vee H_0(R)$$

The weakest precondition for an action $A = g \rightarrow S$ is defined as

$$wp(g \rightarrow S, R) = (g \Rightarrow wp(S, R))$$

The semantics of the **do–od** statement has also been discussed elsewhere in the literature. Hehner [11] advocates the use of recursion as a means of achieving repetitive behavior, whereas Back [2] discusses the **do–od** construct in the presence of a nondeterministic assignment statement. In neither of these cases is Dijkstra's original definition optimal. In our case, **do–od** is mainly used as a means of making parallelism possible, and we find that Dijkstra's definition is the most suitable for our purposes.

3.2.3   Refinement of Statements

A statement S is said to be (*correctly*) *refined* by statement S', denoted $S \sqsubseteq S'$, if for every postcondition Q,

$$wp(S, Q) \Rightarrow wp(S', Q)$$

In other words, refinement means that whatever total correctness criterion S satisfies, S' will also satisfy this criterion (S' can satisfy other total correctness criteria also, which S does not satisfy).

Intuitively, a statement S is refined by a statement S', if (1) whenever S is guaranteed to terminate, S' is also guaranteed to terminate, and (2) any possible outcome of S' for some initial state is also a possible outcome of S for this same initial state. This means that a refinement may either extend the domain of termination of a statement or decrease the nondeterminism of the statement, or both.

The refinement relation is reflexive and transitive. This means that (1) a statement is always refined by itself (reflexivity) and (2) refinements can be chained or ordered such that a statement in the chain is always refined by the statements occupying a place further down the chain (transitivity). This formally supports the stepwise refinement approach. Moreover, the statement constructors considered here are monotonic with respect to the refinement relation. This allows us to refine substatements in the context of other statements.

3.2.4   Refinement of Actions

Refinement between statements can be extended to a notion of refinement between actions as follows. Let A and A' be two actions. Action A is *refined* by action A', $A \sqsubseteq A'$, if for every postcondition Q,

$$wp(A, Q) \Rightarrow wp(A', Q)$$

When considering the definition of the weakest precondition for an action, we have an action A is refined by an action A' if and only if (1) whenever A' is enabled, the body of A is refined by the body of A' (i.e., $\{gA'\}$; $sA \sqsubseteq sA'$) and (2) A is enabled whenever A' is enabled (i.e., $gA' \Rightarrow gA$).

3.3 Action Systems Formalism

We model parallel computations in terms of action systems. In this section, the action systems formalism will be presented. We will describe how the formalism is used to model parallel computation and give several alternative interpretations for action systems.

3.3.1 Parallel Programming with Action Systems

3.3.1.1 Action Systems

An *action system* $\mathcal{A}$ is a statement of the form:

$$\mathcal{A} = |[\textbf{var } x;\ S_0;\ \textbf{do } A_1[]...[]A_m \textbf{ od}]| : z$$

on *state variables* $y = x \cup z$. The *global* variables z are indicated explicitly for notational convenience. Each variable is associated with some domain of values. The set of possible assignments of values to the state variables constitutes the *state space*. The initialization statement S_0 assigns initial values to the state variables. Due to the total correctness semantics we follow, only the input/output behavior of $\mathcal{A}$ with respect to z is of interest.

The behavior of an action system is as follows: the initialization statement is executed first; thereafter, as long as there are enabled actions, one action at a time is nondeterministically chosen and executed. Actions are atomic, i.e., simultaneously enabled actions can be executed in any order, and only their input–output relationship is of interest. Hence, they can be executed in parallel if they do not have any variables in common.

3.3.1.2 Partitioned Action Systems

An action system can be viewed as a parallel program, when a process network is associated with it. Basically this can be done either by (1) assigning each action to a process ("partitioning with respect to actions"), or by (2) assigning each state variable to a process ("partitioning with respect to variables"). The processes then execute the actions concurrently, guaranteeing that the atomicity requirement of actions is respected.

In this chapter, we will only consider partitioning with respect to actions, or rather an extension of that approach. Consider the action system $\mathcal{A}$ above. Let $\aleph = \{A_1, ..., A_m\}$ be the set of actions and let the multiset $\mathcal{P} = \{P_1, ..., P_p\}$ be a partitioning of $\aleph$, i.e., $P_i \subseteq \aleph$, $P_i \neq \emptyset$, for $i = 1, ..., p$ and $\bigcup_{i=1}^{p} P_i = \aleph$. Each action A_i in the partitioned action system $(\mathcal{A}, \mathcal{P})$ is thus associated with at least one process P_j that it can be executed in. An action can be associated with more than one process, and it may be executed in any one of those processes. We have that

$$\textbf{do } A \textbf{ od} \sqsubseteq \textbf{do } A[]A \textbf{ od}$$

Since the partitions do not have to be mutually disjoint, our solution is not strictly a partitioning in the established mathematical sense of the word.

When there are several legal options, an implementation can decide in which of the processes the action will be executed. However, conflicts between variables have to be taken into account when actions are scheduled. Two actions may not be executed at the same time if they have variables in common, since atomicity must be respected. We consider detection of variable conflicts as an implementation issue, the details of which are outside the scope of this chapter, but we assume that there are mechanisms for communication between processes. Degerlund et al. [9] have proposed methods for implementing action systems partitioned with respect to actions and have developed a proof-of-concept scheduler for executing such action systems.

Example 3.1

To illustrate partitioning, consider the following action system for performing *exchange sort*:

$$
\begin{aligned}
S = |[\quad & x.1, ..., x.n := X_i, ..., X_n \\
& \textbf{do} \\
& [] \, x.1 > x.2 \rightarrow x.1, x.2 := x.2, x.1 \qquad\qquad\qquad (E.1) \\
& \quad ... \qquad\qquad\qquad\qquad\qquad\qquad\qquad\qquad\qquad ... \\
& [] \, x.(n-1) > x.n \rightarrow x.(n-1), x.n := x.n, x.(n-1) \quad (E.(n-1)) \\
& \textbf{od} \\
]| \quad & : x.1, ..., x.n \in integer
\end{aligned}
$$

The system contains $n - 1$ sorting actions: $E.1, ..., E.(n - 1)$, each of which swaps pairs of numbers if they are not in ascending order. We especially note that two consecutive actions (e.g., $E.1$ and $E.2$) cannot be executed simultaneously, since they have variables in common. On the other hand, if the actions are not consecutive (e.g., $E.1$ and $E.3$), they have no common variables, and their execution can take place at the same time. The (extended) partitioning with respect to actions can, for example, be done in the following ways:

$$\mathcal{P}_1 = \{\{E.1, ..., E.(n-1)\}\}$$
$$\mathcal{P}_2 = \{\{E.1\}, ..., \{E.(n-1)\}\}$$
$$\mathcal{P}_3 = \{\{E.1, ..., E.(n-1)\}, ..., \{E.1, ..., E.(n-1)\}\}, \quad |P_3| = n - 1$$

In partitioning $\mathcal{P}_1$, there is only one process, and all actions have to be scheduled on that one. In $\mathcal{P}_2$, there are as many processes are there are actions $(n - 1)$, and each action is tied to a particular process. If there were no variable conflicts, all actions could be executed in parallel in their respective processes. However, in our example there are variable conflicts, so this degree of parallelism is not possible since atomicity has to be taken into account. In partitioning $\mathcal{P}_3$, there are $n - 1$ processes, and any action can be executed in any process (as long as variable dependencies are respected).

3.3.2 Derivation of Action Systems

Action systems can be developed in a stepwise manner [5,18] with the target execution platform or process network in mind. Here, the main focus is on grid-like target platforms, where any action can in principle be executed on any processing unit, to the extent permitted by the partitioning. However, the rules we present below apply to a wider class of platforms. Moreover, according to our experience we believe that specific rules for other types of platforms can be developed based on the refinement calculus for action systems.

Because action systems are just a special kind of sequential statements in our specification language, we can use the refinement calculus for stepwise refinement of action systems. However, as the goal is to transform a more or less sequential algorithm into an action system that can be executed in a parallel fashion, we need special kinds of transformation rules that introduce parallelism into the execution.

As an action system is the unit for which we give an interpretation in terms of parallel activity, we need

(i) Rules to turn arbitrary statements into **do–od** constructs

(ii) Rules for merging sequentially composed action systems

(iii) Rules for changing variables in statements

The first class of rules (i) makes it possible to create action systems out of any statement in our language. An example of a class (i) rule is *Sequence-to-loop*. This rule states that the following refinement can be done:

$$\left\{ \bigwedge_{i=1}^{n} g_i \right\}; S_1; \left\{ \bigwedge_{i=2}^{n} g_i \right\}; S_2; ...; \left\{ \bigwedge_{j=n}^{n} g_n \right\}; S_n$$

$$\sqsubseteq$$

$$\textbf{do } g_1 \rightarrow S_1 [] ... [] g_n \rightarrow S_n \textbf{ od}$$

provided that

1. Each $g_i \rightarrow S_i$ terminates with g_i false

2. $g_i \rightarrow S_i$ and $g_j \rightarrow S_j$ can be executed in any order for all $i, j, i \neq j$

3. $g_i \rightarrow S_i$ cannot affect the value of g_j for all $i, j, i \neq j$

Notice that we use the assert statement here to require that every action in the loop is initially enabled. When inserting an assert statement we use the fact that the statement acts as *skip* when the predicate evaluates to true, having no effect on the state. Moreover, we have to guarantee that each action is executed only once in the loop.

Using the rules in class (ii) we can take full advantage of the divide-and-conquer style of program design and introduce parallelism into execution by merging the action systems for subproblems. One of the rules belonging to class (ii) is the *Merge-into-loop* rule. We have that

$$\textbf{do } g_1 \to S_1 \textbf{ od}; \{g_2\}; S_2$$

$$\sqsubseteq$$

$$\textbf{do } g_1 \to S_1[]g_2 \to S_2 \textbf{ od}$$

provided that

1. S_2 cannot affect the value of g_1

2. $g_1 \to S_1$ and $g_2 \to S_2$ can be executed in any order or $g_2 \to S_2$ cannot precede $g_1 \to S_1$

3. $\textbf{do } g_2 \to S_2 \textbf{ od}$ terminates when g_1 is true

The rule also generalizes to n ($n > 2$) actions.

A similar class (ii) rule is *Merge loops*, which combines two sequentially composed **do–od** loops into a single one. We have that

$$\textbf{do } g_1 \to S_1 \textbf{ od}; \textbf{do } g_2 \to S_2 \textbf{ od}$$

$$\sqsubseteq$$

$$\textbf{do } g_1 \to S_1[]g_2 \to S_2 \textbf{ od}$$

provided that

1. S_2 cannot affect the value of g_1

2. $g_1 \to S_1$ and $g_2 \to S_2$ can be executed in any order or $g_2 \to S_2$ cannot precede $g_1 \to S_1$

3. $\textbf{do } g_2 \to S_2 \textbf{ od}$ terminates when g_1 is true

This rule also generalizes to more than two actions.

Even though the rules in (i)–(ii) allow us to create actions and action systems out of arbitrary statements, they do not necessarily introduce any real parallelism into an execution. For this, we need actions which are independent from each other in the sense that they do not have variables in common and that they can be simultaneously enabled. The rules in class (iii) make actions independent of each other and make it possible to change the way in which the program state is represented by variables in the system, in order to replace a centralized representation by a distributed one. Typically, we replace one variable by a number of variables, for example, one for each process in the partitioning. The new variables hold the same information as the original variable, but permit the information to be accessed in a distributed fashion.

This makes the processes less dependent of each other so that more activity can go on in parallel. One of the rules belonging to this class is the *Exchange variables* rule. We have that

$$\{v_1 = v_2\};\ S$$
$$\sqsubseteq$$
$$S[v_2/v_1]$$

where v_1 and v_2 are variables, and S is a statement containing variable v_1. The conditions for applying the rule is that v_1 is free in S and v_2 is not free is S.

A similar rule is the *Exchange expressions* rule. Considering two expressions, e_1 and e_2, it states that if the assertion $\{e_1 = e_2\}$ holds, an expression S containing the expression e_1 can be refined by S', which is the same expression as S except for all occurrences of e_1 having been replaced by e_2.

We furthermore need general rules and methods for turning a program into a form where the above methods can be applied and also for improving the efficiency of the program in general. Many of these rules are not specific to action systems but apply to any kind of statements. An example of a general rule is the *Add new variables* rule:

$$|[\text{var } x;\ S]| \sqsubseteq |[\text{var } x,\ y;\ S_1;\ S;\ S_2]|.$$

Here, we have the condition that S_1 and S_2 do not write to old variables, and S_1 and S_2 need to terminate.

Examples of how the rules presented here can be used in practice are given in the following section.

3.4   Case Study: Integer Factorization

We present the methodology via a case study about integer factorization. Even though we focus on this particular problem, it follows a general derivation pattern that can be used in other examples as well:

1. We first specify the problem using mathematics and discuss the target parallel platform. Thereafter,

2. the mathematical formulae are manipulated using equivalent representations which more reflect the target architecture for the parallel algorithm that we want to develop. When we feel satisfied with the mathematical representation,

3. we turn the formulae directly into an equivalent specification, usually a sequential statement, in our specification language. Thereafter,

4. we use the divide-and-conquer paradigm and identify the independent tasks of the future computation (rules in class (iii)). Finally,

5. the tasks are turned into loops (class (i) rules) and the loops are merged into a single loop (class (ii)) having the form of an action system.

The main intellectual inventions are done at steps (2) and (4) where we actually invent the parallel algorithm. As for our factorization case study, steps (1) and (2) will be described below in Section 3.4.1, whereas steps (3)–(5) will be handled in Section 3.4.2.

3.4.1   Problem Scenario

We will now give a simple example of how action systems can be used to derive a parallel program. Given a number $num \in \{2, 3, ...\}$ the problem we want to solve is finding the greatest number, $factor \in \{1, ..., num - 1\}$, by which num is evenly divisible, i.e. num mod $factor = 0$. Such a number exists for any accepted values of num, since $\forall num \in \{2, 3, ...\} : num$ mod $1 = 0$.

This task can be expressed by the following formula:

$$factor = max(\{i \in \{1, ..., num - 1\} \mid num \text{ mod } i = 0\})$$

3.4.1.1   Topology

We will do the development with a specific target architecture in mind, for which we will gradually adapt the system. We assume that our architecture contains n identical processors for task computation ("slave") and one additional processor for coordination. We consider a star topology where the coordinating processor should be able to communicate directly with each of the slave processors, but where the slaves do not need to be able to communicate directly with each other. We also assume that one process is executing on each processor. In practice, this can be a grid scenario where the slave processes, as well as the coordination process, are run on computers connected to the Internet. Another possibility that is equally plausible is execution on computers in a local area network (LAN), or even on a single computer containing several processors (SMP) or on a computer consisting of one processor containing several computational cores. Our topology is illustrated in Figure 3.1.

3.4.1.2   Tackling the Problem

Our formula states what we want to achieve in a precise mathematical manner. We will use transformation rules to gradually work toward more concrete statements, of which the final one will not only be a true action system but also be adapted to parallel implementation. During the refinement, we do not state specifically *how* the factorization is performed. However, we do have the method of trial division in mind, and we will tackle the problem accordingly. Consider the set of divisors, $\{1, ..., num - 1\}$, by which we will try to divide num. We will split this set into smaller chunks, or subsets, so that these can be distributed to slave processors for trial division. The granularity of the splitting into subsets is given by a parameter $step \in \{1, ..., num - 1\}$, and each of the subsets will contain $step$ elements. If $num - 1$ is not evenly divisible by $step$, some subsets will contain negative numbers

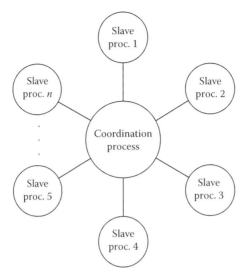

FIGURE 3.1: Process topology used in our example.

that will not affect the final result. We are fully aware of the fact that trial division is not particularly efficient, and that there are more sophisticated factorization algorithms available, such as the *general number field sieve* algorithm. Trial division is, however, very intuitive, and serves our purposes well for showing the general pattern of how parallelism can be introduced when developing software using program refinement.

3.4.1.3 Parallelization Strategy

Our goal is to refine action systems in such a way that n subsets can be tested in parallel on the n processors, i.e., one subset for each processor. We will call such a trial division of n subsets a *round*. Given the parameters num, n, and $step$, the total number of rounds needed will be $r = roof\left(\frac{num-1}{step*n}\right)$. After each round (except the last one), n new subsets can be distributed to the processors. Since actions with common variables are not allowed to execute in parallel, we will refine the system in such a way that the computing actions involved in the same round will not have any variables in common in the final step. Figure 3.2 shows how the splitting takes place when $num = 39$, $n = 3$, and $step = 5$, in which case three rounds will be needed. The last subset (i.e., the one for processor 3 in the last round) contains only negative values and will, in practice, be superfluous. We especially want to emphasize that we work "backward" from the largest number ($num - 1$) down toward 1 (and finally negative numbers if needed for filling the last few subsets in the last round). This means that, for example, processor 1 during round 1 will be assigned the subset containing the largest numbers, i.e., $\{num-step, ..., num-1\}$. The general formulae

	Proc 3	Proc 2	Proc 1
Round 3	−6 to −2	−1 to 3	4 to 8
Round 2	9 to 13	14 to 18	19 to 23
Round 1	24 to 28	29 to 33	34 to 38

FIGURE 3.2: Splitting into computational chunks ($num = 39$, $n = 3$, and $step = 5$).

for deciding the start and stop values for a particular subset as a function of processor and round number are given below:

$$start(round, proc) = num - 1 - (round - 1) * n * step - (proc - 1) * step$$

$$stop(round, proc) = start(round, proc) - step + 1$$

We wish to point out that since we work from higher numbers toward lower ones, the start value is greater than the stop value. In particular, and in full accordance with what we stated above, $start(1, 1) = num - 1$ and $stop(1, 1) = num - step$.

3.4.2 Derivation

We will now define a function for finding the maximum (positive) divisor, by which num is evenly divisible, in an interval $\{low, high\}$ ($1 \leq low \leq high$):

$$search(low, high) = max(\{0\} \cup \{i \in \{low, ..., high\} \mid num \bmod i = 0\}).$$

In case such a divisor does not exist, the value 0 is returned.

We notice that with the parameters $low = 1$, $high = num - 1$, the search function will yield the expression for *factor*. The search function does contain the number 0 in the *max* expression, which our *factor* expression does not. However, with legal values of num, low, and $high$, a (positive) result will always be found, so it is irrelevant whether the number 0 is there or not. Mathematically, the expressions will thus coincide. We can therefore use the *search* function to specify of first version of our statement:

$$\mathcal{A}_0 = |[factor := search(1, ..., num - 1)$$
$$]| : factor \in \mathbb{Z}_+$$

We are now ready to start rewriting the statement to better suit our needs. We will start by splitting the search space into several parts. We split the set $\{1, ..., num - 1\}$ into subsets of the size $step$ numbers. We will also consider rounds, so that n sets are examined in each round. Furthermore, the new variables $found_{11}, ..., found_{rn}$ are introduced to indicate what factor (or 0, if not found) each of the n searches per round has found. Our new version, $\mathcal{A}_1$, is shown below. Note that meta notation

for should be interpreted as a replication of the previous statement, with sequential composition (;) as the delimiter. Later on, we will also use **for** to denote replication where the delimiter is nondeterministic choice ([]). In those cases, we indicate the delimiter by writing [] before the statement to be replicated.

$$\mathcal{A}_1 = |[\text{ \bf var} found_{11}, \ldots, found_{rn} \in \mathbb{Z}_+$$
$$[(found_{ij} := search(stop(i, j), start(i, j))) \text{ \bf for}(1 \leq j \leq n)$$
$$] \text{ \bf for}(1 \leq i \leq r);$$
$$\{search(1, \ldots, num - 1) = max(found_{11}, \ldots, found_{rn})\};$$
$$factor := max(found_{11}, \ldots, found_{rn})$$
$$]| : factor \in \mathbb{Z}_+$$

As compared to $\mathcal{A}_0$, we have replaced the only assignment with $n * r + 1$ sequential assignments. The correctness of the refinement step is proven as follows. The introduction of new variables, as well as statements involving only these variables ($found_{ij} := \ldots$), is justified by the *Add new variables* rule. Having introduced these new statements as shown in $\mathcal{A}_1$, we can show mathematically that the assertion

$$\{search(1, \ldots, num - 1) = max(found_{11}, \ldots, found_{rn})\}$$

holds prior to the *factor* := … statement. Thus, the condition for applying the *Exchange expressions* rule holds. By using this rule to replace the old expression $search(1, \ldots, num - 1)$ with $max(found_{11}, \ldots, found_{rn})$, we achieve an alternative assignment to the factor variable, i.e., $factor := max(found_{11}, \ldots, found_{rn})$. Since we have used two rules (*Add new variables*, *Exchange expressions*), we have actually performed two transformation steps when refining $\mathcal{A}_0$ into $\mathcal{A}_1$.

Next, we introduce partial analysis after each round by adding the new variables $partial_1, \ldots, partial_r$ and logically restructure our block statement:

$$\mathcal{A}_2 = |[\text{ \bf var} found_{11}, \ldots, found_{rn}, partial_1, \ldots, partial_r \in \mathbb{Z}_+$$
$$[(found_{ij} := search(stop(i, j), start(i, j))) \text{ \bf for}(1 \leq j \leq n);$$
$$partial_i := max(found_{i1}, \ldots, found_{in})] \text{ \bf for}(1 \leq i \leq r);$$
$$factor := max(partial_1, \ldots, partial_r)$$
$$]| : factor \in \mathbb{Z}_+$$

Each round will now be analyzed separately and the result of round i will be stored in a variable $partial_i$. After the final round, the variable $factor$ will be assigned the greatest of partial maximum values found after each round, i.e., the maximum value of $partial_1, \ldots, partial_r$. The correctness of this step is proven in a similar manner as the last one, using exactly the same rules.

To illustrate how $\mathcal{A}_2$ works intuitively, Figure 3.3 shows the order in which computations take place as well as what partial results are generated at what stage. The various shades of grey in the figure illustrate what takes place at what time, so that a lighter shade means that the block (subset of numbers) in question is

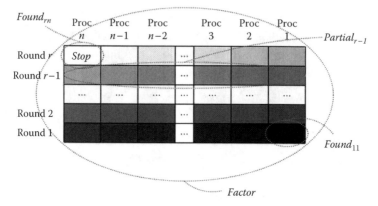

FIGURE 3.3: Computational structure of $\mathcal{A}_2$.

processed at a later time than a darker block. For example, the first subset of numbers (in the lower right-hand corner) will be processed first, and its result (the greatest factor found, or 0 if none is found) will be stored in variable $found_{11}$. Since we have not yet introduced nondeterminism, the subsets (blocks) still cannot be processed in parallel.

Next, we will introduce a **do–od** loop using the *Sequence-to-loop* rule. In this way, we will achieve nondeterministic choice between the search statements within a round. The new variables $comp_1, ..., comp_n$ are added for use in the guards that will be introduced for the search statements. This can be written as

$$\mathcal{A}_3 = \|[\textbf{var}\, found_{11}, ..., found_{rn}, partial_1, ..., partial_r \in \mathbb{Z}_+,$$
$$comp_1, ..., comp_n \in boolean_+$$
$$comp_1, ..., comp_n := true;$$
$$(\,\textbf{do}$$
$$[]comp_j \rightarrow found_{ij} := search(stop(i, j), start(i, j));$$
$$comp_j := false \,\textbf{for}\, (1 \leq j \leq n)$$
$$\textbf{od};$$
$$partial_i := max(found_{i1}, ..., found_{in});$$
$$comp_1, ..., comp_n := true) \,\textbf{for}(1 \leq i \leq r);$$
$$factor := max(partial_1, ..., partial_r)$$
$$]\!| : factor \in \mathbb{Z}_+$$

As for using the *Sequence-to-loop* rule, we have that assertions mentioned in the rule are true at the required points. The guards $comp_1, ..., comp_n$ disable themselves after their respective action has been executed, the actions in question can be executed in any order, and they do not affect each other's guards. Thus, all conditions are fulfilled and the rule can be applied. The intuitive interpretation of the new behavior in $\mathcal{A}_3$ is that the **do–od** loop will terminate after each of the n actions has been running. After this has taken place, a partial maximum value is computed in the same way as in $\mathcal{A}_2$, before the next round starts.

In the next refinement step, we insert the partial maximum computation into the **do–od** loop. Technically, this is justified by using the *Merge-into-loop* rule. Furthermore, we add the new variables $round_1, ..., round_n$ and $roundpart$. Each $round_j$ variable keeps track of which round processor j is currently in. The $roundpart$ similarly indicates the round for which a partial result should next be computed. The resulting $\mathcal{A}_4$ looks as follows:

$\mathcal{A}_4 = |[$ **var** $found_{11}, ..., found_{rn}, partial_1, ..., partial_r \in \mathbb{Z}_+,$
$\qquad comp_1, ..., comp_n \in boolean, round_1, ..., round_n, roundpart \in \mathbb{Z}_+$
$\qquad comp_1, ..., comp_n := true; round_1, ..., round_n := 1; roundpart := 1;$
$\qquad ($ **do**
$\qquad\qquad [] round_j = i \wedge comp_j \rightarrow found_{ij} := search(stop(i, j), start(i, j));$
$\qquad\qquad\qquad comp_j := false; round_j := round_j + 1$ **for**$(1 \leq j \leq n)$
$\qquad\qquad [] roundpart = i \wedge \neg(comp_1 \vee ... \vee comp_n) \rightarrow$
$\qquad\qquad\qquad partial := max(found_{i1}, ..., found_{in}); comp_1, ..., comp_n := true;$
$\qquad\qquad\qquad roundpart := roundpart + 1$
$\qquad$ **od** $)$ **for**$(1 \leq i \leq r);$
$\qquad factor := max(partial_1, ..., partial_r)$
$]| : factor \in \mathbb{Z}_+$

We have now achieved r **do–od** loops, each of which contains actions for the computations of one round. In order to conform to the requirements of an action system, there should however only be one (outer) **do–od** loop. Therefore, we merge the loops by using the *Merge loops* rule repeatedly, after verifying that its conditions of use hold, until we achieve $\mathcal{A}_5$:

$\mathcal{A}_5 = |[$ **var** $found_{11}, ..., found_{rn}, partial_1, ..., partial_r \in \mathbb{Z}_+,$
$\qquad comp_1, ..., comp_n \in boolean, round_1, ..., round_n, roundpart \in \mathbb{Z}_+$
$\qquad comp_1, ..., comp_n := true; round_1, ..., round_n := 1; roundpart := 1;$
$\qquad$ **do**
$\qquad\qquad [] ([] round_j = i \wedge comp_j \rightarrow$
$\qquad\qquad\qquad found_{ij} := search(stop(i, j), start(i, j)); comp_j := false;$
$\qquad\qquad\qquad round_j := round_j + 1$ **for**$(1 \leq j \leq n)$
$\qquad\qquad [] roundpart = i \wedge \neg(comp_1 \vee ... \vee comp_n) \rightarrow$
$\qquad\qquad\qquad partial := max(found_{i1}, ..., found_{in}); comp_1, ..., comp_n := true;$
$\qquad\qquad\qquad roundpart := roundpart + 1)$ **for**$(1 \leq i \leq r)$
$\qquad$ **od**$;$
$\qquad factor := max(partial_1, ..., partial_r)$
$]| : factor \in \mathbb{Z}_+$

We are almost finished at this stage, but we still have to insert the final maximum computation into the **do–od** loop. After verifying that the conditions for using the *Merge-into-loop* are fulfilled, we apply the rule and achieve our final system $\mathcal{A}_6$:

$\mathcal{A}_6 = |[\textbf{var} \ found_{11}, ..., found_{rn}, partial_1, ..., partial_r \in \mathbb{Z}_+,$
$\qquad comp_1, ..., comp_n \in boolean_+, round_1, ..., round_n, roundpart \in \mathbb{Z}_+$
$\qquad comp_1, ..., comp_n := true; \ round_1, ..., round_n := 1; \ roundpart := 1;$
$\quad \textbf{do}$
$\qquad [] \ ([] \ round_j = i \wedge comp_j \rightarrow$
$\qquad\qquad found_{ij} := search(stop(i, j), start(i, j));$
$\qquad\qquad comp_i := false;$
$\qquad\qquad round_j := round_j + 1 \ \textbf{for}(1 \leq j \leq n) \qquad\qquad (Search_{ij})$
$\qquad\quad [] \ roundpart = i \wedge \neg(comp_1 \vee ... \vee comp_n) \rightarrow$
$\qquad\qquad partial := max(found_{i1}, ..., found_{in});$
$\qquad\qquad comp_1, ..., comp_n := true;$
$\qquad\qquad roundpart := roundpart + 1) \ \textbf{for}(1 \leq i \leq r) \qquad (Partial_i)$
$\qquad [] \ \bigwedge_{1 \leq j \leq n} round_j > r \wedge roundpart > r \rightarrow$
$\qquad\qquad factor := max(partial_1, ..., partial_r) \qquad\qquad (Final)$
$\quad \textbf{od}$
$]| : factor \in \mathbb{Z}_+$

In this final version, $\mathcal{A}_6$, we note especially three facts: first, $\mathcal{A}_6$ is a correct refinement of our initial statement $\mathcal{A}_0$, i.e., $\mathcal{A}_0 \sqsubseteq \mathcal{A}_6$. This is a direct result of the transitivity property of the refinement relation. Second, the system fully adheres to the action system requirements, which includes that there is a single **do–od** loop, consisting of actions separated by the nondeterministic choice delimiter ([]). Third, search actions ($Search_{ij}$, $1 \leq i \leq r$, $1 \leq j \leq n$) corresponding to the same round (i.e., with the same value of i) do not contain any shared variables. This implies that they can be scheduled for simultaneous execution, as long as we use a proper partitioning. A suitable partitioning would be

$$\mathcal{P}_1 = \{\{Search_{11}, ..., Search_{r1}\}, ..., \{Search_{1n}, ..., Search_{rn}\},$$
$$\{Partial_1, ..., Partial_r, Final\}\}$$

In $\mathcal{P}_1$, search actions with the same second index (corresponding to processor number in terms of search subsets) are part of the same partition. In this way, each of these partitions contains one action corresponding to each computations round. There is also one partition that contains the actions responsible for computing the partial results after each round, as well as the final result. Partitioning $\mathcal{P}_1$ can be conveniently mapped to the topology we had in mind when developing the action system. This mapping is illustrated in Figure 3.4.

The system can also be partitioned in an alternative way as follows:

$$\mathcal{P}_2 = \{\{Search_{11}, ..., Search_{rn}\}, ..., \{Search_{11}, ..., Search_{rn}\},$$
$$\{Partial_1, ..., Partial_r, Final\}\}, \ |\mathcal{P}_2| = n + 1$$

Here, all search actions are included in all partitions, except for a partition that is dedicated for coordination. This partitioning can be mapped onto our topology as shown in Figure 3.5. In this way, specific actions are not tied to specific processors,

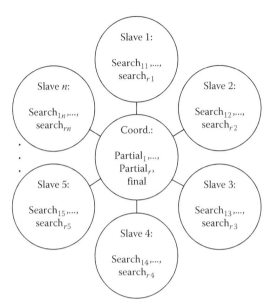

FIGURE 3.4: Partitioning $\mathcal{P}_1$ of the final system.

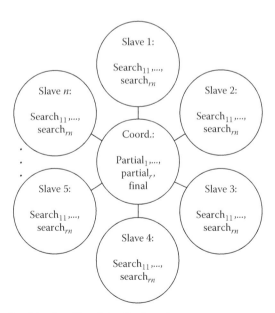

FIGURE 3.5: Partitioning $\mathcal{P}_2$ of the final system.

but the actions can be executed on any available processor (except for the coordinating one). As mentioned in Section 3.3.1, we consider the practical details concerning communication between processes to be outside the scope of this chapter, but we assume an implementation providing the required interprocess communication mechanisms. Such a framework for implementing action systems has been discussed by Degerlund et al. [9]. Furthermore, Boström and Waldén [7] have proposed a specification language for facilitating development of grid systems. Their language, Distributed B, is based on Event B, which in turn is similar to the action systems formalism.

3.5   Concluding Remarks

In this chapter, we have presented a design methodology for the derivation of provably correct parallel algorithms expressed as action systems. We have shown how the combination of refinement calculus and action systems formalism gives a foundation for a design methodology. The method is based on systematic construction of programs by stepwise refinement. Each step in the derivation can be formally proven correct. The main emphasis is on a methodology that allows the programmer to start with a traditional, purely sequential algorithm and gradually turn it into a parallel version. We argue that the transformation rules discussed allow the designer to uncover the potential parallelism in the program in a more innovative and general manner than what a parallelizing compiler can do.

Our presentation was mainly targeting grid- and cluster-type parallel platforms. The main difficulty of grid programming [13], task parallelization of the algorithm, is handled here via the program transformations. The main novelty here is the focus on correctness and formal program transformation when parallelizing a system. The starting point can be a problem expressed using mathematical formulae (as in the case study of Section 3.4), or an already existing algorithm, that may be sequential, and which is turned into a parallel program using transformations.

The presented method is rather general and the target platform can be defined freely via the variable or action partitioning as seen via the derivations. As was observed, we started from the mathematical definitions of the problems at hand and used abstract, high-level constructs to give the initial specifications. In this way our language resembles the programming language Fortress [19] which also allows mathematical notation. Moreover, our approach is independent of programming language compared to many other systems where more concrete Java or C/C++ programming is required [1,17,20]. Indeed, we can define classes of action systems, which can be implemented on different execution platforms and using different programming languages [5,9,18]. Furthermore, the focus here was on total correctness reasoning about parallel algorithms. Action systems are also used to model reactive behavior [3]. Related to this are formalisms such as TLA+ [14] and input/output automata, as well as Event B [15].

References

[1] M. Aldinucci, M. Coppola, M. Danelutto, N. Tonellotto, M. Vanneschi, and C. Zoccolo. High level grid programming with ASSIST. *Computational Methods in Science and Technology*, 12(1):21–32, 2006.

[2] R.J.R. Back. Proving total correctness of nondeterministic programs in infinitary logic. *Acta Informatica*, 15(3):233–249, 1981.

[3] R.J.R. Back. Refinement calculus. Part II: Parallel and reactive programs. In *Stepwise Refinement of Distributed Systems. Models, Formalisms, Correctness (REX Workshop Proceedings)*, Mook, the Netherlands, *Lecture Notes in Computer Science*, 430:67–93, 1990.

[4] R.J.R. Back and R. Kurki-Suonio. Decentralization of process nets with centralized control. In *PODC '83: Proceedings of the Second Annual ACM Symposium on Principles of Distributed Computing*, New York, ACM Press, 1983, pp. 131–142.

[5] R.J.R. Back and K. Sere. Stepwise refinement of action systems. *Structured Programming 12*, pp. 17–30, 1991.

[6] R.J.R. Back and J. von Wright. *Refinement Calculus: A Systematic Introduction*. Springer-Verlag, New York, 1998.

[7] P. Boström and M. Waldén. An extension of Event B for developing grid systems. In *ZB 2005: Formal Specification and Development in Z and B*, Guildford, U.K., *Lecture Notes in Computer Science*, 3455:142–161, 2005.

[8] K.M. Chandy and J. Misra. *Parallel Program Design: A Foundation*. Addison-Wesley Publishing Company, Reading, MA 1988.

[9] F. Degerlund, M. Waldén, and K. Sere. Implementation issues concerning the action systems formalism. In David S. Munro, Hong Shen, Quan Z. Sheng, Henry Detmold, Katrina E. Falkner, Cruz Izu, Paul D. Coddington, Bradley Alexander, and Si-Qing Zheng, editors, *Proceedings of the Eighth International Conference on Parallel and Distributed Computing Applications and Technologies (PDCAT'07)*, IEEE Computer Society Press, 2007, pp. 471–479.

[10] E. Dijkstra. *A Discipline of Programming*. Prentice Hall International, Englewood Cliffs, NJ, 1976.

[11] E. Hehner. do considered od: A contribution to the programming calculus. *Acta Informatica*, 11(4):287–304, 1979.

[12] C.A.R. Hoare. *Communicating Sequential Processes*. Prentice-Hall International, Englewood Cliffs, NJ, 1985.

[13] D. Laforenza. Grid programming: some indications where we are headed. *Parallel Computing*, 28(12):1733–1752, 2002.

[14] L. Lamport. *Specifying Systems: The TLA+ Language and Tools for Hardware and Software Engineers.* Addison-Wesley Longman, Boston, MA 2002.

[15] C. Métayer, J.-R. Abrial, and L. Voisin. Event-B Language. RODIN Deliverable D7. EU Project RODIN (IST-511599), 2005. http://rodin.cs.ncl.ac.uk/deliverables.htm.

[16] R. Milner. *A Calculus of Communicating Systems. Lecture Notes in Computer Science,* 92, 1980.

[17] T.-A. Nguyen and P. Kuonen. Programming the grid with POP-C++. *Future Generation Computer Systems,* 23(1):23–30, 2007.

[18] K. Sere. *Stepwise derivation of parallel algorithms.* PhD thesis, Åbo Akademi University, Åbo, Finland, 1990.

[19] G. Steele. The Fortress parallel programming language, 2006. Presentation slides from 2006 GSPx multicore applications conference, http://research.sun.com/projects/plrg/Publications/GSPx-Lecture2006 public.pdf.

[20] R. van Nieuwpoort, J. Maassen, T. Kielmann, and H.E. Bal. Satin: Simple and efficient Java-based grid programming. *Scalable Computing: Practice and Experience,* 6(3):19–32, 2005.

Part II

Distributed Systems

Chapter 4

Analysis of Distributed Systems with mCRL2

Jan Friso Groote, Aad Mathijssen, Michel A. Reniers, Yaroslav S. Usenko, and Muck van Weerdenburg

Contents

4.1 Introduction

The language mCRL2 [14] is a process algebra with data and time suitable for the specification and verification of a broad range of (typically real-time distributed) systems.

Specific characteristics of the mCRL2 process algebra are local communication, multiactions, and communication-independent parallelism. Local communication allows one to specify communications within subcomponents of the system without affecting the rest of the system. This makes the language particularly suitable for component-based and hierarchical systems. Global communication, which is typically used in other languages, only allows communication to be defined on a global level for the system as a whole. Multiactions enable the specification of (not necessarily related) actions that are to be executed together. Most process algebras only allow a single action to be executed atomically thus forcing an order on the execution of actions. By allowing multiple actions to be executed together, certain systems (such as low-level hardware) can be modeled in a more straightforward manner. These multiactions also allow the separation of parallelism and communication. When two actions can execute at the same time, a multiaction with those actions is the result. Communication can then be applied to these multiactions to make certain actions communicate with each other.

Besides the process algebra, mCRL2 also has a higher-order data language. Processes and actions can be parameterized with this data, which is essential for the description of most practical systems. Besides several built-in data types (e.g., integers, lists, and sets) the user can define his or her own abstract data types in a straightforward manner. Typical functions, such as equality, are automatically generated for such user-defined data types. As the data language is higher-order, functions are first-class citizens and can therefore be used just as easily as other data.

For practical use we have developed a toolset for mCRL2. This toolset contains numerous tools for manipulation of specifications, state space exploration and generation, simulation, verification, and visualization. To be able to work efficiently a specification is transformed into a specific form called a linear process. Such a linear process can be seen as a compact symbolic representation of the (possibly infinite) state space of the system. Many optimization and verification techniques have been developed for these linear processes (see Section 4.3). A typical practice in verification via model checking is to generate a state space of a specification and use that state space to check certain properties. The mCRL2 toolset supports this practice, but also offers an alternative. Given a specification and a (modal) formula expressing a desired (temporal) property, the toolset can generate a parameterized boolean equation system (PBES) [19,21]. Such a PBES can then be solved by specific provers of the toolset. The use of PBESs is particularly interesting for systems that are too large or complex for model checking via explicit state spaces. The mCRL2 toolset has been successfully used in various academic and industrial case studies; see Refs. [27,36,43], or Ref. [29] for an overview.

mCRL2 is the successor of μCRL [15,16] and its timed version timed μCRL [13,37]. The languages are very similar: any process description in (timed) μCRL can be translated straightforwardly into a description of the same process in mCRL2. For a discussion on the differences and their motivation we refer the reader to Section 4.5 and Refs. [14,20].

In Section 4.2, we introduce the mCRL2 language. In Section 4.3, we give an in-depth overview of the toolset. We further illustrate the possibilities offered by the mCRL2 language and toolset by means of some examples in Section 4.4. In Section 4.5, we place mCRL2 in its historical context.

4.2 mCRL2 Language

We define the syntax of the mCRL2 process language together with an intuitive explanation of the semantics. After that, we discuss the data language that is used by the process language.

4.2.1 Process Language

This section gives an overview of the mCRL2 process language. The precise semantics—in terms of timed transition systems and by means of structural operational semantics (SOS)—can be found in Ref. [20, Section 3]. The axioms of mCRL2 can be found in Ref. [20, Section 2].

4.2.1.1 Multiactions

The primary notion in the mCRL2 process language is an *action*, which represents an elementary activity or a communication of some systems. The following example illustrates how action names send, receive, and error can be declared. Actions can be *parameterized* with data. For example:

act error:
     send : $\mathbb{B}$:
     receive : $\mathbb{B} \times \mathbb{N}$:

This declares parameterless action name error, action name send with a data parameter of sort $\mathbb{B}$ (booleans), and action name receive with two parameters of sort $\mathbb{B}$ and $\mathbb{N}$ (natural numbers), respectively. For the above action name declaration, error, send(*true*), and receive(*false*, 6) are valid actions. The means offered by mCRL2 to define sorts and operations on sorts will be discussed in Section 4.2.2.

In general, we write a, b, ... to denote action names and $\vec{d}, \vec{e}, \ldots$ to denote vectors of data parameters. In the notation $a(\vec{d})$, we assume that the vector of data parameters $\vec{d}$ is of the type that is specified for the action name a. An action without data parameters can be seen as an action with an empty vector of data parameters. We often write $\alpha, \beta, \ldots$ for multiactions.

Actions are allowed to occur simultaneously in mCRL2. In this case, we speak about multiactions. A multiaction is a bag of actions that is constructed according to the following BNF grammar:

$$\alpha \;::=\; \tau \;|\; a(\vec{d}) \;|\; \alpha \,|\, \beta.$$

The term τ represents the multiaction containing no (observable) actions; it is called the hidden or internal action. The term $a(\vec{d})$ represents a multiaction that contains only (one occurrence of) the action $a(\vec{d})$ and the term $\alpha \mid \beta$ represents a multiaction containing the actions from both the multiaction α and β.

As multiactions are bags we have the usual equalities on multiactions, viz. commutativity and associativity of $|$. Note that τ acts as the unit element of $|$. That is, $\alpha \mid \tau = \alpha = \tau \mid \alpha$. Using the actions declared previously the following are considered multiactions: τ, error $|$ error $|$ send(*true*), send(*true*) $|$ receive(*false*, 6), and $\tau \mid$ error.

4.2.1.2 Basic Operators

Process expressions, denoted by $p, q, \ldots$, describe when certain multiactions can be executed. For example, "*a* is followed by either *b* or *c*." We make this notion more formal by introducing operators. The most basic expressions are as follows:

- *Multiactions* (α, β, etc.) as described above.

- *Deadlock* or inaction δ, which does not execute any multiactions, but only displays delay behavior.

- *Alternative composition*, written as $p + q$. This expression nondeter-ministically chooses to execute either p or q. The choice is made upon performance of the first multiaction of p or q.

- *Sequential composition*, written $p \cdot q$. This expression first executes p and upon termination of p continues with the execution of q.

- *Conditional operator*, written $c \rightarrow p \diamond q$, where c is a data expression of sort $\mathbb{B}$. This process expression behaves as an if-then-else construct: if c is *true* then p is executed, else q is executed. The else part is optional. This operator is used to express that data can influence process behavior.

- *Process references*, written $\mathsf{P}(\vec{d})$, $\mathsf{Q}(\vec{d})$, etc., are used to refer to processes declared by *process definitions* of the form $\mathsf{P}(\vec{x} : \vec{D}) = p$. This process definition declares that the behavior of the process reference $\mathsf{P}(\vec{d})$ is given by $p[\vec{d}/\vec{x}]$, i.e., p in which all free occurrences of variables $\vec{x}$ are replaced by $\vec{d}$.

- *Summation operator*, written as $\sum_{x:D} p$, where x is a variable of sort D and p is a process expression in which this variable may occur. In this case, we say that x is bound in p. The corresponding behavior is a nondeterministic choice among the processes $p[d/x]$ for all data expressions d of sort D. If $\{d_0, d_1, \ldots, d_n, \ldots\}$ are the data expressions of sort D, then $\sum_{x:D} p$ can be expressed as $p[d_0/x] + p[d_1/x] + \cdots + p[d_n/x] + \cdots$.

- *At operator*, written $p \triangleleft t$, where t is a data expression of sort $\mathbb{R}$ (real numbers). The expression $p \triangleleft t$ indicates that any first multiaction of p happens at time t (in case $t < 0$, $p \triangleleft t$ is equal to $\delta \triangleleft 0$, the latter representing an immediate deadlock).

When writing process expressions we usually omit parentheses as much as possible. To do this, we define precedence rules for the operators. The precedence of the operators introduced so far, in decreasing order, is as follows: ς, $\cdot$, $\rightarrow$, $\sum$, $+$. Furthermore, $\cdot$ and $+$ are associative. So, instead of writing $(a \cdot (b \cdot c)) + (d + e)$ we usually write $a \cdot b \cdot c + d + e$.

Often processes have some *recursive* behavior. A coffee machine, for example, will normally not stop (terminate) after serving only one cup of coffee. To facilitate this, we use process references and process definitions:

act   coin, break, coffee;

proc   Wait = coin $\cdot$ Serve;

         Serve = break $\cdot$ δ + coffee $\cdot$ Wait;

This declares process references (often just called processes) Wait and Serve. Process Wait can do a coin action, after which it behaves as process Serve. Process Serve can do a coffee action and return to process Wait, but it might also do a break action, which results in a deadlock.

A complete process specification needs to have an *initial process*. For example:

init   Wait;

Parameterized processes can be declared as follows:

proc $P(c : \mathbb{B}, n : \mathbb{N})$ = error $\cdot$ $P(c, n)$

                 + send(c) $\cdot$ $P(\neg c, n + 1)$

                 + receive(c, n) $\cdot$ $P(\textit{false}, \textit{max}(n - 1, 0))$;

This declares the process $P(c, n)$ with data parameter c of sort $\mathbb{B}$ and n of sort $\mathbb{N}$. Note that the sorts of the data parameters are declared on the left-hand side of the definition. In the process references on the right-hand side the *values* of the data parameters are specified.

Summation is used to *quantify* over data types. Summations over a data type are particularly useful to model the receipt of an arbitrary element of a data type. For example, the following process is a description of a single-place buffer, repeatedly reading a natural number using action name read, and then delivering that value via action name send.

act   read, send : $\mathbb{N}$;

proc   Buffer = $\displaystyle\sum_{n:\mathbb{N}}$ read(n) $\cdot$ send(n) $\cdot$ Buffer;

init   Buffer;

Time can be added to processes using the operator ς. We give a few examples of the use of this operator. To start with, we specify a simple clock:

act   tick;

proc $C(t : \mathbb{R})$ = tick ς $(t + 1)$ $\cdot$ $C(t + 1)$;

init $C(0)$;

For a positive value u of sort $\mathbb{R}$, the process $\mathsf{C}(u)$ exhibits the single infinite trace $\mathsf{tick} \triangleleft (u + 1) \cdot \mathsf{tick} \triangleleft (u + 2) \cdot \mathsf{tick} \triangleleft (u + 3) \cdot \cdots$.

As a different example, we show a model of a *drifting* clock (taken from Ref. [49]). This is a clock that is accurate within a bounded interval $[1 - \eth, 1 + \eth]$, where $\eth < 1$.

proc $DC(t : \mathbb{R}) = \displaystyle\sum_{\epsilon : \mathbb{R}} (1 - \eth \leq \epsilon \wedge \epsilon \leq 1 + \eth) \rightarrow \mathsf{tick} \triangleleft (t + \epsilon) \cdot DC(t + \epsilon);$

init $DC(0);$

4.2.1.3   Parallel Composition

Besides the basic operators, which are typically used to specify the behavior of core components in the system, we also have *parallel composition* (or *merge*) to compose processes. We write $p \parallel q$ for the parallel composition of p and q. This means that the multiactions of p and q are *interleaved* and *synchronized*. The parallel composition binds stronger than the summation operator, but weaker than the conditional operator.

A simple example is $\mathsf{a} \parallel \mathsf{b}$, which will either first execute a followed by b, or first b followed by a, or a and b together. That is, $\mathsf{a} \parallel \mathsf{b}$ is equivalent to $\mathsf{a} \cdot \mathsf{b} + \mathsf{b} \cdot \mathsf{a} + \mathsf{a} \mid \mathsf{b}$. Another example is $\mathsf{a} \cdot \mathsf{b} \parallel (\mathsf{c} + \mathsf{d})$, which is equivalent to $\mathsf{a} \cdot (\mathsf{b} \cdot (\mathsf{c} + \mathsf{d}) + \mathsf{c} \cdot \mathsf{b} + \mathsf{d} \cdot \mathsf{b} + \mathsf{b} \mid \mathsf{c} + \mathsf{b} \mid \mathsf{d}) + \mathsf{c} \cdot \mathsf{a} \cdot \mathsf{b} + \mathsf{d} \cdot \mathsf{a} \cdot \mathsf{b} + \mathsf{a} \mid \mathsf{c} \cdot \mathsf{b} + \mathsf{a} \mid \mathsf{d} \cdot \mathsf{b}$.

4.2.1.4   Additional Operators

Now that we are able to put various processes in parallel, we need ways to restrict the behavior of this composition and to model the interaction between processes. For this purpose, we introduce the following operators:

- *Restriction operator* $\nabla_V(p)$ (also known as *allow*), where V is a set consisting of (nonempty) bags of action names specifying exactly which multiactions from p are allowed to occur. Restriction $\nabla_V(p)$ disregards the data parameters of the multiactions in p when determining if a multiaction should be allowed, e.g., $\nabla_{\{\mathsf{b} \mid \mathsf{c}\}}(\mathsf{a}(0) + \mathsf{b}(true, 5) \mid \mathsf{c}) = \mathsf{b}(true, 5) \mid \mathsf{c}$.

  Note that the empty multiaction τ is not allowed as an element of the set V, so τ is always allowed (for any V, $\nabla_V(\tau) = \tau$).

- *Blocking operator* $\partial_B(p)$ (also known as *encapsulation*), where B is a set of action names that are *not* allowed to occur. Blocking $\partial_B(p)$ disregards the data parameters of the actions in p when determining should be blocked, e.g., $\partial_{\{\mathsf{b}\}}(\mathsf{a}(0) + \mathsf{b}(true, 5) \mid \mathsf{c}) = \mathsf{a}(0)$.

  Note that τ is not allowed as an element of B, as B is a set of action names (not a bag of multiaction names).

- *Renaming operator* $\rho_R(p)$, where R is a set of renamings of the form $a \rightarrow b$, meaning that every occurrence of action name a in p is replaced by action name b. This set R is required to be a function, i.e., every action name may only occur once as a left-hand side of $\rightarrow$ in R. Renaming disregards the data parameters, but when a renaming is applied the data parameters are retained, e.g., $\rho_{\{a \rightarrow b\}}(a(0) \mid b + a \mid c) = b(0) \mid b + b \mid c$.

- *Communication operator* $\Gamma_C(p)$, where C is a set of allowed communications of the form $a_0 \mid \cdots \mid a_n \rightarrow c$, with $n \geq 1$ and a_i and c action names. For each communication $a_0 \mid \cdots \mid a_n \rightarrow c$, multiactions of the form $a_0(\vec{d}) \mid \cdots \mid a_n(\vec{d}) \mid \alpha$ (for some $\vec{d}$) in p are replaced by $c(\vec{d}) \mid \alpha$. Note that the data parameters are retained in action c. For example, $\Gamma_{\{a \mid b \rightarrow c\}}(a(0) \mid b(0)) = c(0)$, but also $\Gamma_{\{a \mid b \rightarrow c\}}(a(0) \mid b(1)) = a(0) \mid b(1)$. Furthermore, $\Gamma_{\{a \mid b \rightarrow c\}}(a(1) \mid a(0) \mid b(1)) = a(0) \mid c(1)$.

  The left-hand sides of the communications in C should be disjoint (e.g., $C = \{a \mid b \rightarrow c, a \mid d \rightarrow e\}$ is not allowed) to ensure that the communication operator uniquely maps each multiaction to another.

With these additional operators we can, for example, enforce communication between processes. A typical example is as follows:

act send, read, c;

init $\nabla_{\{c\}}(\Gamma_{\{send \mid read \rightarrow c\}}(send \parallel read));$

Since $send \parallel read$ is equivalent to $send \cdot read + read \cdot send + send \mid read$, the communication operator replaces the occurrence of $send \mid read$ by c giving $send \cdot read + read \cdot send + c$. The restriction operator only allows execution of the c alternative. That is, the above process is equivalent to just c.

Passing a data value from one process to another can also be achieved using the communication and restriction operations. For instance:

$$\nabla_{\{c\}}(\Gamma_{\{s \mid r \rightarrow c\}}(s(3) \cdot p \parallel \sum_{n : \mathbb{N}} r(n) \cdot Q(n))) = c(3) \cdot \nabla_{\{c\}}(\Gamma_{\{s \mid r \rightarrow c\}}(p \parallel Q(3)))$$

In this case, the value 3 is passed from one parallel component to the other to become the value of the variable n in process $Q(n)$.

4.2.1.5 Abstraction

An important notion in process algebra is that of *abstraction*. Usually the requirements of a system are defined in terms of *external* behavior (i.e., the interactions of the system with its environment), while one wishes to check these requirements on an implementation of the system which also contains *internal* behavior (i.e., the interactions between the components of the system). So it is desirable to be able to abstract from the internal behavior of the implementation. For this purpose the following constructs are available:

- *Internal action* or silent step τ, which is the empty multiaction that indicates that some (unknown) internal behavior happens.

- *Hiding operator* $\tau_I(p)$, which hides (or renames to τ) all actions with an action name in I from p. Hiding $\tau_I(p)$ disregards the data parameters of the actions in p when determining if an action should be hidden.

Assume we have two simple processes P and Q that are to be run in parallel. Furthermore, we wish that P executes an a and Q a b and that the a is followed by b. We could then write the following specification.

act a, b, r, s, i;
proc P = a · s;
     Q = r · b;
init $\nabla_{\{a,b,i\}}(\Gamma_{\{r|s \rightarrow i\}}(P \parallel Q))$;

The system described by this specification is equivalent to the process a · i · b. That is, first P executes its a action, then P and Q synchronize (producing i as a result) such that Q can execute its b action.

As action i is only used to synchronize P and Q, we are not really interested to see it in the resulting process. Instead, we can hide the result of this synchronization as in the following process:

init $\tau_{\{i\}}(\nabla_{\{a,b,i\}}(\Gamma_{\{a|b \rightarrow i\}}(P \parallel Q)))$;

This process is equivalent to a · τ · b. Under certain equivalences (see Refs. [45,46]) such as weak and branching bisimilarity, this process is also equivalent to a · b (as the τ is unobservable).

4.2.2 Data Language

The mCRL2 data language is a functional language based on *higher-order abstract data types* [30,31,34], extended with concrete data types: *standard data types* and sorts constructed from a number of *type formers*.

4.2.2.1 Basic Data Type Definition Mechanism

Basically, mCRL2 contains a simple and straightforward data type definition mechanism. Sorts (types), constructor functions, maps (functions), and their definitions can be declared. Sorts declared in such a way are called user-defined sorts. For instance, the following declares the sort A with constructor functions c and d. Also functions f, g, and h are declared and (partially) defined:

sort A;

cons $c, d : A$;

map $f : A \times A \to A$;

   $g : A \to A$;

   $h : A$;

var $x : A$;

eqn $f(c, x) = c$;

   $f(d, x) = x$;

   $g(c) = c$;

   $h = d$;

In the equations, *variables* are used to represent arbitrary data expressions. A sort is called a *constructor sort* when it has at least one constructor function. For example, A is a constructor sort. Constructor sorts correspond to inductive data types.

4.2.2.2 Standard Data Types

mCRL2 has the following standard data types:

- Booleans ($\mathbb{B}$) with constructor functions *true* and *false* and operators $\neg$, $\wedge$, $\vee$, and $\Rightarrow$. It is assumed that *true* and *false* are different.

- Unbounded positive integers ($\mathbb{N}^+$), natural numbers ($\mathbb{N}$), integers ($\mathbb{Z}$), and real numbers ($\mathbb{R}$) with relational operators $<$, $\leq$, $>$, $\geq$, unary negation $-$, binary arithmetic operators $+$, $-$, $*$, **div**, **mod**, and arithmetic operations *max*, *min*, *abs*, *succ*, *pred*, *exp*. These functions are only available for the appropriate sorts, e.g., **div** and **mod** are only defined for a denominator of sort $\mathbb{N}^+$. Also conversion functions $A2B$ are provided for all sorts A, $B \in \{\mathbb{N}^+, \mathbb{N}, \mathbb{Z}, \mathbb{R}\}$.

The user of the language is allowed to add maps and equations for standard data types. This also enables the user to specify *inconsistent* theories where the equation *true* = *false* becomes derivable. In such a case, the data specification loses its meaning.

4.2.2.3 Type Formers

There are a number of operators to construct types, such as structured types, function types, and types for lists, sets and bags.

A *structured type* represents a sort together with constructor, projection, and recognizer functions in a compact way. For instance, a sort of machine states can be declared by:

   struct *off* | *standby* | *starting* | *running(mode* : $\mathbb{N}$) | *broken?is_broken*;

This declares a sort with constructor functions *off*, *standby*, *starting*, *running*, and *broken*, projection function *mode* from the declared sort to $\mathbb{N}$ and recognizer *is_broken* from this sort to $\mathbb{B}$. So $mode(running(n)) = n$, and for all constructors c different from *running*, $mode(c)$ is left unspecified; so $mode(c)$ is a natural number, but we do not know which one. Also, $is\_broken(broken) = true$, and for all constructors d different from *broken*, $is\_broken(d) = false$.

Second, we have the *function type* former. The sort of functions from A to B is denoted $A \rightarrow B$. Note that function types are first-class citizens: functions may return functions. It is assumed that parentheses associate to the right in function notations, e.g., $A \rightarrow B \rightarrow C$ means $A \rightarrow (B \rightarrow C)$.

We also have a *list* type former. The sort of (finite) lists containing elements of sort A is declared by $List(A)$ and has constructor operators $[\,] : List(A)$ and $\triangleright :$ $A \times List(A) \rightarrow List(A)$. Other operators include $\triangleleft$, $+\!\!+$ (concatenation),. (element at), *head*, *tail*, *rhead*, and *rtail* together with list enumeration $[\,e_0, \ldots, e_n\,]$. The following expressions of sort $List(A)$ are all equivalent: $[\,c, d, d\,], c \triangleright [\,d, d\,], [\,c, d\,] \triangleleft d$, and $[\,] +\!\!+ [\,c, d\,] +\!\!+ [\,d\,]$.

Possibly infinite *sets* and *bags* where all elements are of sort A are denoted by $Set(A)$ and $Bag(A)$, respectively. The following operations are provided for these sort expressions: set enumeration $\{d_0, \ldots, d_n\}$, bag enumeration $\{d_0 : c_0, \ldots, d_n : c_n\}$ (c_i is the multiplicity or count of element d_i), set/bag comprehension $\{x : s \mid c\}$, element test $\in$, bag multiplicity *count*, set complement $\bar{s}$, and infix operators $\subseteq$, $\subset$, $\cup$, $-$, $\cap$ with their usual meaning for sets and bags. Also conversion functions *Set2Bag* and *Bag2Set* are provided, where the latter one forgets the multiplicity of the bag elements.

4.2.2.4 Sort References

Sort references can be declared. For instance, B is a synonym for A in

sort $B = A$;

Using sort references it is possible to define *recursive sorts*. For example, a sort of binary trees with numbers as their leaves can be defined as follows:

sort $T = $ **struct** $leaf(\mathbb{N}) \mid node(T, T)$;

This declares sort T with constructors $leaf : \mathbb{N} \rightarrow T$ and $node : T \times T \rightarrow T$, without projection and recognizer functions.

4.2.2.5 Standard Functions

For all sorts the equality operator $\approx$, inequality $\not\approx$, conditional *if*, and quantifiers $\forall$ and $\exists$ are provided. For the user-defined data types, the user has to provide equations giving meaning to the equality operator. For the standard data types and the type formers this operation is defined as expected. The inequality operator, the conditional and the quantifiers are defined for all sorts as expected.

So, for instance, with n a variable of sort $\mathbb{N}$, the expression $n \approx n$ is equal to *true* and $n \not\approx n$ is equal to *false*. Using the above declaration of sort A and map f, $if(true, c, d)$ is equal to c and $\forall_{x:A} (f(x, c) \approx c)$ is equal to *true*.

4.2.2.6 Additional Language Constructs

Expressions of sort $\mathbb{B}$ may be used as *conditions* in equations, for instance:

var $\quad x, y : A;$
eqn $\quad x \approx y \ \rightarrow \ f(x, y) = x;$

Furthermore, *lambda abstractions* and *where clauses* can be used. For example:

map $\quad h, h' : A \rightarrow A \rightarrow A;$
var $\quad x, y : A;$
eqn $\quad h(x) \qquad\qquad = \lambda_{y':A}(\lambda_{z:A} f(z, g(z)))(g(f(x, y')));$
$\qquad\quad h'(x)(y) \qquad\quad = f(z, g(z)) \ \textbf{whr} \ z = g(f(x, y)) \ \textbf{end};$

Here functions h and h' are equivalent, i.e., it is possible to show that $h = h'$ (using extensionality).

4.3 Verification Methodology and the mCRL2 Toolset

Different aspects of mCRL2 models can be verified. One might be interested in the equivalence of two mCRL2 models according to some notion of process equivalence. Or in validity of temporal properties, like absence of deadlock, with respect to an mCRL2 model. In this section, we present the methodology and the tools that allow to perform such verifications on mCRL2 models.

The mCRL2 toolset [29] has been developed at the Eindhoven University of Technology to support formal reasoning about systems specified in mCRL2. It is based on term rewriting techniques and on formal transformation of process and data expressions. At the time of writing, it allows one to generate, simulate, reduce, and visualize state spaces; to search for deadlocks and particular actions; to perform symbolic optimizations for mCRL2 specifications (e.g., invariant and confluence reductions); and to verify modal formulas that contain data. Figure 4.1 gives an overview of the toolset. The rectangles represent tools, and the ovals represent the forms of data that are manipulated by the tools.

The *linearizer* transforms a restricted though practically expressive subset of mCRL2 specifications to linear process specifications (LPSs). These LPSs are a compact symbolic representation of the labeled transition system of the specification. Due to its restricted form, an LPS is especially suited as input for tools; there is no need for such tools to take into account all the different operators of the complete language (see Ref. [42] for the details of the linearization process).

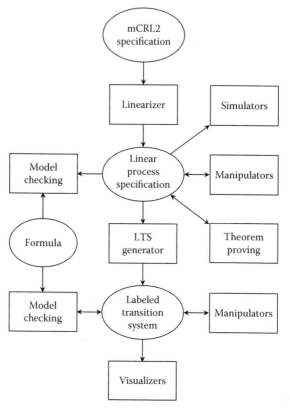

FIGURE 4.1:   mCRL2 toolset.

An LPS contains a single process definition of the *linear form:**

proc $P(x : D) = \sum_{i \in I} \sum_{y_i : E_i} c_i(x, y_i) \rightarrow \alpha_i(x, y_i) \cdot P(g_i(x, y_i));$

init $P(d_0);$

Here, data expressions of the form $d(x_1, \ldots, x_n)$ contain at most free variables from $\{x_1, \ldots, x_n\}$, so d_0 is a closed data expression. Furthermore, I is a finite index set, $\sum_{i \in I} p_i$ is a shorthand for $p_1 + \cdots + p_n$ when $I = \{1, \ldots, n\}$ and $n > 0$, or δ when $n = 0$, and for each $i \in I$:

- $c_i(x, y_i)$ is a boolean expression representing a condition

- $\alpha_i(x, y_i)$ is a multiaction $a_i^1(f_i^1(x, y_i)) \mid \cdots \mid a_i^{n_i}(f_i^{n_i}(x, y_i))$, where $f_i^k(x, y_i)$ (for $1 \leq k \leq n_i$) representing the parameters of action name a_i^k

- $g_i(x, y_i)$ is an expression of sort D representing the next state of the process definition P

---

* Here, for the sake of simplicity, we present an untimed version of the LPS which cannot terminate.

The form of the summand as described above is sometimes presented as the *condition-action-effect* rule. In a particular state d and for some data value e the multiaction $\alpha_i(d, e)$ can be executed if condition $c_i(d, e)$ holds. The effect of the action on the state is given by the fact that the next state is $g_i(d, e)$.

Once an LPS has been generated from an mCRL2 specification, there are a number of possible steps forward. With *simulation* tools one can quickly gain insight into the behavior of the system. It is possible to manually select transitions, but traces can also be automatically generated. Traces—either previously generated with the simulator or obtained via other tools—can be inspected with the simulator.

There are several *LPS manipulation* tools [18] that can simplify an LPS, e.g., by removing unused parameters or by instantiating summation variables. With tools based on *theorem proving* it is possible to check invariants of an LPS or to detect confluence [17].

Model checking tools take an LPS and a modal formula—representing some functional requirements—and create a PBES [19,21] stating whether or not the formula holds for the specification. These formulae are written in a variant of the modal μ-calculus extended with regular expressions [19]; for example, to state that a system is deadlock free we can write $[true^*]\langle true \rangle true$ or, equivalently, $\nu X.\langle true \rangle true \wedge [true]X$. To state that the number *leave* actions never exceeds the number of *enter* actions is expressed by the formula $\nu X(n:\mathbb{N} = 0).[enter]X(n+1) \wedge [leave](n>0 \wedge X(n-1))$. A PBES can be symbolically solved by different tools of the toolset, either via conversion to a boolean equation system (BES) or by direct manipulation of the PBES.

As symbolic model checking is not yet always successful and because certain tools and toolsets only work on labeled transition systems (LTSs), it is possible to use the *LTS generator* to construct the explicit state space of a specification. Besides creating such an LTS, it is also possible to automatically check for the presence of deadlocks or certain actions, and to generate witnessing traces. A variety of exploration techniques are available (such as breadth-first and depth-first searches as well as random simulation).

Besides a wide range of tools on LPSs, the toolset also contains some tools that *manipulate* LTSs, most notably LTS reduction tools that can minimize an LTS modulo certain equivalences (e.g., trace equivalence, and strong and branching bisimilarity).

With *LTS visualization* tools one can gain insight into systems whose size ranges from very small to quite large. In Section 4.4, there are several examples of images that have been created using these tools. Each tool has a different way of visualizing an LTS: either by using automatic positioning algorithms or by clustering states based on state information. These visualization tools have proven to be useful in detecting simple properties as well as more complex properties such as symmetry.

In Section 4.4, we demonstrate in slightly more detail how we can use the toolset to validate systems.

4.4   Examples of Modeling and Verification in mCRL2

In this section, we demonstrate the language and the toolset with several simple examples.

4.4.1   Dining Philosophers

A classical example of a concurrent system is the well-known *dining philosophers problem* [10]. We illustrate how to model this problem in mCRL2 and give a procedure to find deadlocks and traces leading to these deadlocks using the tools from the mCRL2 toolset.

The dining philosophers problem tells the story of a group of philosophers sitting a table at which each philosopher has their own plate. In between each pair of neighboring plates there is precisely one fork. The dish they are served requires two forks to be eaten. In other words, each pair of neighboring entities (philosophers) shares one resource (fork). For simplicity, we only consider three philosophers. This situation is depicted in Figure 4.2.

An mCRL2 model of the problem is presented below.

sort $PhilId = $ **struct** $p_1 \mid p_2 \mid p_3$;
       $ForkId = $ **struct** $f_1 \mid f_2 \mid f_3$;

map $lf, rf : PhilId \rightarrow ForkId$;
eqn $lf(p_1) = f_1$; $lf(p_2) = f_2$; $lf(p_3) = f_3$;
      $rf(p_1) = f_3$; $rf(p_2) = f_1$; $rf(p_3) = f_2$;

act   get, put, up, down, lock, free : $PhilId \times ForkId$;
      eat : $PhilId$;

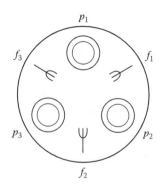

FIGURE 4.2:   Dining table for three philosophers.

proc Phil(p : *PhilId*) = (get($p, lf(p)$) ∥ get($p, rf(p)$)) · eat(p)·
 (put($p, lf(p)$) ∥ put($p, rf(p)$)) · Phil(p);

Fork(f : *ForkId*) = $\sum_{p:Phil}$ up(p, f) · down(p, f) · Fork(f);

init $\partial_{\{get,put,up,down\}}$ ($\Gamma_{\{get|up\rightarrow lock,put|down\rightarrow free\}}$ (

Phil(p_1) ∥ Phil(p_2) ∥ Phil(p_3) ∥ Fork(f_1) ∥ Fork(f_2) ∥ Fork(f_3)

));

In this model the sort *PhilId* contains the philosophers. The nth philosopher is denoted by p_n. Similarly, the sort *ForkId* contains the forks (f_n denotes the nth fork). The functions lf and rf designate the respective left and right fork of each philosopher.

Actions get(p_n, f_m) and put(p_n, f_m) are performed by the philosopher process. They model that the philosopher p_n gets or puts down fork f_m. The corresponding actions up(p_n, f_m) and down(p_n, f_m) are performed by the fork process. They model the fork f_m being taken or put down by philosopher p_n. The action eat(p_n) models philosopher p_n eating. For communication purposes we have added actions lock and free. These will represent a fork actually being taken, respectively, put down by a philosopher.

The process Phil(p_n) models the behavior of the nth philosopher. It first takes the left and right forks (in any order; possibly even at the same time), then eats, then puts both forks back (again in any order) and repeats its own behavior.

The process Fork(f_n) models the behavior of the nth fork (i.e., the fork on the left of the nth philosopher). For any philosopher p it can perform up(p, f_n) meaning that the fork is being taken by philosopher p. Then it performs down(p, f_n) meaning that the same philosopher puts the fork down. Finally it repeats its own behavior again.

The whole system consists of three Phil and three Fork processes in parallel. By enforcing communication between actions get and up and between put and down, we ensure that forks agree on being taken or put down by the philosophers. The result of these communications are lock and free, respectively. Note that the communication operator only ensures that communication happens when possible. The blocking operator makes sure that nothing else happens (i.e., by disallowing the loose occurrences of the actions get, put, up, and down).

This model of the problem can be analyzed by generating the underlying LTS (consisting of 93 states and 431 transitions) and searching for the deadlocks while doing this. After linearizing the specification and exploring the LTS we can find that the trace lock(p_1, f_1) · lock(p_2, f_2) · lock(p_3, f_3) leads to a deadlock. (The LTS generator has a special option to detect deadlocks and report traces to them.)

By using a simulator one can confirm that this trace indeed leads to a deadlock (either by manually executing this trace, or—especially in the case of longer traces—by loading the trace generated by the LTS generator). The detected deadlock represents the situation when each of the philosophers has taken one fork and waits for another one without being able to put the first one back. Note that due to the nature of mCRL2, this deadlock can also be reached by a single multiaction. That is, by performing the multiaction $\mathsf{lock}(p_1, f_1) \mid \mathsf{lock}(p_2, f_2) \mid \mathsf{lock}(p_3, f_3)$ denoting that all philosophers take their left fork at exactly the same time. (In cases where this is not desired the restriction operator can be used to allow only single actions.)

The typical first attempt to a solution is stated as follows: let the philosophers pick the forks in a fixed order (first left, then right). That is, replace the process definition of Phil by the following:

proc $\mathsf{Phil}(p : PhilId) = \mathsf{get}(p, lf(p)) \cdot \mathsf{get}(p, rf(p)) \cdot \mathsf{eat}(p)\cdot$
$\qquad\qquad\qquad\qquad \mathsf{put}(p, lf(p)) \cdot \mathsf{put}(p, rf(p)) \cdot \mathsf{Phil}(p);$

The LTS of this specification consists of 35 states and 97 transitions. However, this solution is incorrect: the trace $\mathsf{lock}(p_1, f_1) \mid \mathsf{lock}(p_2, f_2) \mid \mathsf{lock}(p_3, f_3)$ is still valid in the LTS and thus leads to a deadlock.

One of the solutions is to cross the arms of one of the philosophers (or just tell him to pick the forks in reverse order). We change the definition of Phil into the following:

proc $\mathsf{Phil}(p : PhilId) = (p \approx p_1) \rightarrow \mathsf{get}(p, rf(p)) \cdot \mathsf{get}(p, lf(p)) \cdot \mathsf{eat}(p)\cdot$
$\qquad\qquad\qquad\qquad\qquad \mathsf{put}(p, rf(p)) \cdot \mathsf{put}(p, lf(p)) \cdot \mathsf{Phil}(p)$
$\qquad\qquad\qquad \diamond \quad \mathsf{get}(p, lf(p)) \cdot \mathsf{get}(p, rf(p)) \cdot \mathsf{eat}(p)\cdot$
$\qquad\qquad\qquad\qquad\qquad \mathsf{put}(p, lf(p)) \cdot \mathsf{put}(p, rf(p)) \cdot \mathsf{Phil}(p);$

This results in a LTS consisting of 36 states and 104 transitions without any deadlock states.

In Figure 4.3, the global structure of the LTSs of the last two solutions are visualized using a technique from Ref. [47]. Based on their distance from the initial state, states are clustered and positioned in a 3D structure similar to a cone tree [38], with the emphasis on symmetry. This symmetry is apparent in Figure 4.3a where the three symmetric legs indicate that the behavior of each of the three philosophers is similar. In Figure 4.3b there are two asymmetric legs, reflecting the asymmetry of crossing the arms of one of the philosophers.

4.4.2 Alternating Bit Protocol

We now provide a model of a simple communication protocol, namely the alternating bit protocol (ABP) [3,26], which ensures successful transmission of data through lossy channels. The protocol is depicted in Figure 4.4. It consists of four components, S, R, K, and L, that are connected via communication links 1–6.

The processes K and L are channels that transfer messages from sender to receiver and from receiver to sender, respectively. The data can be lost in transmission, in which case the processes K and L deliver an error. The sender process S and receiver process R must take care that despite this loss of data, transfer between communication links 1 and 4 is reliable, in the sense that messages received over link 1 are sent over link 4 exactly once in the same order in which they were received.

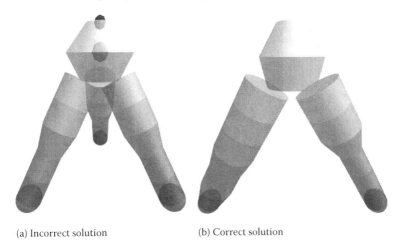

(a) Incorrect solution   (b) Correct solution

FIGURE 4.3: Visualization of the structure of the LTSs of two solutions to the dining philosophers problem. In (a), all three philosophers pick the forks in a fixed order, resulting in a deadlock to which a trace is shown. In (b), one philosopher picks the forks in reverse order, giving a deadlock free solution.

In order to model the protocol, we assume some sort D with data elements to be transferred. The desired external behavior of the ABP is a simple one-place buffer, specified by the following process:

act $r_1, s_4 : D$;

proc $B = \sum_{d:D} r_1(d) \cdot s_4(d) \cdot B$;

Actions $r_1(d)$ and $s_4(d)$ represent the receiving and sending of a message over communication links 1 and 4, respectively. Note that the external behavior is actually the simplest conceivable behavior for a data transfer protocol.

In order to develop the sender and the receiver we must have a good understanding of the exact behavior of the channels K and L. As will be explained below, the process K forwards pairs consisting of a message and a bit, and the process L only forwards bits. We use booleans to represent bits. The processes choose internally whether data are delivered or lost using the action i. If it is lost an error message $\perp$ is delivered:

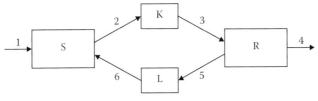

FIGURE 4.4: Alternating bit protocol.

sort $Error = $ **struct** $\perp$;

act $r_2, s_3 : D \times \mathbb{B}$;

      $r_5, s_6 : \mathbb{B}$;

      $s_3, s_6 : Error$;

      i;

proc $K = \displaystyle\sum_{d:D}\sum_{b:\mathbb{B}} r_2(d,b) \cdot (i \cdot s_3(d,b) + i \cdot s_3(\perp)) \cdot K$;

      $L = \displaystyle\sum_{b:\mathbb{B}} r_5(b) \cdot (i \cdot s_6(b) + i \cdot s_6(\perp)) \cdot L$;

Note that the action i cannot be omitted. If it would be removed, the choice between delivering the correct data or the error is made while interacting with the receiver of the message. The receiver can henceforth determine whether the data will be lost or not. This is not what we want to model here. We want to model that whether or not data are lost is determined internally in K and L. Because the factors that cause the message to be lost are outside our model, we use nondeterministic choice to model data loss.

We model the sender and receiver using the protocol proposed in Refs. [3,26]. The first aspect of the protocol is that the sender must guarantee that despite data loss in K, data eventually arrive at the receiver. For this purpose, it iteratively sends the same messages to the sender. The receiver sends an acknowledgment to the sender whenever it receives a message. If a message is acknowledged, the sender knows that the message is received so it can proceed with the next message.

A problem of this protocol is that a datum may be sent more than once, and the receiver has no way of telling whether the datum stems from a single message which is resent, or whether it stems from two messages that contain the same data. In order to distinguish between these options extra control information must be added to the message. A strong point made in Refs. [3,26] is that adding a single bit already suffices for the job. For consecutive messages the bit is alternated for each subsequent new datum to be transferred. If a datum is resent, the old bit is used. This explains the name "alternating bit protocol."

After receiving a message at the receiver over link 3, its accompanying bit is sent back in the acknowledgment. When the bit differs from the bit associated with the last message, the receiver knows that this concerns a new datum and forwards it over link 4. Upon reception of $\perp$, the receiver does not know whether this regards an old or new message, and it sends the old bit to indicate that resending is necessary.

Upon receipt of an unexpected bit or an error message, the sender knows that the old datum must be resent. Otherwise, it can proceed to read new data over link 1 and forward it.

First, we specify the sender S in the state that it is going to send out a datum with the bit b attached to it:

act $s_2 : D \times \mathbb{B}$;

     $r_6 : \mathbb{B}$;

     $r_6 : Error$;

proc $S(b : \mathbb{B}) = \sum_{d:D} r_1(d) \cdot T(d, b)$;

     $T(d : D, b : \mathbb{B}) = s_2(d, b) \cdot (r_6(b) \cdot S(\neg b) + (r_6(\neg b) + r_6(\bot)) \cdot T(d, b))$;

In state $S(b)$, the sender reads a datum d over link 1. Next, the system proceeds to state $T(d, b)$, in which it sends this datum over link 2, with the bit b attached to it. Then, the sender expects to receive the acknowledgment b over link 6, ensuring that the pair (d, b) has reached the receiver unscathed. If the correct acknowledgment b is received, then the system proceeds to state $S(\neg b)$, in which it is going to send out a datum with the bit $\neg b$ attached to it. If the acknowledgment is either the wrong bit $\neg b$ or the error message $\bot$, then the system sends the pair (d, b) over link 2 once more.

Next, we specify the receiver R in the state that it is expecting to receive a datum with the bit b attached to it:

act $r_3 : D \times \mathbb{B}$;

     $r_3 : Error$;

     $s_5 : \mathbb{B}$;

proc $R(b : \mathbb{B}) = \sum_{d:D} r_3(d, b) \cdot s_4(d) \cdot s_5(b) \cdot R(\neg b)$

     $+ \left(\sum_{d:D} r_3(d, \neg b) + r_3(\bot) \right) \cdot s_5(\neg b) \cdot R(b)$;

In state $R(b)$ there are two possibilities.

1. If the receiver reads a pair (d, b) over link 3, then this constitutes new information, so the datum d is sent over link 4, after which acknowledgment b is sent to the sender over link 5. Next, the receiver proceeds to state $R(\neg b)$, in which it is expecting to receive a datum with the bit $\neg b$ attached to it.

2. If the receiver reads a pair $(d, \neg b)$ or an error message $\bot$ over link 3, then this does not constitute new information. Then, the receiver sends acknowledgment $\neg b$ to the sender over link 5 and remains in state $R(b)$.

The desired concurrent system is obtained by putting $S(true)$, $R(true)$, K, and L in parallel, blocking send and read actions over the internal links. That is, the ABP is defined by the following specification:

proc $\text{ABP} = \nabla_V(\Gamma_C(S(true) \parallel K \parallel L \parallel R(true)))$;

init ABP;

Here $C = \{r_2 \mid s_2 \rightarrow c_2, r_3 \mid s_3 \rightarrow c_3, r_5 \mid s_5 \rightarrow c_5, r_6 \mid s_6 \rightarrow c_6\}$ and $V = \{r_1, s_4, c_2, c_3, c_5, c_6, i\}$.

FIGURE 4.5:   Visualization of the LTS structure of the ABP with $|D| = 2$.

If we set the number of different data elements to be transferred to 2, i.e., $|D| = 2$, we can linearize the specification, generate the LTS, and visualize the global structure of the system as shown in Figure 4.5 (using the techniques from Ref. [47]). The structure is completely symmetric: the left and right part precisely correspond to the transfer of the two different data elements.

The interesting question here is whether the scheme with alternating bits works correctly. Or, in other words, do the buffer B and the alternating bit protocol ABP behave the same when only the action names r_1 and s_4 are visible? This question is concisely stated by asking whether the following equation holds in branching bisimulation semantics:

$$B = \tau_{\{c_2, c_3, c_5, c_6, i\}}(\mathsf{ABP}).$$

Using the toolset this can easily be shown to hold by generating the LTS of both processes and automatically comparing their modulo branching bisimilarity.

4.4.3   Sliding Window Protocol

In the alternating bit protocol, the sender sends out a datum and then waits for an acknowledgment before it sends the next datum. In situations where transmission of data is relatively time consuming, this procedure tends to be unacceptably slow. In sliding window protocols (SWPs) [9] (see also Ref. [41]), this situation has been resolved as the sender can send out multiple data elements before it requires an acknowledgment. This protocol is so effective that it is one of the core protocols of the Internet.

The most complex SWP described in Ref. [41] was modeled in 1991 using techniques as described in Ref. [8]. This model revealed a deadlock. When confronted with this, the author of Ref. [41] indicated that this problem remained undetected for a whole decade, despite the fact that the protocol had been implemented a number of times. There is an evidence that this particular deadlock occurs in actual implementations of Internet protocols, but this has never been systematically investigated. In recent editions of Ref. [41] this problem has been fixed.

We concentrate on a variant of the SWP which is unidirectional, to keep the model to its essential minimum. The essential feature of the SWP is that it contains buffers in the sender and the receiver to keep copies of the data in transit. This is needed to be able to resend this data if it turns out after a while that the data did not arrive correctly. Both buffers have size n. This means that there can be at most $2n$ data elements under way.

This suggests that the external behavior of the SWP is a bounded first-in-first-out (FIFO) queue of length $2n$. Such a queue, that receives data from a nonempty data domain D over link 1 and sends them over link 4, behaves as $\mathsf{FIFO}([], 2n)$, where process FIFO is defined as follows:

act $\quad \mathsf{r}_1, \mathsf{s}_4 : D;$

proc $\quad \mathsf{FIFO}(l : List(D), m : \mathbb{N}^+)$

$$= (\#l < m) \to \sum_{d:D} \mathsf{r}_1(d) \cdot \mathsf{FIFO}(l \rhd d, m)$$
$$+ \ (\#l > 0) \to \mathsf{s}_4(head(l)) \cdot \mathsf{FIFO}(tail(l), m);$$

Note that action $\mathsf{r}_1(d)$ can be performed until the list l contains m elements, because in that situation the queue will be filled. Furthermore, $\mathsf{s}_4(head(l))$ can only be performed if l is not empty.

We now give a model of the SWP which implements the bounded FIFO queue on top of two unreliable channels K and L. The setup is similar to that of the ABP, and is depicted in Figure 4.6.

The channels differ from those of the ABP because they do not deliver an error in case of data loss. An indication of an error is necessary for the ABP to work correctly, but is unrealistic. A better model of a channel is one where data are lost without an explicit indication. In this case, we still assume that the channels can only carry a single data item, and have no buffer capacity. But the channels can be replaced by others, for instance with bounded, or unbounded capacity. As long as the

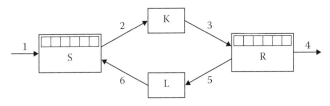

FIGURE 4.6: Sliding window protocol.

channels now and then transfer a message, the sliding window protocol can be used to transform these unreliable channels into a FIFO queue.

act $r_2, s_3 : D \times \mathbb{N}$;

        $r_5, s_6 : \mathbb{N}$;

        i;

proc $K = \sum\limits_{d:D,k:\mathbb{N}} r_2(d, k) \cdot (i \cdot s_3(d, k) + i) \cdot K$;

       $L = \sum\limits_{k:\mathbb{N}} r_5(k) \cdot (i \cdot s_6(k) + i) \cdot L$;

The sender and the receiver in the SWP both maintain a buffer of size n containing the data being transmitted. The buffers are represented by a function from $\mathbb{N}$ to D indicating which data element occurs at which position. Only the first n places of these functions are used. In the receiver, we additionally use a buffer of booleans of length n to recall which of the first n positions in the buffer contain valid data.

sort $DBuf = \mathbb{N} \to D$;

       $BBuf = \mathbb{N} \to \mathbb{B}$;

The SWP uses a numbering scheme to number the messages that are sent via the channels. It turns out that if the sequence numbers are issued modulo $2n$, messages are not confused and are transferred in order. Each message with sequence number j is put at position $j|_n$ (a shorthand for $j \bmod n$) in the buffers.

We use the following auxiliary functions to describe the protocol. The function *empty* below represents a boolean buffer that is false everywhere, indicating that there is no valid data in the buffer. We use notation $q[i:=d]$ to say that position i of buffer q is filled with datum d. Similarly, $b[i:=c]$ is buffer b where c is put at position i. The most involved function is $nextempty_{mod}(i, b, m, n)$. It yields the first position in buffer b starting at $i|_n$ that contains *false*. If the first m positions from $i|_n$ of b are all *true*, it yields the value $(i+m)|_{2n}$. The variable m is used to guarantee that the function $nextempty_{mod}$ is well defined if b is *true* at all its first n positions.

map   *empty* : $BBuf$;

       $\_[\_:=\_] : DBuf \times \mathbb{N} \times D \to DBuf$;

       $\_[\_:=\_] : BBuf \times \mathbb{N} \times \mathbb{B} \to BBuf$;

       $nextempty_{mod} : \mathbb{N} \times BBuf \times \mathbb{N} \times \mathbb{N}^+ \to \mathbb{N}$;

var $d : D$; $i, j, m : \mathbb{N}$; $n : \mathbb{N}^+$; $q : DBuf$; $c : \mathbb{B}$; $b : BBuf$;

eqn $empty = \lambda_{j:\mathbb{N}} false$;

       $q[i:=d] = \lambda_{j:\mathbb{N}} if(i \approx j, d, q(j))$;

       $b[i:=c] = \lambda_{j:\mathbb{N}} if(i \approx j, c, b(j))$;

       $nextempty_{mod}(i, b, m, n) =$

           $if(b(i|_n) \wedge m>0, nextempty_{mod}((i+1)|_{2n}, b, m-1, n), i|_{2n})$;

Below we model the sender process S. The variable ℓ contains the sequence number of the oldest message in sending buffer q, and m is the number of items in the sending buffer. If a datum arrives over link 1, it is put in the sending buffer q, provided there is space. There is the possibility to send the kth datum over link 2 with sequence number $(\ell+k)|_{2n}$, and there is a possibility that an acknowledgment arrives over link 6. This acknowledgment is the index of the first message that has not yet been received by the receiver.

act $s_2 : D \times \mathbb{N};$

        $r_6 : \mathbb{N};$

proc $S(\ell, m : \mathbb{N}, q : DBuf, n : \mathbb{N}^+)$

$$= (m < n) \rightarrow \sum_{d:D} r_1(d) \cdot S(\ell, m+1, q[((\ell+m)|_n):=d], n)$$

$$+ \sum_{k:\mathbb{N}} (k < m) \rightarrow s_2(q((\ell+k)|_n), (\ell + k)|_{2n}) \cdot S(\ell, m, q, n)$$

$$+ \sum_{k:\mathbb{N}} r_6(k) \cdot S(k, (m-k+\ell)|_{2n}, q, n);$$

The receiver is modeled by the process $R(\ell', q', b, n)$ where ℓ' is the sequence number of the oldest message in the receiving buffer q'. A datum can be received over link 3 from channel K and is only put in the receiving buffer q' if its sequence number k is in the receiving window. If sequence numbers and buffer positions would not be considered modulo $2n$ and n, this could be stated by $\ell' \leq k < \ell' + n$. The condition $(k - \ell')|_{2n} < n$ states exactly this, taking the modulo boundaries into account.

The second summand in the receiver says that if the oldest message position is valid (i.e., $b(\ell'|_n)$ holds), then this message can be delivered over link 4. Moreover, the oldest message is now $(\ell'+1)|_{2n}$ and the message at position $\ell'|_n$ becomes invalid.

The last summand says that the index of the first message that has not been received at the receiver is sent back to the sender as an acknowledgment that all lower numbered messages have been received.

act $r_3 : D \times \mathbb{N};$

        $s_5 : \mathbb{N};$

proc $R(\ell' : \mathbb{N}, q' : DBuf, b : BBuf, n : \mathbb{N}^+)$

$$= \sum_{d:D, k:\mathbb{N}} r_3(d, k) \cdot (((k-\ell')|_{2n} < n)$$

$$\rightarrow R(\ell', q'[(k|_n) := d], b[(k|_n) := true], n)$$

$$\diamond R(\ell', q', b, n))$$

$$+ \; b(\ell'|_n) \rightarrow s_4(q'(\ell'|_n)) \cdot R((\ell'+1)|_{2n}, q', b[(\ell'|_n) := false], n)$$

$$+ \; s_5(nextempty_{mod}(\ell', b, n, n)) \cdot R(\ell', q', b, n);$$

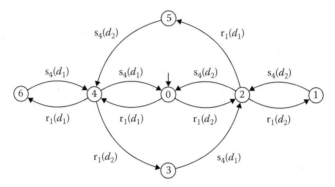

FIGURE 4.7:   Minimized LTS of the SWP with window size 1 and $|D| = 2$.

The behavior of the SWP is characterized by

proc $SWP(q, q' : DBuf, n : \mathbb{N}^+)$
$$= \nabla_V(\Gamma_C(S(0, 0, q, n) \parallel K \parallel L \parallel R(0, q', empty, n)));$$

Here $C = \{r_2|s_2 \to c_2, r_3|s_3 \to c_3, r_5|s_5 \to c_5, r_6|s_6 \to c_6\}$ and $V = \{c_2, c_3, c_5, c_6, i, r_1, s_4\}$.

The contents of q and q' can be chosen arbitrarily without affecting the correctness of the protocol. This is stressed by not instantiating these variables. The sliding window protocol behaves as a bounded FIFO buffer for any $n > 0$ and $q, q' : DBuf$:

$$FIFO([], 2n) = \tau_I(SWP(q, q', n))$$

where $I = \{c_2, c_3, c_4, c_5, i\}$.

In Figure 4.7 the LTS of the SWP with window size 1 and 2 data elements is presented after minimization modulo branching bisimilarity. It is not difficult to see that the 2-place FIFO queue has a bisimilar LTS.

Proving correctness of the SWP for any window size and for an arbitrary data domain is a more difficult problem that involves techniques based on theorem proving, like the cones and foci method [22]. In Ref. [2] it is proven that the equivalence between the SWP and the FIFO queue holds also in this case. Due to the tricky nature of modulo calculation, this proof is quite involved.

4.5   Historical Context

Work on process algebras can be traced back to observations by Milner and Bekič around 1973 [4,32]. They observed that actions are the basic notion to describe behavior. This led to the development of three main process algebraic formalisms (CCS [33], ACP [5], and CSP [23]). These formalisms did not contain data and primarily served as vehicles to study the mathematical and semantic properties of such languages.

A major next step was the extension of process algebras to overcome the lack of practical expressivity that process algebras have. The most well known is LOTOS [24,44]. It contains equational abstract data types, a module mechanism, interrupts, and mechanisms to yield return values. LOTOS became an international standard in 1990. A language that is very similar to LOTOS is PSF [28]. A later member to be added to the family of these process algebras is the CSP variant CSPm [39], although it does not use abstract data types.

In an attempt to specify a universal process algebraic language (the Common Representation Language, CRL), it was realized that these extended process algebras became too complex, which would hamper further development. In response a micro CRL (μCRL [15,16]) was defined. This language has essentially the same expressivity of its larger brothers (LOTOS, PSF, and CRL), but does not contain unnecessary decorum.

For approximately one decade the language μCRL has been used to develop theory, proof methodology, algorithms, and an analysis toolset. This led to a variety of axiomatization and expressivity results [25]. The cones and foci theorem [22], confluence, and τ-priorization results [17] enabled the verification of more complex protocols, culminating in the proof-checked algebraic verification of the sliding window protocol [2]. In response to modeling of several industrial case studies, a toolset for μCRL was developed [7], which had an interesting side effect work on visualizing huge state spaces [47] and PBES, to verify modal formulas with data and time [19,21].

Although work using μCRL was progressing it was felt that the abstract data types became more and more of a nuisance. The problem was not with expressivity as most practically relevant systems could easily be expressed using the abstract data types. The problem was that abstract data types are less suitable as a means to build up knowledge. Using abstract data types it is required to specify data types such as booleans and natural numbers for each specification. This means that when studying a specification, these user-defined standard data types must be read and understood, because they can (and will) differ at subtle points from user-defined standard data types defined by others. Furthermore, no standard meta results on the data types can be accumulated (e.g., associativity of addition, distributivity of modulo operations), because they do not hold for all possible equational specifications of these operators. The use of abstract data types also led to rather long μCRL specifications, generally starting with a dozen pages of data types. A disadvantage that we only saw after defining and using μCRL is the lack of sets and functions. This meant that most data types were encoded in lists (e.g., arrays are modeled as lists), which is not efficient for state space generation, awkward to read and unpleasant to prove properties about. It is worth noting that the clarity of abstract equational data types has substantial advantages. They are easier to learn, a perfect means of communication when it comes to discussions about details and relatively straightforward to implement efficiently [48].

With these arguments, it was decided to extend μCRL with concrete data types and give the resulting language the somewhat uninspiring name mCRL2. Under influence of, among others, the work by Dijkstra and associates [11], it was felt that a proper specification language should not only contain basic data types but also commonly used mathematical objects such as functions, sets and quantifiers. Furthermore, in the style of abstract data types, it was considered necessary that these data types should not be a finite approximation, but the "real thing." So, numbers have an infinite range, sets can both be finite and infinite and quantifiers can both have finite and infinite domains. Among the process algebraic inspired languages, mCRL2 is unique in this respect (other "rich" formalisms are Cold [12], Z [1], and languages used in proof checkers such as Coq [6] and PVS [35]).

Besides a major extension of the data language, mCRL2 differs in some aspects from the μCRL process language, such as the availability of multiactions, and the use of local communication and communication-independent parallellism (as explained in Section 4.1).

A similar movement from μCRL to mCRL2 can be observed in the development of E-LOTOS [40] from LOTOS. The most notable difference is that E-LOTOS is developed with the implementability of the language as guiding principle, whereas mCRL2 has been developed with specifiability in mind. So, typically mCRL2 contains quantifiers and sums over infinite data types. E-LOTOS has no quantifiers and restricts sums to finite data types. mCRL2 has infinite sets and natural numbers with unbounded range (which has a performance penalty). E-LOTOS maps the numbers on machine-based numbers with a finite range. Conversely, E-LOTOS allows exceptions, while loops, and assignments that are not available in mCRL2.

What are the future developments for mCRL2? We expect to see a steady improvement of the language and the tools. More important is the integration of all behavioral aspects in unified formalisms: e.g., behavior, data, modal logics, time, performance, throughput, continuous interaction, and feedback in one mathematical framework. But what we consider the most important development is that attention will shift from the language to its use. We expect articles on specification style—e.g., design by confluence to reduce the state spaces to be analyzed, design with synchronized increases of clocks—to enable abstract interpretation techniques. But we also expect that attention will move to the design and study of system behavior per se using mCRL2 (or any similar formalism) as a self-evident means.

References

[1] J.-R. Abrial, S. A. Schuman, and B. Meyer. A specification language. In *On the Construction of Programs*, pp. 343–410. Cambridge University Press, Cambridge, MA, 1980.

[2] B. Badban, W. Fokkink, J. F. Groote, J. Pang, and J. van de Pol. Verification of a sliding window protocol in μCRL and PVS. *Formal Aspects of Computing*, 17(3):342–388, 2005.

[3] K. A. Bartlett, R. A. Scantlebury, and P. T. Wilkinson. A note on reliable full-duplex transmission over half-duplex links. *Communications of the ACM*, 12(5):260–261, 1969.

[4] H. Bekič. Towards a mathematical theory of processes. In C. B. Jones, editor, *Programming Languages and Their Definition. Lecture Notes in Computer Science*, 177:168–206, 1984.

[5] J. A. Bergstra and J. Willem Klop. Process algebra for synchronous communication. *Information and Control*, 60(1–3):109–137, 1984.

[6] Y. Bertot and P. Castéran. *Interactive Theorem Proving and Program Development. Coq'Art: The Calculus of Inductive Constructions*. Texts in Theoretical Computer Science. An EATCS Series. Springer, Berlin and New York, 2004.

[7] S. C. C. Blom, W. Fokkink, J. F. Groote, I. van Langevelde, B. Lisser, and J. van de Pol. μCRL: A toolset for analysing algebraic specifications. In G. Berry, H. Comon, and A. Finkel, editors, *Proc. Computer Aided Verification (CAV 2001)*, Paris, France, *Lecture Notes in Computer Science*, 2102: 250–254, 2001.

[8] J. J. Brunekreef. Sliding window protocols. In S. Mauw and G. J. Veltink, editors, *Algebraic Specification of Communication Protocols*. Cambridge University Press, New York, 1993.

[9] V. G. Cerf and R. E. Kahn. A protocol for packet network intercommunication. *IEEE Transactions on Communications*, 22:637–648, 1974.

[10] E. W. Dijkstra. Hierarchical ordering of sequential processes. *Acta Informatica*, 1:115–138, 1971.

[11] W. H. J. Feijen and A. J. M. van Gasteren. *On a Method of Multiprogramming*. Springer, New York, 1999.

[12] L. M. G. Feijs and H. B. M. Jonkers. *Formal Specification and Design*. Cambridge University Press, New York, 2005.

[13] J. F. Groote. The syntax and semantics of timed μCRL. Technical Report SEN-R9709, CWI, Amsterdam, 1997.

[14] J. F. Groote, A. Mathijssen, M. van Weerdenburg, and Y. S. Usenko. From μCRL to mCRL2: Motivation and outline. In *Proc. Workshop Essays on Algebraic Process Calculi (APC 25)*, Bertinoro, Romagna, Italy, *Electronic Notes in Theoretical Computer Science*, 162:191–196, 2006.

[15] J. F. Groote and A. Ponse. The syntax and semantics of μCRL. In A. Ponse, C. Verhoef, and S. F. M. van Vlijmen, editors, *Algebra of Communicating Processes, Utrecht 1994*, Workshops in Computing, Springer, 1995, pp. 26–62.

[16] J. F. Groote and M. A. Reniers. Algebraic process verification. In J. A. Bergstra, A. Ponse, and S. A. Smolka, editors, *Handbook of Process Algebra*, chapter 17, pp. 1151–1208. Elsevier Science Publishers B.V., Amsterdam, 2001.

[17] J. F. Groote and M. P. A. Sellink. Confluence for process verification. *Theoretical Computer Science*, 170(1–2):47–81, 1996.

[18] J. F. Groote and B. Lisser. Computer assisted manipulation of algebraic process specifications. *SIGPLAN Notices*, 37(12):98–107, 2002.

[19] J. F. Groote and R. Mateescu. Verification of temporal properties of processes in a setting with data. In A. M. Haeberer, editor, *Proc. Algebraic Methodology and Software Technology (AMAST 1998)*, Amazonia, Brazil, *Lecture Notes in Computer Science*, 1548:74–90, 1999.

[20] J. F. Groote, A. Mathijssen, M. Reniers, Y. Usenko, and M. van Weerdenburg. The formal specification language mCRL2. In Ed Brinksma, David Harel, Angelika Mader, Perdita Stevens, and Roel Wieringa, editors, *Methods for Modelling Software Systems (MMOSS)*, number 06351, in *Dagstuhl Seminar Proceedings*, Dagstuhl, Germany, 2007. Internationales Begegnungs- und Forschungszentrum fuer Informatik (IBFI), Schloss Dagstuhl, Germany.

[21] J. F. Groote and T. A. C. Willemse. Parameterised boolean equation systems (extended abstract). In P. Gardner and N. Yoshida, editors, *Proc. CONCUR 2004*, London, UK, *Lecture Notes in Computer Science*, 3170:308–324, 2004.

[22] J. F. Groote and J. Springintveld. Focus points and convergent process operators: A proof strategy for protocol verification. *Journal of Logic and Algebraic Programming*, 49(1–2):31–60, 2001.

[23] C. A. R. Hoare. *Communicating Sequential Processes*. Prentice Hall, Englewood Cliffs, NJ, 1985.

[24] ISO. *ISO 8807: Information Processing Systems—Open Systems Interconnection—LOTOS—A Formal Description Technique Based on the Temporal Ordering of Observational Behaviour*. International Standards Organization, Geneva, Switzerland, 1989.

[25] B. Luttik. On the expressiveness of choice quantification. *Annals of Pure and Applied Logic*, 121(1):39–87, 2003.

[26] W. C. Lynch. Reliable full duplex file transmission over half-duplex telephone lines. *Communications of the ACM*, 11(6):407–410, 1968.

[27] A. Mathijssen and A. J. Pretorius. Verified design of an automated parking garage. In L. Brim, B. R. Haverkort, M. Leucker, and J. van de Pol, editors, *Proc. FMICS and PDMC 2006*, Bonn, Germany, *Lecture Notes in Computer Science*, 4346:165–180, 2007.

[28] S. Mauw and G.J. Veltink. A process specification formalism. *Fundamenta Informaticae*, XIII:85–139, 1990.

[29] mCRL2 website. http://www.mcrl2.org/.

[30] K. Meinke. Universal algebra in higher types. *Theoretical Computer Science*, 100(2):385–417, 1992.

[31] K. Meinke. Higher-order equational logic for specification, simulation and testing. In *The 1995 Workshop on Higher-Order Algebra, Logic and Term Rewriting (HOA 1995)*, Paderborn, Germany, *Lecture Notes in Computer Science*, 1074:124–143, 1996.

[32] R. Milner. Processes: A mathematical model of computing agents. In H.E. Rose and J.C. Shepherdson, editors, *Proc. Logic Colloquium*, North-Holland, 1973, pp. 158–173.

[33] R. Milner. *A Calculus of Communicating Systems. Lecture Notes in Computer Science*, 92, 1980.

[34] B. Möller, A. Tarlecki, and M. Wirsing. Algebraic specification of reachable higher-order algebras. In *Recent Trends in Data Type Specification. Lecture Notes in Computer Science*, 332:154–169, 1988.

[35] S. Owre, J. M. Rushby, and N. Shankar. PVS: A prototype verification system. In D. Kapur, editor, *Proc. Int. Conf. on Automated Deduction (CADE 1992)*, Saratoga Springs, New York, *Lecture Notes in Computer Science*, 607:748–752, 1992.

[36] B. Ploeger and L. Somers. Analysis and verification of an automatic document feeder. In Y. Cho, R. L. Wainwright, H. Haddad, S. Y. Shin, and Y. W. Koo, editors, *Proc. ACM Symposium on Applied Computing (SAC 2007)* 2007, Seoul, Korea, pp. 1499–1505.

[37] M. A. Reniers, J. F. Groote, M. B. van der Zwaag, and J. van Wamel. Completeness of timed μCRL. *Fundamenta Informaticae*, 50(3–4):361–402, 2002.

[38] G. G. Robertson, J. D. Mackinlay, and S. K. Card. Cone trees: Animated 3D visualizations of hierarchical information. In *CHI '91: Proceedings of the SIGCHI Conference on Human Factors in Computing Systems*, New York, ACM, 1991, pp. 189–194.

[39] A. W. Roscoe, C. A. R. Hoare, and R. Bird. *The Theory and Practice of Concurrency*. Prentice Hall PTR, Upper Saddle River, NJ, 1997.

[40] M. Sighireanu. Contribution à la définition et à l'implémentation du langage "Extended LOTOS." PhD thesis, Université Joseph Fourier, Grenoble, France, 1999.

[41] A. S. Tanenbaum. *Computer Networks*. Prentice Hall, Upper Saddle River, NJ 1981.

[42] Y. S. Usenko. Linearization in μCRL. PhD thesis, Technische Universiteit Eindhoven (TU/e), 2002.

[43] M. van Eekelen, S. ten Hoedt, R. Schreurs, and Y. S. Usenko. Analysis of a session-layer protocol in mCRL2. Verification of a real-life industrial implementation. In *Proc. FMICS 2007, Lecture Notes in Computer Science, 4916*, Springer, 2008. Also appeared as a Radboud University Nijmegen Technical Report ICIS-R07014.

[44] P. van Eijk and M. Diaz, editors. *Formal Description Technique LOTOS: Results of the Esprit Sedos Project.* Elsevier, New York, 1989.

[45] R. J. van Glabbeek and W. P. Weijland. Branching time and abstraction in bisimulation semantics. *Journal of the ACM*, 43(3):555–600, 1996.

[46] R. J. van Glabbeek. The linear time—branching time spectrum II. In E. Best, editor, *Proc. CONCUR 1993*, Hildesheim, Germany, *Lecture Notes in Computer Science*, 715: 66–81, 1993.

[47] F. van Ham, H. van de Wetering, and J. J. van Wijk. Interactive visualization of state transition systems. *IEEE Transactions on Visualization and Computer Graphics*, 8(4):319–329, 2002.

[48] M. van Weerdenburg. An account of implementing applicative term rewriting. *ENTCS*, 174(10):139–155, 2007.

[49] T. A. C. Willemse. Semantics and verification in process algebras with data and timing. PhD thesis, Technische Universiteit Eindhoven (TU/e), 2003.

Chapter 5

Business Process Specification and Analysis

Uwe Nestmann and Frank Puhlmann

Contents

5.1   Introduction

The theory behind business processes is investigated in a research area known as business process management (BPM). In contrast to business administration, BPM focuses on the technical issues of designing, managing, enacting, analyzing, adapting, and mining business processes [5]. The theories of BPM have been carried over from its preceding research area, workflow management (WfM). WfM covered a subset of BPM, namely static business processes within companies [6,16]. While ease of adaptation and flexibility was often a major design goal [2,28,30], the

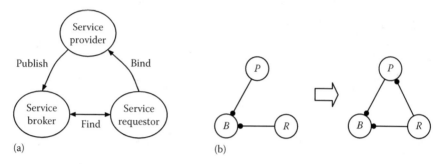

FIGURE 5.1:   SOA vs. π-calculus. (a) SOA and (b) channel passing in π-calculus.

concepts investigated could all be represented using static process theory such as Petri nets or CSP [13,21].

With the rise of the Internet and service-oriented architectures (SOA) [10], however, the focus shifted away from static workflows that are executed in isolation. Instead, dynamic collaborations between different business partners moved into the center of attention. The key players in SOA are three entities. First, there exist so called service providers, which want to provide their services to customers. Second, service brokers allow service providers to register their—typically Web—services so that they can be found by the third kind of entity, service requesters. The whole setting is completely dynamic—service providers can (de)-register their services during the runtime of the system and service requesters can search for and bind to all currently available services.

The key concept of SOA—dynamic binding—is depicted in Figure 5.1a. Interestingly, it directly resembles the triangle of an example written over a decade earlier, in a report introducing the π-calculus as a theory of mobile processes [17], shown in Figure 5.1b. Technically, the π-calculus added label-passing mobility [12] to the well-known static process theory CCS [18]. Since the theory has already moved from static to mobile systems years ago, the application of this new kind of theory to BPM is overdue. In this chapter, the authors present the state-of-the-art research on the application of the π-calculus to BPM.

This chapter is written in a tutorial style. We first introduce the preliminaries—key concepts of BPM, the π-calculus syntax and semantics, and a mapping algorithm from graphical business processes to π-calculus agents. This algorithm relies on a set of patterns introduced in the following section. Afterward, we introduce the basic principles of interaction specification. The chapter is concluded by showing how bisimulation techniques can be applied to analysis of the formalized business processes.

5.2   Preliminaries

This section introduces the required preliminaries, starting with the key concepts of BPM and the grammar and semantics of the π-calculus variant that is applied. Afterward, the concepts of BPM are represented in π-calculus terms.

5.2.1 Key Concepts

As already stated, BPM has inherited a rich theoretical and practical foundation from WfM. These includes several perspectives [11] as well as a catalogue of existing workflow patterns [4]. Regarding the perspectives, we consider the behavioral perspective, which defines the routing between different activities. The other perspectives—including, e.g., functional and data views—are only treated as needed. Regarding the workflow patterns—i.e., recurring pieces of behavior found in business processes—we consider a selected subset. If required, all patterns can be looked up in Ref. [25].

We start with the key ingredient of every business process, activities, followed by further concepts required for this chapter.

DEFINITION 5.1   Activity
   An activity *is a piece of work to be done.*

An activity might be, for instance, a manual activity like phoning someone, writing a letter, etc., or an automated activity, like invoking a script or computer program. An activity can also be a decision, e.g., between two further activities, or another situation like waiting for previous activities to finish, e.g., a bus driver waiting for at least three passenger to enter the bus. In the π-calculus formalization, an activity is represented by a τ-prefix.

DEFINITION 5.2   Activity Instance
   An activity instance *is a concrete realization of an activity.*

Examples of an activity instance are actually phoning Mr. Smith, actually waiting for three passengers, etc. In the π-calculus formalization, we usually abstract from the concrete activity instance and denote it as a τ-transition.

DEFINITION 5.3   Control Flow
   A control flow *defines temporal execution dependencies between activities.*

An example is writing a letter and thereafter sending it. Control flow relations are written as tuples of activities, e.g., (*Write Letter, Send Letter*). We assume transitivity of control flow relations, but not symmetry and reflexivity. After having defined a sequence of activities by control flow as well as activity itself, we can refine the definition of a process. In the π-calculus formalization, control flow is represented by the possibility of an interaction between two components.

DEFINITION 5.4   Process
   A process *is a set of activities related by control flow.*

An example is a credit broker process that finds the lowest interest rates for a given credit request. It might consist of the activities: (A) *Receive Credit Request*, (B) *Process Credit Request*, and (C) *Show Results*. The dependencies are straightforward: A has to happen before B and B has to be finished before C. Accordingly, the control flow relations are given by (A, B) and (B, C). Note that (A, C) is given by the transitivity of control flow.

DEFINITION 5.5   Process Instance

A process instance *is the concrete realization of a business process.*

Examples of a process instance are the actual processing of an insurance claim from Mr. Smith or buying a house including several steps.

DEFINITION 5.6   Interaction Flow

Interaction flow *defines temporal dependencies between two activities of different processes.*

An example is sending and receiving a letter. Interaction flow relations are written as pairs of activities from different processes (*Send Letter, Receive Letter*). In the π-calculus formalization, interaction flow is possible via free names.

DEFINITION 5.7   Interaction

An interaction *is given by a set of processes related by interaction flow.*

An example interaction is given by the credit broker process from Definition 5.4 and a customer process. The customer has the activities (D) *Ask for Credit Offer* and (E) *Read Credit Offer* with the single relation D before E, formally: (D, E). The credit broker and the customer need to synchronize their processes using interaction flow from activity D to A, and from C to E. Thus, the interaction flows are given by the tuples (D, A) and (C, E) and the complete interaction is given by the processes of the credit broker and the customer as well as the interaction flows.

The structure of a business process is captured in a process graph. A process graph resembles the graph structure of business process notations, such as BPMN [8] that is applied in this chapter. An example is shown in Figure 5.2. The business process starts at the left-hand side with a *start event*, denoted as a circle. Afterward, the control flow is split into two parallel branches using a parallel split gateway (the interpretation of the "+" sign, here, is quite different compared to its interpretation in process algebra). A part (of a bicycle) is registered and painted concurrently. Afterward, the control flow is joined with a synchronizing merge gateway, denoted with an "O" inside. This kind of merge waits for all activated, preceding branches to finish before continuing the control flow further downstream. If that happens, the painting is (quality) checked. If the quality check fails, the part is painted again, while otherwise the business process terminates with an *end event*. The condition is

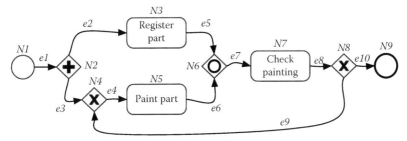

FIGURE 5.2: Simple business process.

evaluated using an exclusive choice gateway denoted with an "X" inside. If a part is painted again, the synchronizing merge gateway pays for its considerable implementation cost, since in the second iteration it only waits for the paint part activity to finish before continuing the control flow. Formally, a process graph is defined as a triple.

DEFINITION 5.8 Process Graph

A process graph $G = (\mathbf{N}, \mathbf{E}, \mathrm{T})$ is a triple representing a business process:

- **N** *is a nonempty set of nodes,*

- $\mathbf{E} \subseteq (\mathbf{N} \times \mathbf{N})$ *is a set of directed edges between nodes, and*

- $\mathrm{T} : \mathbf{N} \rightarrow \{ACT, XOR, AND, DISC, OR, RIM\}$ *is a total function assigning a type to each node.*

In addition to the function T*, we introduce some notation to relate nodes with their connecting edges. Let* $N \in \mathbf{N}$*. Then* $\bullet N = \{(X, N) \in \mathbf{E}\}$ *denotes the set of incoming edges for node* N*, whereas* $N \bullet = \{(N, X) \in \mathbf{E}\}$ *denotes the set of outgoing edges of node* N*. Moreover,* $\mathcal{R}(\mathbf{E})$ *denotes the transitive closure of* $\mathbf{E}$*.*

The different node types are precisely described in Section 5.3, except for *RIM*, described in Section 5.4.3. The example of Figure 5.2 is given by the following sets **N**, **E**, and T:

$$\mathbf{N} = \{N1, \ldots, N9\}$$
$$\mathbf{E} = \{(N1, N2), (N2, N3), (N2, N4), \ldots, (N8, N9)\}$$
$$\mathrm{T} = \{(N1, ACT), (N2, AND), (N3, ACT), (N4, XOR), (N5, ACT), (N6, OR),$$
$$(N7, ACT), (N8, XOR), (N9, ACT)\}$$

For presentational convenience, we restrict the types of process graphs considered to a subset known as *structurally sound process graphs*. This restriction is taken over from workflow theory [1] and simplifies the theory without restricting its expressive power.

DEFINITION 5.9 *Structural Soundness*

Let $G = (\mathbf{N}, \mathbf{E}, \mathrm{T})$ be a process graph. G is structurally sound, *if the following properties hold:*

- *There is exactly one node $init_G \in \mathbf{N}$ with no incoming edges, i.e., $\bullet init_G = \emptyset$ and $\forall N \in \mathbf{N} \setminus init_G : \bullet N \neq \emptyset$; $init_G$ is called* the initial node of G.

- *There is exactly one node $final_G \in \mathbf{N}$ with no outgoing edges, i.e., $final_G \bullet = \emptyset$ and $\forall n \in \mathbf{N} \setminus final_G : n \bullet \neq \emptyset$; $final_G$ is called* the final node of G.

- *Every node $N \in \mathbf{N} \setminus \{init_G, final_G\}$ is located on a path from the initial to the final node of G, i.e., $(init_G, N) \in \mathcal{R}(\mathbf{E})$ and $(N, final_G) \in \mathcal{R}(\mathbf{E})$.*

Furthermore, we require that the type of the initial and the final node is always ACT: $\mathrm{T}(init_G) = ACT$ and $\mathrm{T}(final_G) = ACT$. The example of Figure 5.2 fulfills the structural soundness property.

5.2.2 Pi-Calculus Syntax and Semantics

In contrast to prior published work on the π-calculus and BPM, we consider the asynchronous variant, in which sending processes are not blocked in case no matching receiver is available. It has the advantage of supporting more efficient implementations as well as providing simplified reasoning [19] about various properties of (business) processes.

DEFINITION 5.10 *Pi-Calculus*

The set of process terms $P \in \mathcal{P}$, also called agent terms, *is generated by the grammar*

$$P ::= \overline{x}\langle \tilde{y} \rangle \mid M \mid P|P \mid \nu z\ P \mid A(x_1, \ldots, x_n)$$
$$M ::= \mathbf{0} \mid x(\tilde{z}).P \mid \tau.P \mid M + M$$

assuming infinitely many names $x, y, z \in \mathcal{N}$ and process constants $A \in \mathcal{K}$, also called agent identifiers, *at our disposal. Tuples $x_1 \ldots, x_n$ of finite length are usually denoted as $\tilde{x}$, with $|\tilde{x}| = n$. Actions $\alpha \in \mathcal{A}$ are of the form:*

$$\alpha ::= \tau \mid x(\tilde{z}) \mid \nu \tilde{z}\ \overline{x}\langle \tilde{y} \rangle$$

where $\tilde{z} \subseteq \tilde{y}$. The asynchronous π-calculus is defined as the labeled transition system $(\mathcal{P}, \{\overset{\alpha}{\longrightarrow}\}_{\alpha \in \mathcal{A}})$ where the transition relation $\overset{\alpha}{\longrightarrow}$ is determined as the smallest relation generated by the rules in Figure 5.3; the definition also refers to the notion of structural congruence $\equiv$, *determined as the smallest congruence generated by the laws of Figure 5.4.*

The syntax and both the transition relation and the structural congruence are standard except for SC-INP-GARB and SC-OUT-GARB that are added as special-purpose

$$\text{OUT} \; \frac{}{\overline{x}\langle \tilde{y}\rangle \xrightarrow{\overline{x}\langle \tilde{y}\rangle} 0} \qquad \text{INP} \; \frac{}{x(\tilde{z}).P \xrightarrow{x\tilde{y}} \{\tilde{y}/\tilde{z}\}P} \; \text{ IF } \{\tilde{y}\} \subseteq \mathcal{N} \text{ AND } |\tilde{y}| = |\tilde{z}|$$

$$\text{TAU} \; \frac{}{\tau.P \xrightarrow{\tau} P} \qquad\qquad \text{SUM} \; \frac{M_1 \xrightarrow{\alpha} P}{M_1 + M_2 \xrightarrow{\alpha} P}$$

$$\text{PAR} \; \frac{P_1 \xrightarrow{\alpha} P_1'}{P_1 \mid P_2 \xrightarrow{\alpha} P_1' \mid P_2} \; \text{ IF } bn(\alpha) \cap fn(P_2) = \emptyset \qquad \text{RES} \; \frac{P \xrightarrow{\alpha} P'}{vz\,P \xrightarrow{\alpha} vz\,P'} \; \text{ IF } z \notin n(\alpha)$$

$$\text{COMM} \; \frac{P_1 \xrightarrow{\overline{x}\langle\tilde{y}\rangle} P_1' \quad P_2 \xrightarrow{x\tilde{y}} P_2'}{P_1 \mid P_2 \xrightarrow{\tau} P_1' \mid P_2'} \qquad \text{OPEN} \; \frac{P \xrightarrow{v\tilde{z}\,\overline{x}\langle\tilde{y}\rangle} P'}{va\,P \xrightarrow{va.\tilde{z}\,\overline{x}\langle\tilde{y}\rangle} P'} \; \text{ IF } x \neq a \in \{\tilde{y}\}\setminus\{\tilde{z}\}$$

$$\text{STRUCT} \; \frac{Q \equiv P \quad P \xrightarrow{\alpha} P' \quad P' \equiv Q'}{Q \xrightarrow{\alpha} Q'}$$

FIGURE 5.3: Asynchronous π-calculus transition rules.

rules to clean up useless remaining parts of business process formalizations. The knowledgeable reader may notice that, due to our use of the structural congruence $\equiv$, we do not need a CLOSE transition rule in addition to our simple COMM rule to derive all possible τ-transitions that arise from successful communication. The structural laws also justify the use of n-ary counterparts for summation $\sum_{i=1}^{n}$ and parallel composition $\prod_{i=1}^{n}$ (for finite n). We sometimes use the notation $\{\pi\}_1^n$ to denote a sequence of n identical input actions π.

SC-ALPHA	P_1	$\equiv$	P_2, if $P_1 = P_2$
SC-SUM-ASSOC	$M_1 + (M_2 + M_3)$	$\equiv$	$(M_1 + M_2) + M_3$
SC-SUM-COMM	$M_1 + M_2$	$\equiv$	$M_2 + M_1$
SC-SUM-INACT	$M + 0$	$\equiv$	M
SC-COMP-ASSOC	$P_1 \mid (P_2 \mid P_3)$	$\equiv$	$(P_1 \mid P_2) \mid P_3$
SC-COMP-COMM	$P_1 \mid P_2$	$\equiv$	$P_2 \mid P_1$
SC-COMP-INACT	$P \mid 0$	$\equiv$	P
SC-RES	$vz\, vw\, P$	$\equiv$	$vw\, vz\, P$
SC-RES-INATCT	$vz\, 0$	$\equiv$	0
SC-RES-COMP	$vz\,(P_1 \mid P_2)$	$\equiv$	$P_1 \mid vz\, P_2$, if $z \notin fn(P_1)$
SC-UNFOLD	$A(\tilde{y})$	$\equiv$	$P\{\tilde{y}/\tilde{x}\}$, if $A(\tilde{x}) \overset{def}{=} P$
SC-INP-GARB	$vz\, z.P$	$\equiv$	0
SC-OUT-GARB	$vz\, \prod_{i=1}^{n} \overline{z}$	$\equiv$	0

FIGURE 5.4: Axioms of structural congruence.

The term constructors $x(\tilde{z}).P$ and $\nu\tilde{z}\,P$ (short for $\nu z_1, \cdots, \nu z_n\, P$) are binding occurrences of $\tilde{z}$ within the scope of P; we silently assume the names of $\tilde{z}$ in a binding occurrence to be pairwise distinct. The occurrence of a name that is not bound is called free. The sets of free and bound names of an agent P are denoted, respectively, as $fn(P)$ and $bn(P)$. We use the standard notation $P\{\tilde{y}/\tilde{x}\}$ to denote the simultaneous substitution of names in $\tilde{y}$ for all (free) occurrences of respective names in $\tilde{x}$ in the agent term P, assuming equal lengths of the tuples $\tilde{x}$ and $\tilde{y}$.

For notational convenience, instead of replication $!P$, we use instantiations $A(\tilde{y})$ of process constants A with concrete parameters $\tilde{y}$, where we silently assume that suitable defining equations of the form $A(\tilde{x}) \stackrel{\text{def}}{=} P$ are provided. Often, we omit the parameter list $\tilde{x}$ from such a definition, assuming that all free names of the right-hand side of the equation are included in this list. Also for notational convenience, we permit ourselves to use nonname-passing send and receive actions of the form a and $\overline{a}$.

One of the advantages of using asynchronous π-calculus instead of the standard π-calculus is that naive bisimulation schemes already give rise to congruences. Here, we present the early [bi]simulation scheme; its main advantage is its simple definition when building it, as we do, upon an operational semantics that employs so-called early transitions. Note further that, since we do not exploit the additional equalities of asynchronous bisimulation techniques, we use their synchronous counterparts, again in favor of simplicity of the definitions [29].

DEFINITION 5.11   *Simulation, Bisimulation*
     Let $\mathcal{R} \subseteq \mathcal{P} \times \mathcal{P}$.

1. *$\mathcal{R}$ is a* simulation *if $P\,\mathcal{R}\,Q$ implies that*

   if $P \xrightarrow{\alpha} P'$ with $bn(\alpha) \cap fn(P, Q) = \emptyset$,
   then there is Q' with $Q \xrightarrow{\alpha} Q'$ such that $P'\,\mathcal{R}\,Q'$.

   Agent P is similar to (i.e., is simulated by) agent Q, denoted as $P \precsim Q$, if $P\,\mathcal{R}\,Q$ for some simulation $\mathcal{R}$. $\precsim$ is called similarity.

2. *$\mathcal{R}$ is a* bisimulation *if both $\mathcal{R}$ and $\mathcal{R}^{-1}$ are simulations. Agents P and Q are* bisimilar, *denoted as $P \sim Q$, if $P\,\mathcal{R}\,Q$ for some bisimulation $\mathcal{R}$. $\sim$ is called* bisimilarity.

So-called weak variants of the above simulation-based notions are obtained by replacing $Q \xrightarrow{\alpha} Q'$ with $Q \xRightarrow{\hat{\alpha}} Q'$ in the above definition, where $\Longrightarrow$ represents zero or more τ transitions, i.e., $\xrightarrow{\tau}{}^{*}$, $\xRightarrow{\alpha}$ as $\Longrightarrow\xrightarrow{\alpha}\Longrightarrow$, and $\xRightarrow{\hat{\alpha}}$ as $\xRightarrow{\alpha}$ if $\alpha \neq \tau$ and $\Longrightarrow$ if $\alpha = \tau$. Weak bisimilarity is denoted as $\approx$. Weak similarity is denoted as $\precapprox$.

5.2.3   Trios

To support interactions between—and reasoning about—business processes, we introduce a nonstandard notion of contexts. Instead of the standard case, where a context arises from some process term P by replacing a (single) occurrence of $\mathbf{0}$ within P by a so-called hole $[\cdot]$, we provide contexts that arise from P by replacing (possibly several) occurrences of τ-actions by a so-called trio $\{\alpha; \beta; \gamma\}$.* The three items α, β, γ serve as placeholders for specific kinds of actions:

> α : a single arbitrary input prefix of the form $x(\tilde{z})$
>
> β : one of the three special forms $i, \overline{o}$ or $i.\overline{o}$
>
> γ : an arbitrary output action of the form $\nu\tilde{z}\ \overline{y}\langle\tilde{v}\rangle$

The first and the last actions—α and γ—will be used to plug in communication actions inside business processes that are represented as π-calculus terms. While α is capable of receiving via an input action, γ has two purposes. First, it is able to create new names via a restriction. Second, it can emit existing names as well as the freshly created names via an output prefix. Section 5.4 explains the application in detail. The β-action will be used to analyze correctness properties of business processes given by π-calculus terms. Its purpose is explained Section in 5.5.

For each of the items, we will provide possibly separate instantiations, including the possibility of a trivial instantiation, indicated by the use of ϵ. The use of ";" in a trio indicates a sequential flow relation among the items, modulo the asynchrony of outputs. Sometimes, when the actual naming of the items in a trio is less important or obvious, we abbreviate the occurrences of a trio $\{\alpha; \beta; \gamma\}$ by a simple $\{\cdot\}$. For example, $a.\{\alpha; \beta; \gamma\}.\overline{b}$ could be abbreviated by $a.\{\cdot\}.\overline{b}$.

Usually, we denote agents by means of agent identifier definitions $A(\tilde{x})\overset{\text{def}}{=}P$. To denote an agent identifier A whose defining process P contain trios $\{\alpha; \beta; \gamma\}$, we require to refer to them explicitly as $A\{\alpha; \beta; \gamma\}(\tilde{x})$, or simply $A\{\alpha; \beta; \gamma\}$. To denote a (partial) instantiation of such a term, for example replacing β by $\overline{o}$, we then simply write $A\{\alpha; \overline{o}; \gamma\}$. If the defining process P was $a.\{\alpha; \beta; \gamma\}.\overline{b}$, then its corresponding partial instantiation would become $a.\{\alpha; \overline{o}; \gamma\}.\overline{b}$.

As soon as all of the items of a trio have been instantiated, we can transform the whole trio-enhanced term into a standard agent term according to the mapping $\mathbf{T}(\cdot)$ that is spelled out in Figure 5.5 in detail for the case of $\mathbf{T}(\{\alpha; \beta; \gamma\}.P)$; for the remaining term cases, $\mathbf{T}(\cdot)$ is defined purely homomorphically. The mapping requires a bit of care, due to the occurrence of asynchronous outputs in a conceptual sequence of actions, possibly mixed with restrictions on names. Note that, as for the standard plugin operation of a process inside a context, we also keep the plugin operation of instantiations for the trio components at the meta level. Although a trio may occur multiple times in a term, it is always instantiated with the same data for each occurrence in the same term.

For example, the agent

$$B\{x(y); \epsilon; \nu v\ \overline{y}\langle v\rangle\}, \quad \text{for } B\{\alpha; \beta; \gamma\} \overset{\text{def}}{=} a.\{\alpha; \beta; \gamma\}.\overline{b}$$

---

* The name is borrowed from Ref. [20], but has obviously different meaning, here. Note that α, β, γ only serve to distinguish the different items within a trio but not to distinguish different occurrences of trios within the same term. The latter are excluded by definition.

$$\mathbf{T}(\lbrack\epsilon; \epsilon; \epsilon\rbrack.P) \stackrel{\mathrm{def}}{=} \tau.\,\mathbf{T}(P)$$

$$\mathbf{T}(\lbrack x(\tilde{y}); \epsilon; \epsilon\rbrack.P) \stackrel{\mathrm{def}}{=} x(\tilde{y}).\,\mathbf{T}(P)$$

$$\mathbf{T}(\lbrack\epsilon; i; \epsilon\rbrack.P) \stackrel{\mathrm{def}}{=} i.\,\mathbf{T}(P)$$

$$\mathbf{T}(\lbrack\epsilon; \overline{o}; \epsilon\rbrack.P) \stackrel{\mathrm{def}}{=} \tau.(\overline{o} \mid \mathbf{T}(P))$$

$$\mathbf{T}(\lbrack\epsilon; i.\overline{o}; \epsilon\rbrack.P) \stackrel{\mathrm{def}}{=} i.(\overline{o} \mid \mathbf{T}(P))$$

$$\mathbf{T}(\lbrack\epsilon; \epsilon; \nu\tilde{z}\,\overline{x}\langle\tilde{y}\rangle\rbrack.P) \stackrel{\mathrm{def}}{=} \nu\tilde{z}\;\tau.(\overline{x}\langle\tilde{y}\rangle \mid \mathbf{T}(P))$$

$$\mathbf{T}(\lbrack x(\tilde{y}); \epsilon; \nu\tilde{z}\,\overline{y}\langle\tilde{v}\rangle\rbrack.P) \stackrel{\mathrm{def}}{=} \nu\tilde{z}\;x(\tilde{y}).(\overline{y}\langle\tilde{v}\rangle \mid \mathbf{T}(P))$$

$$\mathbf{T}(\lbrack x(\tilde{y}); i; \epsilon\rbrack.P) \stackrel{\mathrm{def}}{=} x(\tilde{y}).i.\,\mathbf{T}(P)$$

$$\mathbf{T}(\lbrack x(\tilde{y}); \overline{o}; \epsilon\rbrack.P) \stackrel{\mathrm{def}}{=} x(\tilde{y}).(\overline{o} \mid \mathbf{T}(P))$$

$$\mathbf{T}(\lbrack x(\tilde{y}); i.\overline{o}; \epsilon\rbrack.P) \stackrel{\mathrm{def}}{=} x(\tilde{y}).i.(\overline{o} \mid \mathbf{T}(P))$$

$$\mathbf{T}(\lbrack\epsilon; i; \nu\tilde{z}\,\overline{y}\langle\tilde{v}\rangle\rbrack.P) \stackrel{\mathrm{def}}{=} i.\nu\tilde{z}\;(\overline{y}\langle\tilde{v}\rangle \mid \mathbf{T}(P))$$

$$\mathbf{T}(\lbrack\epsilon; \overline{o}; \nu\tilde{z}\,\overline{y}\langle\tilde{v}\rangle\rbrack.P) \stackrel{\mathrm{def}}{=} \tau.(\overline{o} \mid \nu\tilde{z}\;(\overline{y}\langle\tilde{v}\rangle \mid \mathbf{T}(P)))$$

$$\mathbf{T}(\lbrack\epsilon; i.\overline{o}; \nu\tilde{z}\,\overline{y}\langle\tilde{v}\rangle\rbrack.P) \stackrel{\mathrm{def}}{=} i.(\overline{o} \mid \nu\tilde{z}\;(\overline{y}\langle\tilde{v}\rangle \mid \mathbf{T}(P)))$$

$$\mathbf{T}(\lbrack x(\tilde{y}); i; \nu\tilde{z}\,\overline{y}\langle\tilde{v}\rangle\rbrack.P) \stackrel{\mathrm{def}}{=} x(\tilde{y}).i.\nu\tilde{z}\;(\overline{y}\langle\tilde{v}\rangle \mid \mathbf{T}(P))$$

$$\mathbf{T}(\lbrack x(\tilde{y}); \overline{o}; \nu\tilde{z}\,\overline{y}\langle\tilde{v}\rangle\rbrack.P) \stackrel{\mathrm{def}}{=} x(\tilde{y}).(\overline{o} \mid \nu\tilde{z}\;(\overline{y}\langle\tilde{v}\rangle \mid \mathbf{T}(P)))$$

$$\mathbf{T}(\lbrack x(\tilde{y}); i.\overline{o}; \nu\tilde{z}\,\overline{y}\langle\tilde{v}\rangle\rbrack.P) \stackrel{\mathrm{def}}{=} x(\tilde{y}).i.(\overline{o} \mid \nu\tilde{z}\;(\overline{y}\langle\tilde{v}\rangle \mid \mathbf{T}(P)))$$

FIGURE 5.5:   Inductive expansion of trios into agent terms.

becomes

$$\mathbf{T}(a.\lbrack x(y); \epsilon; \nu v\,\overline{y}\langle v\rangle\rbrack.\overline{b}) = a.x(y).\nu v\;(\overline{y}\langle v\rangle \mid \overline{b})\ .$$

5.2.4   Mapping Algorithm

While a process graph defines a static structure for a business process, a process instance is generated by a mapping onto π-calculus terms. Technically, we create a composition made up of different components for each node of the process graph. The components intra-act via restricted names that represent the edges between the nodes of the process graph. Since the nodes of the process graph might have different semantics according to their types, we use a pattern-based approach to determine the terms for the components. A selected subset of these patterns is introduced in the next section. For now, we just assume their existence.

ALGORITHM 5.1   *Mapping Process Graphs to Agents*

A process graph $G = (\mathbf{N}, \mathbf{E}, T)$ is mapped to π-calculus agents as follows:

1. *Each node $N \in \mathbf{N}$ is assigned to a unique π-calculus agent identifier. The set of unique agent identifiers is denoted as $\mathcal{N}$, where the elements are denoted as Ni for notational convenience.*

2. *Each edge $E \in \mathbf{E}$ is assigned to a unique π-calculus name. The set of unique names is denoted as $\mathcal{E}$, where the elements are denoted as $e_1, \ldots, e_n$ for notational convenience.*

3. *π-calculus agents are defined according to the type of the corresponding node. Each agent is supplied with a private trio that allows us to tie it properly within suitable interaction and analysis contexts. The applicable set of process patterns is given in Section 5.3.*

4. *An agent $N \stackrel{\text{def}}{=} (ve_1, \ldots, e_n)(\prod_{i=1}^{|\mathbf{N}|} Ni)$ representing a process instance is defined by instantiating the trios of each Ni.*

For simplicity, we use identical identifiers Ni for the nodes of $\mathbf{N}$ and their corresponding agent definitions. An example of the algorithm is given in Section 5.3.4, after the process patterns have been introduced.

5.3   Processes

This section introduces a formal semantics for the different node types of a process graph. The semantics is based on a subset of the workflow patterns [4]. The workflow patterns provide a catalog—with descriptions in natural language—of recurring behaviors in business processes, that are referred to as process patterns. A complete formalization of all Workflow Patterns can be found in Ref. [25].

5.3.1   Basic Control Flow Patterns

Each node of a process graph is given a semantics according to a simple schema: incoming edges represent preconditions that must (partially) be fulfilled to activate the node. Outgoing edges represent postconditions that (partially) hold after the node has been executed. Pre- and postconditions are represented by restricted names. In the following pattern descriptions, we use a_i as the set of indexed names that represent preconditions. Similar, b_i is set of indexed names that represent postconditions. Naturally, these placeholder names have to be replaced with the actual names created in Algorithm 5.1. All agents are defined as $N_X\llbracket \cdot \rrbracket$, where $N_X\llbracket \cdot \rrbracket$ is also just a placeholder for the actual definition.

Pattern 1 (Sequence)   A node $N \in \mathbf{N}$ with at most one incoming and at most one outgoing edges is denoted as a *sequence* if it represents an activity. Formally

$$N_{\text{Seq}}\llbracket \cdot \rrbracket \stackrel{\text{def}}{=} a.\llbracket \cdot \rrbracket.(\overline{b} \mid N_{\text{Seq}}\llbracket \cdot \rrbracket).$$

If $\bullet n = \emptyset$, a and $N_{\text{Seq}}\llbracket \cdot \rrbracket$ are removed, i.e., the corresponding node is $init_G$:

$$N_{\text{Seq}}\llbracket \cdot \rrbracket \stackrel{\text{def}}{=} \llbracket \cdot \rrbracket.\overline{b}.$$

The removal of $N_{\text{Seq}}\{\cdot\}$ in the right-hand side is required, since otherwise an infinite recursion arises. If $N\bullet = \emptyset$, b is removed. The pattern applies if $T(N) = ACT$, $|\bullet N| \leq 1$, and $|N \bullet| \leq 1$.

Pattern 2 (Parallel split) A node $N \in \mathbf{N}$ with more than one outgoing edge is denoted as a *parallel split* if all outgoing edges are activated in parallel. Formally

$$N_{\text{ParSplit}}\{\cdot\} \stackrel{\text{def}}{=} a.\{\cdot\}.\left(\prod_{i=1}^{|N\bullet|}(\overline{b_i}) \mid N_{\text{ParSplit}}\{\cdot\}\right).$$

The pattern applies if $T(N) = AND$, $|\bullet N| = 1$, and $|N \bullet| \geq 2$.

Pattern 3 (Synchronization) A node $N \in \mathbf{N}$ with more than one incoming edge is denoted as a *synchronization* if all incoming edges are activated. Formally

$$N_{\text{Sync}}\{\cdot\} \stackrel{\text{def}}{=} \{a_i\}_{i=1}^{|N\bullet|}.\{\cdot\}.(\overline{b} \mid N_{\text{Sync}}\{\cdot\}).$$

The pattern applies if $T(N) = AND$, $|\bullet N| \geq 2$, and $|N \bullet| = 1$.

Pattern 4 (Exclusive choice) A node $N \in \mathbf{N}$ with more than one outgoing edge is denoted as an *exclusive choice* if only one outgoing edge is activated. Formally

$$N_{\text{ExclChoice}}\{\cdot\} \stackrel{\text{def}}{=} a.\{\cdot\}.\left(\sum_{i=1}^{|N\bullet|}\tau.(\overline{b_i} \mid N_{\text{ExclChoice}}\{\cdot\})\right).$$

The pattern applies if $T(N) = XOR$, $|\bullet N| = 1$, and $|N \bullet| \geq 2$.

Pattern 5 (Simple merge) A node $N \in \mathbf{N}$ with more than one incoming edge is denoted as a *simple merge* if only one incoming edges is activated. Formally

$$N_{\text{SimplMerge}}\{\cdot\} \stackrel{\text{def}}{=} \sum_{i=1}^{|\bullet N|} a_i.\{\cdot\}.(\overline{b} \mid N_{\text{SimplMerge}}\{\cdot\}).$$

The pattern applies if $T(N) = XOR$, $|\bullet N| \geq 2$, and $|N \bullet| = 1$.

5.3.2 Advanced Control Flow Patterns

We discuss advanced control flow patterns in this section. One of them, the synchronizing merge pattern, requires a special treatment of the contained transitions that is discussed in detail.

Pattern 6 (Discriminator) A node $N \in \mathbf{N}$ with more than one incoming edge is denoted as a *discriminator* if all edges are activated, but the node activates the downstream edge after the first incoming edge has been activated.

$$N_{\text{Disc}}\{\cdot\} \stackrel{\text{def}}{=} vh \prod_{i=1}^{|\bullet N|}(a_i.\overline{h}) \mid h.(\{\cdot\}.\overline{b} \mid \{h\}^{|\bullet N|-1}.N_{\text{Disc}}\{\cdot\}).$$

The pattern applies if $T(N) = DISC$, $|\bullet N| \geq 2$, and $|N \bullet| = 1$.

Pattern 7 (Multi choice) A node $N \in \mathbf{N}$ with more than one outgoing edge is denoted as *multi choice* if at least one and at most all edges are activated. Formally

$$N_{\text{MultChoice}}\langle\cdot\rangle \stackrel{\text{def}}{=} \nu c \; a.\langle\cdot\rangle. \left(\prod_{i=1}^{|N\bullet|} \left(\underbrace{\tau.\overline{b_i}}_{\text{enable}} + \underbrace{c.\mathbf{0}}_{\text{cancel}} \right) \; \middle| \; \prod_{i=1}^{|N\bullet|-1} \overline{c} \; \middle| \; N_{\text{MultChoice}}\langle\cdot\rangle \right).$$

This pattern activates at least one postcondition b_i via the *enable* part. This is due to the right-hand component, that only allows the *cancel* part to be executed $n-1$ times. The pattern applies if $T(N) = OR$, $|\bullet N| = 1$, and $|N\bullet| \geq 2$.

Pattern 8 (Synchronizing merge) A node $N \in \mathbf{N}$ with more than one incoming edge is denoted as *synchronizing merge* if at least one and at most all incoming edges are activated. Formally

$$N_{\text{SyncMerge}}\langle\cdot\rangle \stackrel{\text{def}}{=} (\nu done, c, ack)$$

$$\left(\prod_{i=1}^{|\bullet N|} \left(\underbrace{a_i.(\overline{done} \mid \overline{ack})}_{\text{enable}} + \underbrace{c.\overline{ack}}_{\text{cancel}} \right) \right.$$

$$\left. \middle| \quad done. \left(\left(\prod_{i=1}^{|\bullet N|-1} \overline{c} \right) \mid \{ack\}^{|\bullet N|}.\langle\cdot\rangle.(\overline{b} \mid N_{\text{SyncMerge}}\langle\cdot\rangle) \right) \right)$$

This pattern waits for at least one precondition via a_i. Afterward, an intraaction via *done* activates the right hand component of the term. The *done*-transition, however, requires a special treatment that is discussed below. The right-hand component provides cancel triggers via c as well as collecting as many *ack* triggers as there are preconditions for N. The *ack* triggers are either sent via the *enable* (if the corresponding precondition true) or *cancel* (if the corresponding precondition is false) part of the left-hand compositions. The pattern applies if $T(n) = OR$, $|\bullet n| \geq 2$, and $|n\bullet| = 1$.

5.3.3 Synchronizing Merge Semantics

The given specification of the synchronizing merge requires a special treatment of the contained transitions, that will be given informally to introduce the key ideas. In particular, the intraaction via the restricted name *done* needs to be captured. After this transition has occurred, the remaining incoming edges denoted by the names a_i are canceled via $\prod \overline{c}.\mathbf{0}$. Afterward, the synchronizing merge passes the control flow to the downstream nodes via $\overline{b}$. The informal semantics of the patterns states that the control flow should only pass on if there is no possibility that further incoming edges could be activated. Consider again Figure 5.2. The synchronizing merge given by the node $N6$ waits for both, $e5$ and $e6$ in the first iteration, whereas later it waits for only $e6$. Hence, the *done*-transition has to be blocked until all possible activated preconditions have been gathered.

Technically, there are three possibilities of deciding whether the *done* intraaction should be allowed. First, the *done* intraaction might be blocked until no other transition is possible. If only *done* intraactions are possible, one has to be chosen nondeterministically. This approach has the lowest computational effort. However, parallel branches of the business process need to be finished first, regardless whether they affect the synchronizing merge. Thus, from a practical point of view, this kind of semantics should be avoided. Second, a state space analysis according to Ref. [3] could be made to determine if any incoming edge of the synchronizing merge could possibly be activated from the current state. This approach has the highest computational effort. From a practical point of view, this kind of semantics should also be avoided since it requires complex calculations. Third, heuristics could be used. Heuristics represent intermediate complexity. They do not always deliver the best result possible, but limit the required computations.

A simple heuristics is given by precalculating the shortest path between all nodes of a process graph before execution time. The resulting matrix is used to look up whether any other active node is on a path to a synchronizing merge node that is currently in the state of deciding whether to activate downstream nodes or wait further. The resulting computational effort is linear with respect to the number of active nodes of a process graph for each active synchronizing merge node.

To derive an algorithm, we introduce a function

$$shortestPath : N \times N \rightarrow Integer$$

that returns the shortest path between two nodes of a process graph. Furthermore, we need a function that maps an interaction between two agents to the node of the process graph that represents the target node:

$$targetNode : Transition \rightarrow N.$$

Having these two functions, we can process the list of available transitions in a given state of the π-calculus system as shown in Algorithm 5.2.

ALGORITHM 5.2   *Heuristic Synchronizing Merge Algorithm*

```
def orJoinHeuristics(transitions)
  # Step (1)
  blocked, allowed = transitions.partition{|t| t.subject == 'done'}
  return blocked if allowed.size == 0
  # Step (2)
  allowed += blocked.select do |b|
    allowed.find do |a|
      shortestPath(targetNode(a), targetNode(b)) < INFINITE
    end
  end
end
```

In the first step, the list of transitions is partitioned into blocked and allowed transitions. Each transition with a *done* subject is placed into the blocked list. If the list

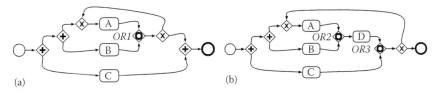

FIGURE 5.6: Business processes with synchronizing merge nodes. (a) Example 1 and (b) example 2.

of allowed transitions is empty, all blocked transitions are allowed and the algorithm ends. In the second step, all blocked transitions are examined. If no path between any node related to the set of allowed transitions to the node of the blocked transition exists, the blocked transition is unblocked by adding it to the set of allowed transitions. Finally, the new set of allowed transitions is returned implicitly.

Figure 5.6 shows business processes containing synchronizing merge nodes. We start with discussing subfigure 5.6a. When applying the simple synchronizing merge algorithm (i.e., block all *done* transitions), an execution trace consisting of A and B will not let the synchronizing merge start the downstream control flow before C finishes. This is, however, not desirable, since C has no influence on the synchronizing merge node. By applying the introduced heuristic algorithm, the execution of C is decoupled from the synchronizing merge *OR1*, since there exists no path from C to *OR1*.

In subfigure 5.6b, the business process has been changed slightly. The back-loop has now been placed after *OR3*. With the same execution trace consisting of A and B, the synchronizing merge node *OR3* is not enabled, because there exists a path from C to *OR3*. Another interesting case arises if two synchronizing merge nodes are in conflict. Consider when A, B, and C have been finished and the only transitions remaining are $done_{OR2}$ and $done_{OR3}$ in the agent definition of the nodes *OR2* and *OR3*. Both transitions are unblocked, since no other transition is currently available. However, only one could—nondeterministically—be selected. We need to stick to nondeterminism, since the "right" transition cannot be detected in the general case (see Ref. [14] for details).

Obviously, the introduced heuristics does not consider the semantics of the nodes. Hence, process graphs with wide-spanning circles will almost always give results close to the simple synchronizing merge algorithm. In many practical cases, however, parallel processing of paths with synchronizing merge nodes is possible. Furthermore, the algorithm never allows a synchronizing merge to synchronize too early. Since the computational effort is acceptable, it provides a good balance between being too restrictive and performing complex state space evaluations.

5.3.4   Process Example

To give an illustration of the mapping Algorithm 5.1, we apply it to the example of Figure 5.2. The agent identifiers as well as the restricted names are directly taken from annotations of the figure, hence the first and the second steps are already done. Following a top-down approach, we continue with the fourth step:

$$N \stackrel{\text{def}}{=} (ve1, \ldots, e10) \left(\prod_{i=1}^{9} Ni \right).$$

This agent can of course only be resolved after the trios of all components Ni have been instantiated. We now give the formal definitions of the different nodes. Instantiations of the involved trios are considered in the interactions and analysis section later on (Sections 5.4 and 5.5). All nodes with the type ACT represent activities and are implemented by Pattern 1 (Sequence):

$$N1\{\cdot\} \stackrel{\text{def}}{=} \{\cdot\}.\overline{e1} \quad \text{and} \quad N9\{\cdot\} \stackrel{\text{def}}{=} e10.\{\cdot\}.N9\{\cdot\}.$$

The agent $N1$ represents the initial node of the process graph. It has no precondition. As stated, the trio $\{\cdot\}$ represents a placeholder (or slot) for inserting interactions as well as modifying the agent terms for analysis later on. Afterward, the agent term $N1$ is able to emit once via the restricted name $e1$ to signal the completion of the node's execution. The agent term $N9$ represents the final node of the process graph. The reactivation of the agent term is handled via recursion.

$$N3\{\cdot\} \stackrel{\text{def}}{=} e2.\{\cdot\}.(\overline{e5} \mid N3\{\cdot\}), \quad N5\{\cdot\} \stackrel{\text{def}}{=} e4.\{\cdot\}.(\overline{e6} \mid N5\{\cdot\}), \quad \text{and}$$
$$N7\{\cdot\} \stackrel{\text{def}}{=} e7.\{\cdot\}.(\overline{e8} \mid N7\{\cdot\}).$$

The other nodes that represent activities are similar; they have both pre- and post-conditions given by restricted names. Node $N2$ is a parallel split gateway as denoted by the type AND. The agent term is given by Pattern 2 (Parallel Split):

$$N2\{\cdot\} \stackrel{\text{def}}{=} e1.\{\cdot\}.(\overline{e2} \mid \overline{e3} \mid N2\{\cdot\}).$$

The agent emits via two restricted names, $e2$ and $e3$. Node $N4$ is a simple merge, as given by the type XOR:

$$N4\{\cdot\} \stackrel{\text{def}}{=} e3.\{\cdot\}.(\overline{e4} \mid N4\{\cdot\}) + e9.\{\cdot\}.(\overline{e4} \mid N4\{\cdot\}).$$

For each intraaction via $e3$ or $e9$, the name $e4$ is emitted once, which gives an exclusive join semantics. The node $N6$ has a more subtle semantics. It is implemented by Pattern 8 (Synchronizing Merge) according to its OR type. This kind of merge waits for all incoming edges that might be activated:

$$N6\{\cdot\} \stackrel{\text{def}}{=} (vdone, c, ack)(e5.(\overline{done} \mid \overline{ack}) + c.\overline{ack} \mid$$
$$e6.(\overline{done} \mid \overline{ack}) + c.\overline{ack} \mid$$
$$done.\overline{c} \mid ack.ack.\{\cdot\}.(\overline{e8} \mid N6\{\cdot\})).$$

As discussed, the pattern requires a special treatment of τ-transitions that occur via the restricted name *done*. The last node missing is *N8*, having the same type as node *N4*, but the converse behavior. Instead of merging control flow, it splits as:

$$N8[\cdot] \stackrel{\text{def}}{=} e8.[\cdot].((\overline{e9} + \overline{e10}) \mid N8[\cdot]).$$

By applying the given algorithm to any kind of graphical process model—that consists of elements matching to the different node types—complex business processes can be formalized into π-calculus terms.

So far, we conclude the section on the representation of internal—static—business processes. All of the patterns introduced have also been discussed in static process theory. The additional value, however, comes in the ability to extend the formal representation from internal business processes to dynamic interactions among them. To achieve this, we make use of the trios, that have been neglected so far.

5.4 Interactions

We now discuss the key ideas behind the extension from internal business processes to interacting services. This is the crucial point in favor of the π-calculus. Instead of providing a breakdown from higher abstraction levels, we need to stay at the π-calculus level, since no graphical representation of dynamic interactions exists today in the area of BPM.

5.4.1 Correlations

A common task between interacting business processes invoking other business processes is response matching. This matchmaking is done using correlations that relate a response with a request. Usually, some kind of correlation identifier is placed inside each request and response. The invoking as well as the responding processes have to take care to correlate the requests based on the identifiers. In the π-calculus, the unique identifier of a request is represented by a restricted name. Since names are unique and can be used as interaction channels, an unambiguous representation of the correlations is given. Consider, for instance, the interacting business processes represented by the agents A and B:

$$A \stackrel{\text{def}}{=} \nu ch \ (\overline{b}\langle ch \rangle \mid ch(r).A' \mid \tau.A) \ \text{ and } \ B \stackrel{\text{def}}{=} \nu r \ b(ch).(\tau.\overline{ch}\langle r \rangle \mid B).$$

Agent A is able to invoke B several times via b, even before a first response is received. B in turn is able to process multiple request initiated via b at the same time. Hence, matching requests and responses have to be correlated. This is done by using ch in A as a correlation identifier. Since ch is unique for each recursive execution of A, the matchmaking is done implicitly via ch.

5.4.2 Static and Dynamic Binding

In contrast to other formalizations, the π-calculus directly expresses dynamic binding of interaction partners as found in SOA. Figure 5.7 depicts how dynamic binding is realized using link passing mobility. The left-hand side shows the three different roles of a SOA, denoted as circles. A service requester (R) knows a service broker (B). The service broker has knowledge about a number of service providers (P). The service broker evaluates the request of the service requester and returns a corresponding link to a service provider. The service requester then uses this link to dynamically bind to the service provider. Hence, the link structure changes over time as shown at the right-hand side of the figure. A simple implementation of a broker having static knowledge of two providers reachable via $p1$ and $p2$ is given by the agent B:

$$B \stackrel{\text{def}}{=} b(ch).((\tau.\overline{ch}\langle p1 \rangle + \tau.\overline{ch}\langle p2 \rangle) \mid B).$$

The agent B is able to emit either the name $p1$ or $p2$ based on an internal decision via the received name ch. A more elaborate implementation might allow providers to register and deregister during the runtime of the broker. The service providers are given by the parameterized agent P:

$$P(p) \stackrel{\text{def}}{=} vresp\ p(req, ch).(\tau.\overline{ch}\langle resp \rangle \mid P(p)).$$

A service requester that is able to dynamically incorporate a service provider according to the interaction behavior of P is given by

$$R \stackrel{\text{def}}{=} vreq\ vch1\ vch2\ (\overline{b}\langle ch1 \rangle \mid ch1(p).(\overline{p}\langle req, ch2 \rangle \mid ch2(resp))).$$

In the first two transitions, R acquires a link to a specific service provider represented by p. Thereafter it uses p to dynamically bind to the service provider. The working system is given by

$$SYS \stackrel{\text{def}}{=} vb\ vp1\ vp2\ (B \mid P1(p1) \mid P2(p2) \mid R).$$

The system is composed out of the requester's agent R as well as other agents, building an environment inside which R is running. This environment can now be changed, e.g., new service providers can be added or removed, all without modifying the service requester.

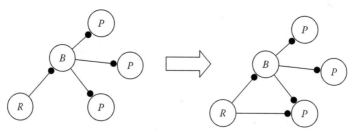

FIGURE 5.7: Dynamic binding in π-calculus.

5.4.3 Selected Interaction Patterns

Common patterns of interaction between business processes have been collected as the service interaction patterns [7]. These patterns are expanded into the agent terms of business processes mapped to π-calculus as described in the previous section. The important role is played by the α and γ parts of the trios found in the pattern formalizations of the previous section. By assigning values to these parts of a trio, each node of a business process is able to communicate with other business processes.

Pattern 9 (Send) The agent definition of a node N that wants to emit data matches the *send* pattern. Technically, an output prefix is placed concurrently with the output prefixes that are used to represent postconditions. The former output prefix might emit restricted names that are also scoped to the latter output prefixes. The latter ones, in turn, might pass them as objects to other input prefixes representing preconditions of other nodes. Formally, the additional output prefix is given by the γ part of a trio. With a partial instantiation of corresponding node's term, e.g.,

$$C[\{\alpha; \beta; \gamma\}.Q] \quad \text{becomes} \quad C[\{\alpha; \beta; \overline{x}\langle y \rangle\}.Q]$$

and with trivial instantiations for α and β, assuming that $C[\cdot]$ and Q do not contain trios themselves:

$$\mathbf{T}(C[\{\epsilon; \epsilon; \overline{x}\langle y \rangle\}.Q]) = C[\tau.(\overline{x}\langle y \rangle \mid Q)].$$

$C[\cdot]$ and Q represent subterms according to Definition 5.10. The pattern can be applied to each trio found in the agent definitions of a process graph.

An example that is shown in Figure 5.8a and formally given by

$$NI\{\cdot\} \overset{\text{def}}{=} e1.\{\cdot\}.\overline{e2}$$

becomes

$$NI\{\alpha; \beta; \overline{ch}\langle data \rangle\} \overset{\text{def}}{=} e1.\{\alpha; \beta; \overline{ch}\langle data \rangle\}.\overline{e2} .$$

If we assume α and β to be ϵ, the instantiation is given as

$$\mathbf{T}(NI\{\alpha; \beta; \overline{ch}\langle data \rangle\}) = e1.\tau.(\overline{ch}\langle data \rangle \mid \overline{e2}).$$

Since no acknowledgment from the receiver is contained, the actual transmission of the message cannot be guaranteed. In practical settings, however, downstream nodes receive results, thus restoring the causalities. The send pattern also manifests a key advantage of the π-calculus, the support for static and dynamic binding. In a statically predefined system, the name ch is already known at design time:

$$SYS \overset{\text{def}}{=} vch \; (NI\{\cdot\} \mid P).$$

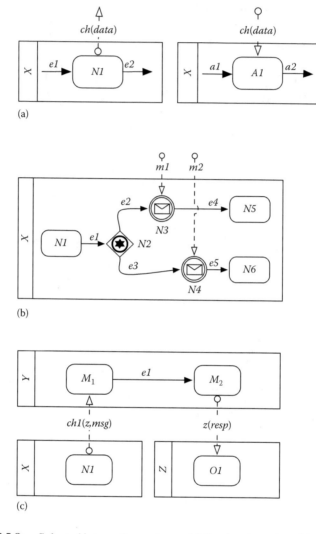

FIGURE 5.8:   Selected interaction patterns. (a) Send and receive, (b) racing incoming messages, and (c) request with a referral.

P denotes an agent representing a service provider accessible via ch, whereas $N1\llbracket\cdot\rrbracket$ is the same as above. Using link passing mobility, the name ch could be acquired at runtime via a statically predefined broker channel:

$$SYS \stackrel{\text{def}}{=} \nu broker \; (N1\llbracket\cdot\rrbracket \mid \nu ch \; (P \mid B))$$

The component P registered the name ch at another component B, representing a broker. Technically, at least one name has to be statically predefined, since otherwise the components would be unable to interact with each other.

Pattern 10 (Receive) The agent definition of a node N that wants to receive data accords to the *receive* pattern. Technically, an input prefix is placed directly after the prefixes that represent the preconditions. Formally, the additional input prefix is given by the α part of a trio. With a partial instantiation of a corresponding receive node's term, e.g.,

$$C[\{\alpha; \beta; \gamma\}.Q] \quad \text{becomes} \quad C[\{x(y); \beta; \gamma\}.Q].$$

Again, $C[\cdot]$ and Q represent subterms according to Definition 5.10. The pattern can be applied to each trio found in the agent definition of a process graph.

An example that is shown in Figure 5.8a and formally given by

$$A1[\cdot] \stackrel{\text{def}}{=} a1.[\cdot].\overline{a2}.\mathbf{0}$$

becomes

$$A1[ch(data); \beta; \gamma] \stackrel{\text{def}}{=} a1.[ch(data); \beta; \gamma].\overline{a2}.$$

Another interaction pattern extends Pattern 4 (Exclusive Choice). Instead of having a nondeterministic or internal decision that routes the control flow, external events represented by incoming messages route the control flow. The pattern is timed, meaning that the event occurring first renders the reception of the other possibilities void.

Pattern 11 (Racing incoming messages) If a node N selects a certain outgoing edge (postcondition) based on incoming messages $m_1 \ldots m_n$, it accords to the *racing incoming messages* pattern. Formally

$$N_{\text{RIM}} \stackrel{\text{def}}{=} a.[\cdot]. \left(\sum_{i=1}^{|\bullet n|} m_i.(\overline{b_i} \mid N_{\text{RIM}}) \right).$$

The pattern applies if $T(n) = RIM$, $|\bullet n| = 1$, and $|n \bullet| \geq 2$.

Interestingly, this pattern cannot be directly associated with existing graphical notations. Actual notations, such as BPMN and UML activity diagrams, use additional nodes to represent the reception of the messages. While, especially in the UML representation, it is assumed that a cancelation of the remaining receive nodes takes place, this cannot be represented in the π-calculus semantics given. The problem occurs inside cycles, where it is not possible to decide whether an incoming message belongs to a certain iteration of the pattern, since all external message subjects are the same for all iterations.

An example is shown in Figure 5.8b, node $N2$. According to the above pattern definition, it is formally denoted as follows:

$$N2[\cdot] \stackrel{\text{def}}{=} e1.[\cdot].(m1.(\overline{e4} \mid N2[\cdot]) + m2.(\overline{e5} \mid N2[\cdot])).$$

The nodes $N3$ and $N4$, including the edges $e2$ and $e3$, have been removed in the formalization.

The last interaction pattern—that is considered in this chapter—makes use again of link passing mobility.

Pattern 12 (Request with a referral) A process choreography made up of (at least) three interacting processes accords to the *request with a referral* pattern if an activity of a process X sends a request to another process Y, that in turn sends the response to a third process Z, where Z has been specified by X. Hence, the channel or link to Z is passed from X to Y.

Since this pattern resembles a combination of the Patterns 9 (Send) and 10 (Receive), there is no distinct formalization. Instead, it needs to be matched to the actual context. Consider for instance Figure 5.8c. It shows a minimum variant of the request with a referral pattern in a proprietary extension of the BPMN. It basically enhances the message flows between the processes with information about the channels to use. Its formal semantics is given by

$$SYS \stackrel{\text{def}}{=} vch1 \; (vz \; (X[\![\epsilon; \epsilon; \overline{ch1}\langle z, msg \rangle]\!] \mid Z[\![z(resp); \epsilon; \epsilon]\!]) \mid Y),$$

with

$$X[\![\cdot]\!] \stackrel{\text{def}}{=} [\![\cdot]\!]$$

$$Y \stackrel{\text{def}}{=} ve1 \; (Y1[\![ch1(z, msg); \epsilon; \epsilon]\!] \mid Y2[\![\epsilon; \epsilon; vresp \; \overline{z}\langle resp \rangle]\!])$$

$$Y1[\![\cdot]\!] \stackrel{\text{def}}{=} [\![\cdot]\!].\overline{e1}\langle z \rangle$$

$$Y2[\![\cdot]\!] \stackrel{\text{def}}{=} e1(z).[\![\cdot]\!].Y2[\![\cdot]\!]$$

$$Z[\![\cdot]\!] \stackrel{\text{def}}{=} [\![\cdot]\!].$$

The activities X and Z of the corresponding processes do not have any pre- or post-conditions, hence they only consist of $[\![\cdot]\!]$. We added data flow between the components of Y to denote the transmission of the received channel in $Y1$ to $Y2$. A formal data flow semantics, however, is out of scope for this chapter.

5.4.4 Interaction Example

To further illustrate the concepts introduced so far, we provide the formal representation of a more elaborate example that includes a simple broker. For the sake of simplicity, we assume the broker to be able to (1) accept registrations of new services at any time via a channel *add* and (2) return any of the registered services nondeterministically upon a request via *get*:

$$B(add, get) \stackrel{\text{def}}{=} add(ch).(B1(get, ch) \mid B)$$

with

$$B1(get, ch) \stackrel{\text{def}}{=} \overline{get}\langle ch \rangle.B1(get, ch).$$

Different services can register at the given broker. We consider a simple bank that is able to process a loan request and return either accept or reject messages. The bank

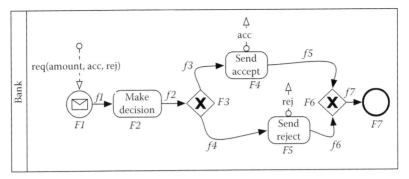

FIGURE 5.9:   Bank example.

must be able to receive loan requests via a channel *req* with three objects, *acc, rej,* and *amount*. While *amount* is only a placeholder for an actual value, the two others are channels which will be used to signal the successful or unsuccessful result of the request. A BPMN diagram of the bank is shown in Figure 5.9. The agent of the bank is given by

$$F(req) \stackrel{\text{def}}{=} (vf1,\ldots,f7)req(amount, acc, rej).\left(\prod_{i=1}^{7} Fi \mid F(req)\right).$$

The above agent has an interesting difference from the global agents given so far. Instead of representing a single instance of the process graph, it supports the execution of multiple instances. Each time a new request is received via *req*, a recursive copy of *F* is created. Hence, multiple requests could be handled at the same time. The components are fairly standard according to the patterns. The only thing that needs to be taken care of is the handling of the data flow. This is, however, not required for this example, since the names *amount, acc, rej* are already correctly restricted to the agents representing the activity instances of the current bank's process instance. We give the agents *F4* and *F5* that send the accept/reject messages:

$$F4\{\epsilon;\ \epsilon;\ \overline{acc}\} \stackrel{\text{def}}{=} f3.\{\cdot\}.(\overline{f5} \mid F4\{\cdot\}) \quad \text{and} \quad F5\{\epsilon;\ \epsilon;\ \overline{rej}\} \stackrel{\text{def}}{=} f4.\{\cdot\}.(\overline{f6} \mid F5\{\cdot\}).$$

Further banks are possible, e.g., ones that always accept a loan request (at higher rates) or require additional security (via an additional interaction). All these banks are able to register at the broker provided. Since they offer a different visible behavior, however, they are not simply interchangeable. Furthermore, it depends on the interaction behavior of a possible customer whether a successful interaction is possible. We consider a sample customer matching to the bank given in Figure 5.10. The customer is defined by the agent

$$C(get) \stackrel{\text{def}}{=} (vc1,\ldots,c8)\left(\prod_{i=1}^{8} Ci\right).$$

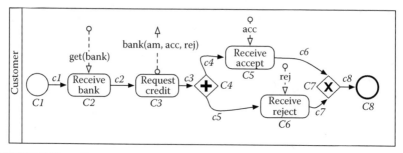

FIGURE 5.10:   Customer example.

The interesting node is *C2*. Via this node, a currently registered bank is selected by the broker and returned to the customer. Afterward, the customer is able to dynamically bind to a new instance of the bank formerly unknown to him. Technically, the free name *get* is used to receive a pointer to a certain bank.

$$C2\{get(bank); \epsilon; \epsilon\} \overset{\text{def}}{=} c1.\{\cdot\}.(\overline{c2}\langle bank\rangle \mid C2\{\cdot\}).$$

In the agent *C3*, representing the downstream node *credit request*, we need to create the names *am* (for amount), *acc*, and *rej* since they are not received earlier in the customer agent:

$$C3\{\epsilon; \epsilon; vam, acc, rej \,\overline{bank}\langle am, acc, rej\rangle\} \overset{\text{def}}{=} c2(bank).\{\cdot\}.(\overline{c3}\langle acc, rej\rangle \mid C3\{\cdot\}).$$

Note that the names *acc* and *rej* are forwarded via *c3*, since they are required later on. In particular, the agent definition of the node *C4* is given by

$$C4\{\epsilon; \epsilon; \epsilon\} \overset{\text{def}}{=} c3(acc, rej).\{\cdot\}.(\overline{c4}\langle acc\rangle \mid \overline{c5}\langle rej\rangle).$$

Here, the names are routed further on depending where they are used. The order of the nodes that require certain names to work with can be determined via a shortest path analysis of the corresponding process graph.

Since the derivation of the other nodes is fairly standard according to the patterns, we omit them and continue with a global system bringing the broker, bank, and customer into one system:

$$SYS \overset{\text{def}}{=} (vadd, get\,)\big(B(add, get) \mid C(get) \mid vreq\,\overline{add}\langle req\rangle.F(req)\big).$$

Consider the prefix of the agent *F* representing the bank. Via the name *add* the request channel of the bank is registered. Afterward, the bank starts its commercial offerings.

5.5 Analysis

This section introduces different properties for the formerly specified business processes and interactions. The analyses will be based on weak simulation and bisimulation. Technically, we will enhance certain agent definitions of process graph nodes with free names. Via these free names, we are able to observe their occurrence. The enhancement will be done using the β part of the trios.

The first property, easy soundness, has been introduced in Ref. [24]. It provides a minimum property which each business process should fulfill. A more elaborate property, lazy soundness [27], provides a sufficient analysis for many application areas. Finally, lazy soundness will be enhanced to interactions.

5.5.1 Easy Soundness

The least property a business process should fulfill is given by easy soundness, informally given by

> A structurally sound process graph representing a business process is easy sound if a result can be provided.

As indicated by the word *can*, we have to use simulation to prove this property for process algebraic formalizations of business processes. In particular, we have to be able to observe the occurrence of the initial and the final node. The idea is depicted in Figure 5.11. A structurally sound process graph is fed into a black box. Each time we press the start button, an instance of the process graph is executed. Each time the final node of process graph is executed, the done bulb flashes. Regarding easy soundness, we have to find at least one process instance where the done bulb flashes, denoting the delivery of the result. An agent fulfilling this invariant is given by

$$S_{\text{EASY}} \overset{\text{def}}{=} i.\tau.\bar{o} . \tag{5.1}$$

The input prefix i denotes the pushbutton, whereas the output prefix $\bar{o}$ resembles the done bulb. Both are in a fixed sequence, i.e., $\bar{o}$ follows always after i. The τ in between denotes the abstraction from complex actions. Since we use weak (bi)-simulations, however, it could also be omitted. To be able to decide whether a business process given by π-calculus agents is weakly similar to S_{EASY}, we have to enhance the agents representing the business process:

FIGURE 5.11: Black box investigation of a structurally sound process graph.

ALGORITHM 5.3   ***Soundness Annotated π-Calculus Mapping***

To annotate a π-calculus mapping D of a process graph $G = (\mathbf{N}, \mathbf{E}, \mathbf{T})$ *for reasoning on soundness, we need to instantiate the trios of the agent definitions. Let Ni iterate over the agent definitions/nodes of all elements of $\mathbf{N}$. Furthermore, $\{i, o\} \cap (fn(D) \cup bn(D)) = \emptyset$. The trios have to be instantiated as follows, where we assume α and γ to be ϵ, since we do not consider interaction flows for soundness:*

- $Ni \langle \epsilon; \epsilon; \epsilon \rangle$ *if* $|\bullet Ni| > 0 \wedge |Ni \bullet| > 0$,

- $Ni \langle \epsilon; i; \epsilon \rangle$ *if* $|\bullet Ni| = 0 \wedge |Ni \bullet| > 0$,

- $Ni \langle \epsilon; \overline{o}; \epsilon \rangle$ *if* $|\bullet Ni| > 0 \wedge |Ni \bullet| = 0$, *and*

- $Ni \langle \epsilon; i.\overline{o}; \epsilon \rangle$ *if* $|\bullet Ni| = |Ni \bullet| = 0$.   □

A formal definition of easy soundness using weak similarity is now given by

DEFINITION 5.12   ***Easy Sound Process Graph***

A *structurally sound process graph G with a semantics given by the soundness annotated π-calculus mapping D of G is easy sound if $S_{EASY} \precsim D$ holds.*

We can prove the business process of the customer from Figure 5.10 to be easy sound by finding a relation $S_{EASY} \precsim C_{EASY}$, with C_{EASY} being syntactically equal to C from Section 5.4.4 and the following instantiations:

$$C1 \langle \epsilon; i; \epsilon \rangle \quad \text{and} \quad C8 \langle \epsilon; \overline{o}; \epsilon \rangle .$$

Since finding the relation by hand is quite difficult, we can convert the business process model into π-calculus terms with a proprietary tool-chain developed by the authors. The result is shown in Figure 5.12. Due to the automated conversation (as described in Ref. [22]), different agent identifiers have been generated. Using the bisimulation checker ABC [9], easy soundness can be proved:

```
abc > wlt S_EASY(i,o) N(i,o)
The two agents are weakly related (4).
Do you want to see the core of the simulation (yes/no) ? no
```

5.5.2   Lazy Soundness

One obvious extension to easy soundness is given by enforcing that all instances of a process graph provide a result:

> A structurally sound process graph representing a business process is lazy sound if in any case a result is provided exactly once.

This property can be proved using weak bisimilarity, where all assumptions from easy soundness also hold. In particular, we need to be able to observe the occurrence

```
agent N1398(e1409,e1400)=e1409.t.('e1400.0 | N1398(e1409,e1400))
agent N1322(e1331,o)=e1331.t.('o.0 | N1322(e1331,o))
agent N1314(e1329,e1328,e1331)=(e1329.N1314_1(e1329,e1328,e1331) +
     e1328.N1314_1(e1329,e1328,e1331))
agent N1314_1(e1329,e1328,e1331)=t.('e1331.0 | N1314(e1329,e1328,e1331))
agent N1304(e1327,e1329)=e1327.t.('e1329.0 | N1304(e1327,e1329))
agent N1303(e1326,e1328)=e1326.t.('e1328.0 | N1303(e1326,e1328))
agent N1283(e1325,e1327,e1326)=e1325.t.('e1327.0 | 'e1326.0 |
     N1283(e1325,e1327,e1326))
agent N1282(e1400,e1325)=e1400.t.('e1325.0 | N1282(e1400,e1325))
agent N534(e1409,i)=i.t.'e1409.0
agent N(i,o)=(^e1409,e1400,e1331,e1329,e1328,e1327,e1326,e1325)
     (N1398(e1409,e1400) | N1322(e1331,o) | N1314(e1329,e1328,e1331) |
     N1304(e1327,e1329) | N1303(e1326,e1328) |
     N1283(e1325,e1327,e1326) | N1282(e1400,e1325) | N534(e1409,i))
```

FIGURE 5.12: Tool-generated agent definitions of The customer from Example 5.10.

of the final node after each occurrence of the initial node. Regarding the black box from Figure 5.11, this means that each time the start button is pressed, we need to be able to observe exactly one flash of the done bulb. In contrast to easy soundness, we cannot try until we observe a flash of the done bulb, but have to consider all possibilities instead. This faces us with two problems: (1) How can we be sure that all paths of the process graph have been traversed, i.e., we do not need to press the start button anymore? (2) How do we know if we do not need to wait any longer for the done bulb to flash, i.e., a deadlock has occurred? If we are able to find a bisimulation between an invariant given by

$$S_{\text{LAZY}} \stackrel{\text{def}}{=} i.\tau.\overline{o} \tag{5.2}$$

and the annotated agent definitions of a process graph, both problems are solved: the former due to the fact that a bisimulation enumerates all possible states, and the latter by the fact that a bisimulation is finite. Since S_{LAZY} exactly resembles S_{EASY}, the same annotation for the agents has to be used.

A formal definition of lazy soundness using weak bisimilarity is given by

DEFINITION 5.13   Lazy Sound Process Graph
   A structurally sound process graph G with a semantics given by the lazy soundness annotated π-calculus mapping D of G is lazy sound if D ≈ S_{LAZY} holds.

The business process from Figure 5.10 can be checked for satisfying lazy soundness; i.e., we need to prove that $S_{\text{LAZY}} \approx N_{\text{EASY}}$ holds. Again, we use ABC and the formalization from Figure 5.12.

```
abc > weq S_LAZY(i,o) N(i,o)
The two agents are not weakly related (4).
Do you want to see some traces (yes/no) ? no
```

As might be expected, the business process does not fulfill the lazy soundness property, since the final node is reached not only once but twice (due to the parallel split gateway in the beginning). This holds, however, only until we take the interactions with possible environments into account, as considered in the next section.

5.5.3 Interaction Soundness

If we consider the bank from Figure 5.9, the final node of the customer will only be reached once, since either the reject or accept activity will be stuck waiting for a response forever. More generally, it can be said that if a business process is lazy sound according to a certain environment, the combination of both is interaction sound:

> A structurally sound process graph representing a business process is
> interaction sound regarding an environment, if the composition of both
> is lazy sound.

Interaction soundness is a so-called compatibility property. By using the π-calculus as a foundation for compatibility, the effects of dynamic binding can be analyzed.

For ease of presentation, we stick to environments made up of arbitrary π-calculus agents according to the grammar from Definition 5.10. These agents might contain parts that are derived from process graphs. Since interactions are also considered, however, we need to update Algorithm 5.3 to comply with interaction flows:

ALGORITHM 5.4   *Interaction Soundness Annotated π-Calculus Mapping*

To annotate a π-calculus mapping D of a process graph $G = (\mathbf{N}, \mathbf{E}, \mathbf{T})$ for reasoning on interaction soundness, we need to partially instantiate the trios of the agent definitions. Let Ni iterate over the agent definitions/nodes of all elements of $\mathbf{N}$. Furthermore, $\{i, o\} \cap (fn(D) \cup bn(D)) = \emptyset$. The trios have to be instantiated as follows, where we assume α and γ to be ϵ, since we do not consider interaction flows for soundness:

- $Ni\{\alpha; \epsilon; \gamma\}$ *if* $| \bullet Ni | > 0 \wedge |Ni \bullet | > 0$

- $Ni\{\alpha; i; \gamma\}$ *if* $| \bullet Ni | = 0 \wedge |Ni \bullet | > 0$

- $Ni\{\alpha; \overline{o}; \gamma\}$ *if* $| \bullet Ni | > 0 \wedge |Ni \bullet | = 0$

- $Ni\{\alpha; i.\overline{o}; \gamma\}$ *if* $| \bullet Ni | = |Ni \bullet | = 0$

We assume the remaining partial instantiations of the trios to be already made during the interaction enhancement of the agent definitions as given in Section 5.4.

Interaction soundness is defined based on environments made up of π-calculus agents as well as an interaction soundness annotated π-calculus mapping. If the mapping of the process graph and the environment have common free names, interactions are possible. If these do not lead to deadlocks, the process graph is interaction sound regarding the investigated environment:

DEFINITION 5.14 Interaction Soundness

Let D be an interaction soundness annotated mapping of a structurally sound process graph and E be an agent with $\{i, o\} \notin fn(E)$ *and* $\tilde{z} = (fn(D) \setminus \{i, o\}) \cup fn(E)$. *The composition S of D and E according to*

$$S \stackrel{\text{def}}{=} (\nu\tilde{z})(D \mid E)$$

is called interaction sound *if* $S \approx S_{\text{LAZY}}$.

As a direct result, each process graph is interaction sound with an environment if $fn(D) \cap fn(E) = \emptyset$, since the possible interactions (i.e., zero) do not disturb the deadlock-freedom of the process graph. Interaction soundness can also be checked via existing bisimulation reasoners. Due to the similarity with lazy soundness and due to space reasons, we omit an example here and refer to Ref. [26] for an extended discussion.

5.6 Conclusion

In this chapter, we gave an introduction to the application of the π-calculus to the area of business process management. We started with the key concepts of BPM and related them to π-calculus expressions. Thereupon, we gave a short summary of selected process patterns that are required to transform process graphs into agents. Afterward, we continued with a discussion on how link passing mobility of the π-calculus might be used to extend the scope from internal business processes to interacting ones. Using link passing mobility, we are able to directly express dynamic binding as found in SOA—the major implementation architecture for today's BPM systems. We concluded the chapter with an insight on how a business process could be analyzed regarding easy and lazy soundness properties.

From a formal point of view, we introduced an extension point—so called trios—that served two purposes. First, they support the flexible integration of communication actions between different business processes. Second, they provide plugin ports for deciding formal properties such as lazy soundness. More elaborate properties, such as weak soundness, require all β-values of the trios to be filled with more complex terms [24].

Another important issue, that has only been sketched due to the page limit, is data flow. Technically, the objects of the input and output actions that represent preconditions also need to be updated with the addition of communication actions. Received values—as well as freshly created names—need to be forwarded to other agents that require them. As stated, the consideration of data flow is out of scope for this chapter.

As a concluding remark, it is manifest that the π-calculus supports the ideas of SOA—that is the key architecture for today's BPM systems—in a direct and precise manner. By making a closer investigation, as has been done in Ref. [23], it becomes clear, however, that certain concepts need to be enhanced, such as the addition of certain kinds of transactions or time [15]. Within this chapter, we provided the key foundations on the application of the π-calculus to business process design and analysis.

References

[1] W. M. P. van der Aalst. Verification of Workflow nets. In P. Azéma and G. Balbo, (Eds.), *Application and Theory of Petri Nets 1997*, Berlin. *Lecture Notes in Computer Science*, 1248:407–426, 1997.

[2] W. M. P. van der Aalst. Exterminating the dynamic change bug: A concrete approach to support workflow change. *Information System Frontiers*, 3(3):297–317, 2001.

[3] W. M. P. van der Aalst and A. H. M. ter Hofstede. YAWL: Yet another workflow language (Revised version). Technical Report FIT-TR-2003-04, Queensland University of Technology, Brisbane, 2003.

[4] W. M. P. van der Aalst, A. H. M. ter Hofstede, B. Kiepuszewski, and A. P. Barros. Workflow patterns. *Distributed and Parallel Databases*, 14(1):5–51, 2003.

[5] W. M. P. van der Aalst, A. H. M. ter Hofstede, and M. Weske. Business Process Management: A Survey. In Wil M. P. van der Aalst, Arthur H. M. ter Hofstede, and M. Weske, (Eds.), *Business Process Management*, Berlin. *Lecture Notes in Computer Science*, 2678:1–12, 2003.

[6] W. M. P. van der Aalst and K. van Hee. *Workflow Management*. MIT Press, Cambridge, MA, 2002.

[7] A. Barros, M. Dumas, and A. ter Hofstede. Service interaction patterns. In W. M. P. van der Aalst, B. Benatallah, and F. Casati, (Eds.), *Business Process Management*, Berlin. *Lecture Notes in Computer Science*, 3649:302–318, 2005.

[8] BPMI.org. *Business Process Modeling Notation*, 1.0 edition, May 2004.

[9] S. Briais. ABC bisimulation checker. Available at http://lamp.epfl.ch/sbriais/abc/abc.html, 2003.

[10] S. Burbeck. The Tao of E-Business Services. Available at http://www-128.ibm.com/developerworks/library/ws-tao/, 2000.

[11] B. Curtis, M. I. Kellner, and J. Over. Process modeling. *Communications of the ACM*, 35(9):75–90, 1992.

[12] U. Engberg and M. Nielsen. A calculus of communication systems with label passing. Technical Report DAIMI PB-208, University of Aarhus, Denmark, 1986.

[13] C. A. R. Hoare. Communicating sequential processes. *Communications of the ACM*, 21(8):666–677, 1978.

[14] E. Kindler. On the semantics of EPCs: A framework for resolving the vicious circle. In J. Desel, B. Pernici, and M. Weske, (Eds.), *Business Process Management*, Berlin. *Lecture Notes in Computer Science*, 3080:82–97, 2004.

[15] C. Laneve and G. Zavattaro. Foundations of Web transactions. In V. Sassone, (Ed.), *Foundations of Software Science and Computational Structures, Lecture Notes in Computer Science, 3441*, Berlin, Springer Verlag, 2005, pp. 282–298.

[16] F. Leymann and D. Roller. *Production Workflow: Concepts and Techniques.* Prentice Hall PTR, Upper Saddle River, NJ, 2000.

[17] R. Milner, J. Parrow, and D. Walker. A calculus of mobile processes pt.1. Laboratory for Foundations of Computer Science Technical Report ECS-LFCS-89-85, School of Informatics, University of Edinburgh, Edinburgh, Scotland, 1989.

[18] R. Milner. *A Calculus of Communicating Systems*, volume 94 of LNCS. Springer Verlag, Berlin, 1980.

[19] U. Nestmann. Welcome to the jungle: A subjective guide to mobile process calculi. In C. Baier and H. Hermanns, (Eds.), *CONCUR 2006—Concurrency Theory, Lecture Notes in Computer Science, 4137*, Berlin, Springer Verlag, 2006, pp. 52–63.

[20] J. Parrow. Trios in concert. In *Proof, Language, and Interaction: Essays in Honour of Robin Milner*, Cambridge, MA, MIT Press, 2000, pp. 623–637.

[21] C. A. Petri. *Kommunikation mit Automaten*. PhD thesis, Institut für Instrumentelle Mathematik, Bonn, 1962.

[22] F. Puhlmann. A tool chain for lazy soundness. In J. Mendling, (Ed.), *Demo Session of the 4th International Conference on Business Process Management, CEUR Workshop Proceedings*, Vol. 203, Vienna, 2006, pp. 9–16.

[23] F. Puhlmann. *On the Application of a Theory for Mobile Systems to Business Process Management*. Doctoral thesis, University of Potsdam, Potsdam, Germany, July 2007.

[24] F. Puhlmann. Soundness verification of business processes specified in the pi-calculus. In R. Meersman and Z. Tari, (Eds.), *OTM 2007, Part I, Lecture Notes in Computer Science, 4803*, Berlin, Springer Verlag, 2007, pp. 6–23.

[25] F. Puhlmann and M. Weske. Using the pi-calculus for formalizing workflow patterns. In W. M. P. van der Aalst, B. Benatallah, and F. Casati, (Eds.), *Business Process Management, Lecture Notes in Computer Science, 3649*, Berlin, Springer Verlag, 2005, pp. 153–168.

[26] F. Puhlmann and M. Weske. Interaction soundness for service orchestrations. In A. Dam and W. Lamersdorf, (Eds.), *Service-Oriented Computing – ICSOC 2006, Lecture Notes in Computer Science, 4294*, Berlin, Springer Verlag, 2006, pp. 302–313.

[27] F. Puhlmann and M. Weske. Investigations on soundness regarding lazy activities. In S. Dustdar, J. L. Fiadeiro, and A. Sheth, (Eds.), *Business Process Management, Lecture Notes in Computer Science, 4102*, Berlin, Springer Verlag, 2006, pp. 145–160.

[28] S. Sadiq, W. Sadiq, and M. Orlowska. Pockets of flexibility in workflow specification. In H. S. Kunii, S. Jajodia, and A. Sølvberg, (Eds.), *Conceptual Modeling - ER 2001 : 20th International Conference on Conceptual Modeling, Lecture Notes in Computer Science, 2224*, Berlin, Springer Verlag, 2001, pp. 513–526.

[29] D. Sangiorgi and D. Walker. *The π-calculus: A Theory of Mobile Processes.* Cambridge University Press, Cambridge, MA, paperback edition, 2003.

[30] M. Weske. *Workflow Management Systems: Formal Foundation, Conceptual Design, Implementation Aspects.* Habilitationsschrift, Fachbereich Mathematik und Informatik, Universität Münster, Münster, 2000.

Chapter 6

Behavioral Specification of Middleware Systems

Nelson Souto Rosa

Contents

6.1 Introduction

Middleware specifications are not trivial to be understood, as the middleware itself is usually very complex. First, middleware systems have to hide the complexity of underlying network mechanisms from the application. Second, the number of services provided by the middleware is becoming larger. Finally, in addition to hiding low-level communication mechanisms, the middleware also has to hide failures, mobility, changes in network traffic conditions, concurrency, and so on. From the point of view of application developers, they very often do not know how the middleware actually works. From the point of view of middleware developers, the complexity poses many challenges that include how to integrate services into a single product, how to satisfy new requirements of emerging applications, or even how to design the middleware itself, as traditional methodologies do not cover the development of this kind of software.

The aforementioned specifications are usually described through APIs. Essentially, service operation signatures are described in Interface Definition Language (IDL) and the behavior of each operation is described by informal prose. For example, CORBA common object services (COS) (e.g., security, transaction) are described in IDL CORBA and informal text. In practical terms, developers who want to implement these services have a hard task to produce a final product by interpreting what the specifications actually describe.

In this context, we present three approaches for formalizing middleware behavior using Language Of Temporal Ordering Specification (LOTOS) [6]. In the first approach, the middleware behavior is specified by adopting software architecture principles. The middleware specification is defined in different levels of abstraction, which are usually adopted by application developers, standards bodies, and middleware developers. Here, LOTOS is used as an Architecture Description Language (ADL) [19] that allows one to formally specify the behavior of middleware software architectures. The second approach presents a framework that helps to formally describe middleware behavior in LOTOS by providing a set of basic abstractions. These abstractions are generic in the sense that may be combined in different ways in order to specify different middleware systems. Finally, the third approach treats the complexity of specifying the behavior of middleware systems, specifically for wireless sensor networks (WSN).

All these approaches based on LOTOS have a common purpose, that is, the formal specification of middleware behavior. However, they differ in relation to the time they may be employed. The software architecture approach is useful when the middleware development process is at the initial stages and software architecture principles may be used to structure the middleware architecture and behavior. The framework, however, is more applicable when the middleware architecture has already been defined and the framework abstractions may be used like semispecifications (in the sense of reusing or extending them) for the elements specified in the architecture. Finally, the last approach incorporates steps of the design of middleware for WSN and is fully based on characteristics of TinyOS and nesC technologies.

LOTOS allows the checking (by using tools) of particular behavioral properties of middleware systems, e.g., deadlock freedom, liveness, and safety. It also makes it possible to check behavioral equivalence, either between the specifications of different middleware models or between two specifications during the refinement process. For the first case, if one desires to replace a transactional middleware with a procedural one, it is possible to check if their behaviors are equivalent. Furthermore, a formal specification eliminates ambiguities in the middleware specification and provides a better understanding of what is actually described. Finally, the formalization creates the possibility of automatic generation of tests.

In terms of related work, formal description techniques have been used for middleware in the specification of the Reference Model–Open Distributed Processing (RM-ODP) [17], in which the behavior of the trader service is formally specified in Enhancements to LOTOS (E-LOTOS) [14]. CORBA has been the main focus of researches on the formalization of middleware systems, which includes the formal specification of the CORBA security service [3] and CORBA naming service [15] using the Z notation, the CORBA event service in high level Petri nets [4,5], and CORBA-distributed objects [20] in a real-time process algebra. Furthermore, more related to CORBA applications, [8] presents a methodology to transform a formal specification written in TRIO [11], a first-order temporal logic, into a high-level design document written in an extension of TRIO, named TRIO/CORBA. Apart from CORBA, Venkatasubramanian [30] proposes a metaarchitectural framework, the Two-Level Actor Model, that may be used to specify the behavior of QoS-enabled middleware elements. Subramonian [26] defines a reusable library of formal models specially developed to capture essential timing and concurrency semantics of middleware building blocks. [12,13] also adopts Petri nets for specifying middleware behavior. Finally, Chattopadhyay [7] concentrates on the formalization of mobile middleware behavior using basic constructs of a predicate calculus.

All those works, however, do not adopt software architecture principles for structuring the service descriptions. In terms of software architecture, a few ADLs like Wright (a CSP-based ADL) [2] include the possibility of describing the behavior of the software architecture. However, there are no tools available for manipulating Wright specifications. Medvidovic [18] has observed the convergence of middleware and software architecture principles in an informal way.

The basic idea of this chapter is to put together our experiences in using LOTOS for specifying middleware behavior. Hence, instead of presenting new developments, it serves as a guide on how to use LOTOS to specify middleware systems that run on both desktops and WSNs. Our initial approach is motivated by two basic facts [22]: the large number of open specifications of both middleware systems and middleware services, and the complexity of those specifications, which are traditionally described through APIs (the operation signatures) and informal prose (the behavior). As an alternative to the informal specification, and by using concepts of software architecture, architectural elements and architectural rules are used to structure middleware behavior specifications. In order to extend this initial approach, we then defined a framework that serves as an initial specification for those interested in formalizing the middleware behavior [23]. The framework consists of a set of reusable building blocks common to different middleware systems. By using the framework, the task of specifying the middleware behavior moves from understanding the complexity of middleware low-level details to combining already-specified and more abstract middleware basic elements. Finally, motivated by practical reasons, and its increasing importance and new challenges imposed on middleware developers, we decided to use LOTOS in the exciting field of WSNs [24]. By simulating the LOTOS specification of a WSN middleware in different scenarios, developers may better understand the middleware behavior prior to implementing and deploying it in the field.

This chapter is organized as following: Section 6.2 gives some background on LOTOS. Section 6.3 presents how software architectural elements (components, connectors, and configuration) may be defined in LOTOS, and then used in Section 6.4 to specify middleware behavior. This approach is applied to CORBA in the case study of Section 6.5. Next, Section 6.6 describes a LOTOS framework for specifying middleware systems whatever their distribution models (e.g., message-oriented middleware [MOM] or object-oriented middleware [OOM]). In Section 6.7, LOTOS is used to specify middleware for WSNs. Finally, Section 6.8 presents some concluding remarks.

6.2 LOTOS Background

A LOTOS [6] specification describes a system through a hierarchy of processes. A process is an entity able to realize nonobservable internal actions and interact with other processes through externally observable actions. The unit of atomic interaction among processes is called an event, which corresponds to synchronous communication that may occur among processes able to interact with one another. Events are atomic, i.e., they happen instantaneously and are not time-consuming. The point where an event interaction occurs is known as a port. Such events may or may not actually involve the exchange of values. A nonobservable action is referred to as an internal action or internal event. A process has a finite set of ports that can be shared.

An essential component of a specification or process definition is its behavior expression. A behavior expression is built by applying an operator (e.g., action prefix operator "$;$") to other behavior expressions. A behavior expression may also include instantiations of other processes, whose definitions are provided in the "where" clause following the expression [6]. The complete list of basic LOTOS behavior expressions is given in Table 6.1. Symbols B, B_1, B_2 in the table stand for any behavior expression, g_i is a gate, and i is an internal action. The term "gate" is used interchangeably with "port".

When using abstract data types (ADTs), it is possible to enrich synchronizations with value passing in such way that a finite list of attributes may be associated to a gate. LOTOS has two types of attributes, namely value declaration ("!E", where E is a value expression) and variable declaration ("?x:t", where x is a name of a variable and t indicates the domain of values over which x ranges). In this way, values may be passed between processes composed in parallel when they offer the

TABLE 6.1: Syntax of behavior expressions in basic LOTOS.

Name	Syntax	Semantics
Inaction	`stop`	It cannot offer anything to the environment, nor it can perform internal actions.
Action prefix		
Unobservable	`i;B`	It is capable of performing action $i(g)$ and transform into process B.
Observable	`g;B`	
Choice	B_1 `[]` B_2	It denotes a process that behaves either like B_1 or like B_2.
Parallel composition		
General case	$B_1 \mid$`[`$g_1,\dots,g_n$`]`$\mid B_n$	A parallel composition expression is able to perform any action that either component expression is ready to perform at a gate (not in $g_1,\dots,g_n$) or any action that both components are ready to perform at a gate in $[g_1,\dots,g_n]$.
Pure interleaving	B_1 `|||` B_2	
Full synchronization	B_1 `||` B_2	
Hiding	`hide `$g_1,\dots,g_n$` in B`	Hiding allows one to transform some observable actions of a process into unobservable ones.
Process instantiation	`p [`$g_1,\dots,g_n$`]`	It is used to express infinite behaviors.
Successful termination	`exit`	`exit` is a process whose purpose is solely that of performing the successful termination.
Sequential composition (enabling)	B_1 `>>` B_2	B_2 is enabled only if and when B_1 terminates successfully.
Disabling	B_1 `[>` B_2	B_1 may or may not be interrupted by the first action of process B_2.

Source: Bolognesi, T. and Brinksma, E., *Comp. Netw. ISDN Syst.*, 14, 25, 1987. With permission.

TABLE 6.2:   Types of interaction.

Process A	Process B	Synchronization Condition	Type of Interaction	Effect
g !E1	g !E2	Value(E1) = Value(E2)	Value matching	Synchronization
g !E	g ?x:t	Value(E) is of sort t	Value passing	After synchronization x = value(E)
g ?x:t	g ?y:u	t = u	Value generation	After synchronization x = y = v, where v is some value of sort t

Source: Bolognesi, T. and Brinksma, E., *Comp. Netw. ISDN Syst.*, 14, 25, 1987. With permission.

same action (gate and value) in a synchronization port. Table 6.2 shows all possible interactions between two processes composed in parallel.

Multiple values may be matched or passed together by writing a series of attributes. For example, "g !E1 !E2" would pass the values E1 and E2 to variables x and y, respectively, in "g ?x:t ?y:u". Another combination seen later in the chapter relies on mixing attribute types: If processes A and B contain "g !E1 !E2" and "g !E1 ?y:u", respectively, this means that the value E2 will pass to variable y when A and B synchronize on the same combination of gate g and value E1.

Here, we present the LOTOS specification of a simple client-server system:

```
(1)  specification ClientServer[request,reply]:noexit
(2)  behaviour
(3)    Client[request,reply] || Server[request,reply]
(4)  where
(5)   process Client[request,reply]: noexit :=
(6)      request; reply; Client[request, reply]
(7)   endproc
(8)  process Server[request,reply]:noexit:=
(9)    hide processRequest in
(10)   request;
(11)   processRequest;
(12)   reply;
(13)   Server [request, reply]
(14)  endproc
(15) endspec
```

The top-level specification (3) is a parallel composition (operator "||") of the processes Client and Server. Hence, every action externally observed is executed by the process Client synchronized with a corresponding observable action from the process `Server`. The process `Client` (5) performs two actions, namely `request` and `reply` (6), and then reinstantiates itself. The action prefix operator ("; ") applies to the process operand on its right, and is parameterized (on the left) with an action identifier. In this case, it defines the temporal ordering of the actions `request` and `reply` (the action `request` occurs before the action `reply`) in the `Client` (6). Informally, the `Server` (8) receives a request (10), processes it by an internal action (11) (hidden by the operator `hide`) and then sends a reply (12) to the process `Client`.

LOTOS specifications may be compared with one another in order to check their behavioral equivalences such as strong, observational, and safety equivalence. All of them may be checked through the CADP Toolbox* [9].

6.3 Software Architecture Middleware Behavior

The definition of software architectures involves three basic abstractions, namely components, connectors, and configurations [25]. A component is a unit of computation or a data store. Components represent a wide range of different elements, from a simple procedure to an entire application, and have an interface used to communicate between the component and the external environment. A connector models interactions among components and the rules that govern those interactions. Some examples of connectors include client–server protocols, variables, buffers, sequence of procedure calls, etc. A connector has an interface that contains interaction points between the connector and the component and other connectors attached to it. Finally, the configuration describes how components and connectors are connected.

These three elements are defined in terms of LOTOS specifications in the subsections below. LOTOS is used as an ADL that allows formally specifying the behavior of middleware software architectures. While most ADLs only describe architectural aspects, LOTOS is used in this case to specify both architectural and behavioral aspects of the software architecture [22].

6.3.1 Components

According to Medvidovic [19], an ADL may include the following elements in order to describe a component: the component's interface (set of interaction points between the component and the external world), component type (for reuse), component semantics (the component behavior), some constraints (a property of or assertion about a part of the system), evolution (ability to specify modification of component properties), and nonfunctional properties (e.g., security and fault-tolerance). While LOTOS may be used to specify the component, component interface, and semantics, it is not able to define component types, evolution characteristics, and

---

* http://www.inrialpes.fr.

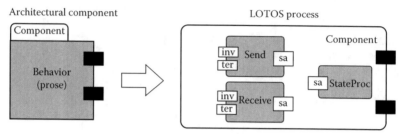

FIGURE 6.1: Structure of a component.

nonfunctional properties. However, desired key characteristics of ADLs are covered by LOTOS. Furthermore, these limitations are also common to other ADLs [19], which usually only address component interface and structural aspects.

A component is modeled in LOTOS through the basic abstraction provided by the language, the process. Figure 6.1 shows informally how an architectural component is structured in LOTOS. The component interface is mapped into a set of ports of the LOTOS process (`inv` and `ter`), whilst the process behavior refers to the component behavior. The component behavior is specified as a parallel composition of three LOTOS processes (3–5) as shown in the following:

```
(1)  process Component [inv, ter]: noexit :=
(2)  hide sa in
(3)   (Send [inv, ter, sa] ||| Receive [inv, ter, sa])
(4)      |[sa]|
(5)   StateProc [sa]
(6)  where
(7)    process Send [inv, ter, sa]: noexit
(8)    := (* Send behaviour *)
(9)    endproc
(10)   process Receive[inv,ter,sa]: noexit
(11)   := (* Receive behaviour *)
(12)   endproc
(13)   process StateProc [sa]: noexit
(14)   := (* StateProc behaviour *)
(15)   endproc
(16) endproc
```

The process `Send` (7) specifies the behavior of requests issued by the component (operations the component needs from the external environment), whilst the process `Receive` (10) refers to requests received by the component (operations the component provides to the external environment). The process `StateProc` (13) represents the component state. The behavior of `Component` is a parallel composition of three processes: `Send`, `Receive`, and `StateProc`. In practical terms, processes `Send` and `Receive` "execute" in parallel (interleaving operator |||), but they have to synchronize (||) with the process `StateProc` that maintains the component state.

It is worth observing that the ports `inv` (invocation) and `ter` (termination) are used for invocations from/to the component and for returning results to/from the component, respectively. For example, a client component in a client–server interaction makes requests through the port `inv` and waits for the reply on the port `ter`. Meanwhile, a server component receives the requests in the port `inv` and returns the result in the port `ter`. Note that LOTOS ports are not directional, i.e., dedicated for input or output, as might be expected. Rather, they are symbolic of interprocess synchronization and communication, defined to occur when matching port names are used with the "!" and "?" attributes.

6.3.2 Connectors

In a similar way to components, the connector specification includes the interface, types, semantics, constraints, evolution, and nonfunctional properties. Despite the use of similar elements for describing the connector, the semantics of a connector are obviously different from the component. As mentioned before, the connector is responsible for defining how two (or more) components interact. Figure 6.2 depicts how an architectural connector is defined in LOTOS.

A connector is also modeled in LOTOS through the basic abstraction provided by the language, the process. The connector is made up of three processes, namely `Source`, `Sink`, and `Choreography`, as follows:

```
(1)process Connector[invT, terT, invR, terR] : noexit :=
(2)  (Source [invR, terT]
(3)     |||
(4)   Sink [invT, terR])
(5)   |[invT, terT, invR, terR]|
(6)   Choreography [invT, terT, invR, terR]
(7) where
(8)    process Source [invR, terT]: noexit
(9)    :=
(10)      (* behaviour specification *)
```

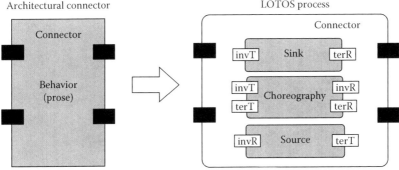

FIGURE 6.2: Structure of a connector.

```
(11)   endproc
(12)   process Sink [invT, terR]: noexit
(13)   :=
(14)      (* behaviour specification *)
(15)   endproc
(16)   process Choreography [invT, terT, invR, terR]: noexit
(17)   :=
(18)      (* behaviour specification *)
(19)   endproc
(20) endproc
```

The process Sink (12–15) refers to the transport of messages the connector receives from the components plugged into it. The process Source (8–11) models the transport of messages that the connector has to send. The process Choreography (17–19) takes responsibility of ordering the messages the connector receives and sends, i.e., it coordinates the way the components plugged into the connector interact. In more practical terms, the process Choreography usually defines the communication protocol the connector implements.

6.3.3   Configuration

The architectural configuration consists of the composition of components and connectors (see Figure 6.3). The configuration is the top-level specification. It is made

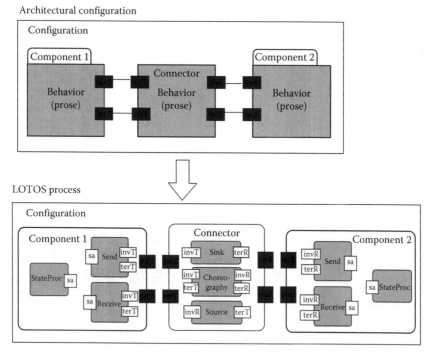

FIGURE 6.3:   Structure of a configuration.

up of the composition of LOTOS processes, i.e., the components and connectors (`Component1 || Connector || Component2`).

A basic architectural rule must be followed to define the configuration: two components cannot be connected directly, i.e., there is a connector between any two components in the configuration. For example, a connector is necessary between the client and the server components in order to explicitly define how those elements interact.

Next, we present how these elements are used to specify middleware software architectures at three abstraction levels.

6.4 Specifying Middleware Software Architecture in LOTOS

Since middleware systems neither perform any application-specific computation nor store long-term data, they are naturally modeled as connectors. Unlike connectors that only define the communication between components, the middleware provides services in addition. In the software architecture discipline, however, only components (not connectors) are traditionally decomposed into smaller elements or provide services. In our approach, connectors may also be decomposed.

As mentioned previously, in order to define the middleware software architecture in LOTOS, we have adopted an approach in which the architecture is viewed at three different levels of abstractions: a simple connector that enables the interaction between distributed applications, a composite connector made up of services, and a distributed composite connector. In fact, these levels represent refinement steps in the design of a middleware platform.

The middleware software architecture has also been defined following some basic principles:

- Each service provided by the middleware (e.g., security, event, naming) defines a component within the composite connector. Additionally, each service may be defined as a composition of fine-grained components. For example, the CORBA security service is made up of a principal authenticator and a component responsible for the cryptography. Both are accessible remotely.

- Communication service, whatever the middleware model or product, is the only mandatory service. Whether the middleware has additional services or not, depends enormously on the middleware (or standard) specification.

- Each service in the distributed composite connector is defined through two parts, namely client and server (or sender and receiver). The underlying communication layers (e.g., transport and network layers) are also defined as a connector.

The following subsections present how the middleware software architecture is defined in each abstraction level by adopting the abstractions defined in the previous section.

6.4.1   Simple Connector

The middleware as a simple connector is the highest abstraction view. At this level, the middleware specification is commonly used/understood by application developers who are not interested in details of how the middleware actually works. In fact, the application developer views the middleware as a communication element that transports messages between components.

Figure 6.4 shows the middleware as a simple connector (without services) that simply defines how the components 1 and 2 interact from the point of view of an external observer. In a typical case, the middleware receives an invocation from Component1 that contains both the name of the requested service and the operation being requested. Next, the middleware passes both of them to Component2 and waits for the reply that comes from Component2. Finally, the middleware passes the reply containing the result to Component1.

As an example of a trace produced by the CADP Toolbox, Table 6.3 shows a simulation of a simple connector middleware where Component2 offers a service ServiceStd. This trace reports the externally visible behavior: just the traffic on the top-level ports, plus the execution of the requested operation.

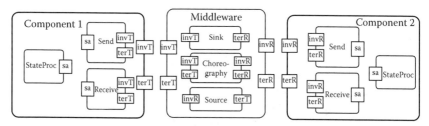

FIGURE 6.4:   Middleware as a simple connector.

TABLE 6.3:   Trace generated by CADP.

CADP Trace	Comment
< *initialstate* >	
invT !'ServiceStd' !'op1'	Component1 invokes (through the Middleware) the operation 'op1' of service 'ServiceStd'.
invR !'ServiceStd' !'op1'	Middleware forwards the invocation to Component2.
i (IOP1 [62])	Component2 processes the request ("i" is an internal action).
terR !'ServiceStd' !'ok'	Component2 returns the result of request ('ok') to Middleware.
terT !'ServiceStd' !'ok'	Middleware forwards the result to Component1.
< *goalstate* >	

It is worth noting that, at this level of abstraction, the description of the middleware behavior is very simple/abstract and it is not possible to know how a request is actually passed between `Component1` and `Component2`. In practical terms, the behavior of individual middleware products may not be differentiated (by an external observer) when the middleware is viewed as a simple connector. The only exception occurs if two middleware systems have different communication models, e.g., OOM and MOM. For example, in the previous trace, the middleware transfers both the requests and the replies. However, in a MOM, the request (service and operations) is simply replaced by a message and a reply is not necessary.

6.4.2 Composite Connector

The middleware as a composite connector is typically adopted in open specifications such as CORBA, Java Message Service [28] and Java Transaction Service [27]. Unlike the application developer's view of the middleware as simply a communication element, in this case the middleware is viewed as a collection of services such as security, events, time, transactions, etc. Hence, the middleware adds value to the communication through these services.

In Figure 6.5, the number of available services and the way they may be composed depend on the particular middleware being specified. For example, when a request gets to the middleware, it may firstly pass to the security service and then to the transaction service before being forwarded to the remote component. Hence, an important point of this specification is the ordering of composition of the middleware services. For this particular purpose, we adopt the LOTOS constraint-oriented specification style [31–33] in which the `Choreography` constrains both the component interactions and the way the services are composed. The process `Choreography` of Figure 6.5 could be specified as follows to invoke internal services 1 and 2 in that order:

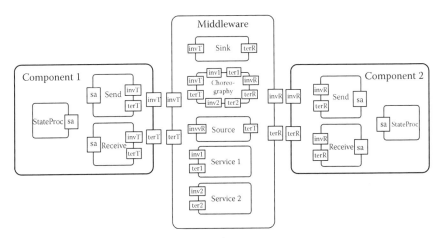

FIGURE 6.5: Middleware as a composite connector.

```
(1) process Choreography[invT, terT, invR, terR, inv1,
                         ter1, inv2, ter2]: noexit :=
(2)    invT ?s:SERVICE ?op:OPER;
(3)      inv1 !Service1 !Op1S1;
(4)        ter1 !Service1 ?r:RESULT;
(5)          inv2 !Service2 !Op1S2;
(6)            ter2 !Service2 ?r:RESULT;
(7)              invR !s !op;
(8)                terR ?s:SERVICE ?r:RESULT;
(9)                  terT !s !r;
(10)                   Choreography[invT, terT, invR, terR, inv1,
                                    ter1, inv2,ter2]
(11) endproc
```

Types SERVICE, OPER, and RESULT, and the constants Service1, Service2, Op1S1, Op1S2 have been defined previously in the ADT definition and are not shown in the specification. According to the constraints imposed by Choreography, after the request gets to the middleware (2), it is passed to Service1 (3–4) followed by Service2 (5–6). Then, the request is sent to Component2 (7) where it is processed and sent back to Component1 (8–9). At this level of abstraction, the behavior of distinct middleware systems is still not differentiated by an external observer like a client (observational equivalence), i.e., a client only knows that it passes a request to the middleware and then receives a reply without perceiving internal actions carried out by the middleware. Whatever the services found in the middleware, the client receives a reply without knowing the actual path taken by the request inside the middleware. However, when strong equivalence is checked (internal actions are considered) it reveals that the behaviors of two middleware systems that provide distinct services are not the same, as their internal actions are different.

6.4.3   Distributed Composite Connector

Finally, the last view of middleware software architectures concerns middleware developers, i.e., a detailed view of the middleware. The middleware is now defined as a distributed composite connector, which is decomposed into two parts, referred to as middleware transmitter and middleware receiver (see Figure 6.6).

The middleware transmitter receives a request from Component1 and passes it to the middleware receiver through another connector (the network). The middleware receiver receives the request from the network, passes it to Component2, and waits for the reply that must be sent to the Component1 through the network and the middleware transmitter. The top-level specification (1) of the software architecture presented in Figure 6.6 is shown in the following:

```
(1) specification Configuration [invT, terT, invR,
                                 terR]: noexit
(2) behaviour
(3) hide reqTN,repTN, reqRN,repRN in
(4)  (
```

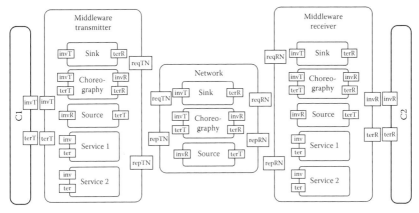

FIGURE 6.6: Middleware as a distributed composite connector.

```
(5)    Component1 [invT, terT]
(6)       |[invT, terT]|
(7)    MiddlewareTransmitter [invT, terT, reqTN, repTN]
(8)    )
(9)       |[reqTN, repTN]|
(10)   Network [reqTN, repTN, reqRN, repRN]
(11)      |[reqRN, repRN]|
(12)   (
(13)    MiddlewareReceiver [invR, terR, reqRN, repRN]
(14)       |[invR, terR]|
(15)    Component2 [invR, terR]
(16)   )
(17)  where
(18)   (* behaviour *)
(19)  endspec
```

The middleware on the transmitter side, the process `MiddlewareTransmitter`
(7), and the middleware on the receiver side, the process `MiddlewareReceiver`
(13), do not have the same behavior. This is an interesting point to be observed as
middleware products are different on both sides. This fact has a direct impact on
how the middleware services are composed. Additionally, a service (or some of its
components) may be present in the server and absent in the client. Hence, the process
`Choreography` and the set of services on both sides may be different.

6.5 Case Study: CORBA

The widely known CORBA standard has been adopted for building middleware
products. According to the CORBA specification [21], in addition to the communication service Object Request Broker (ORB), 15 distributed services may also

be found: persistence, externalization, events, transactions, properties, concurrency, relationships, time, licensing, trader, naming, query, collections, lifecycle, and security. These services are usually not all implemented in a single product, but some of them such as the naming, lifecycle, and communication services are usually available in any CORBA-complaint product.

Three points must be observed in the CORBA software architecture. First, the CORBA standard defines that the COS services may be either inside or outside the ORB. In this particular architecture, we adopt the second approach. Second, stubs, skeletons and portable object adapter (POA) have been incorporated into the ORB and are not explicit elements in the software architecture. Finally, it is worth observing that the CORBA specification is object oriented, but its architecture is not, i.e., the CORBA architecture is a set of components (not objects). Hence, the adoption of a nonobject-oriented language like LOTOS for specifying architectural aspects of CORBA does not create any inconvenience.

6.5.1   CORBA as a Simple Connector

The behavior of CORBA as a simple connector is very similar to that shown in Figure 6.4. At this level of abstraction, CORBA receives a request from the client and sends it to the server. After being processed, the reply is sent back to the client. The behavior of the simple connector CORBA is specified as the temporal ordering of events executed in the CORBA interface. The CORBA interface is made up of the dynamic invocation, stub, ORB, static skeleton, dynamic skeleton, and POA interfaces.

The LOTOS specification of CORBA as a simple connector is shown in the following (including details of the choreography of CORBA that defines the way the invocations to CORBA are ordered):

```
process CORBAasSimpleConnector[invT, terT, invR, terR] :
                               terR] noexit:=
        (Source [invR, terR] ||| Sink [invT, terT])
            |[invT, terT, invR, terR]|
        Choreography [invT, terT, invR, terR]
    where
        process Source [invR, terR] : noexit
        :=
            (* behaviour of process Source *)
        endproc
        process Sink [invT, terT] : noexit
        :=
            (* behaviour of process Sink *)
        endproc
        process Choreography [invT, terT, invR, terR] : noexit
        :=
            invT ?s:SERVICE ?op:OPER;
                invR !s !op;
```

```
        terR ?s:SERVICE ?r:RESULT;
          terT !s !r;
              Choreography [invT, terT, invR, terR]
    endproc
endproc
```

The next section presents CORBA as a composite connector that provides a more detailed view of CORBA.

6.5.2 CORBA as a Composite Connector

CORBA as a composite connector is defined as a collection of services according to Section 6.4.2. Figure 6.7 (some services have been omitted in the figure for clarity) presents the CORBA software architecture, which is composed by COS (components) and ORBCore (connector). The ORBCore is defined as a connector for three main reasons: it implements the communication service between applications; it is the communication channel between COS services; and, two or more components cannot communicate directly (architectural constraint presented in Section 6.4.2).

The top-level specification is a parallel composition of 14 different services (components), the process ORBCore, the process Choreography, the process Sink, and the process Source:

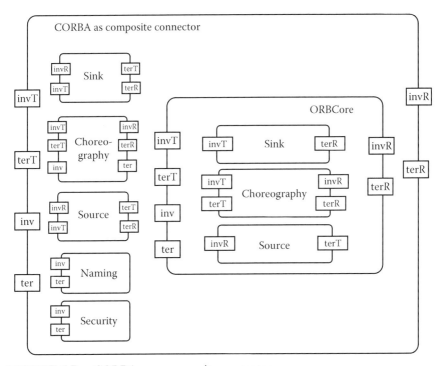

FIGURE 6.7: CORBA as a composite connector.

```
process CORBAasCompositeConnector[invT, terT, invR, terR,
                                  inv, ter]: noexit :=
  (Source[invT, terT, invR, terR] ||| Sink[invT, terT,
                                                invR, terR])
           |[invT, terT, invR, terR]|
    Choreography [invT, terT, invR, terR, inv, ter]
           |[inv, ter]|
  ((Naming [inv, ter] ||| Event [inv, ter] |||
    Persistent[inv, ter] ||| LifeCycle [inv, ter] |||
    Concurrency [inv, ter] ||| Externalization [inv, ter] |||
    Relationship[inv, ter] ||| Transaction [inv, ter] |||
    Query [inv, ter] ||| Licensing [inv, ter] |||
    Property [inv, ter] ||| Time [inv, ter] |||
    Security [inv, ter] ||| Trading [inv, ter])
         |[inv, ter]|
    ORBCore [invT, terT, invR, terR, inv, ter])
where
    (* behaviour *)
endproc
```

As stated in Section 6.4.2, the LOTOS process Choreography takes responsibility for ordering the actions performed by the middleware and defining the way the services are composed. For example, an ordering constraint related to the naming service (referred to as COSNaming) may define that every distributed service must be registered in the naming service before being used by clients. Additionally, clients must obtain an interface reference to the service in order to use it. The next specification presents a possible choreography for the CORBA software architecture shown in Figure 6.7.

```
(1) Process Choreography[invT, terT, invR, terR,
                            inv, ter]:noexit :=
(2)  invR !COSnaming !register;
(3)   inv !COSnaming !register;
(4)    ter !COSnaming ?r:RESULT;
(5)     terR !COSnaming !r;
(6)      invT !COSnaming !lookup;
(7)       inv !COSnaming !lookup;
(8)        ter !COSnaming ?r:RESULT;
(9)         Loop [invT, terT, invR, terR]
(10) where
(11)   process Loop [invT, terT, invR, terR,
                      inv, ter]: noexit :=
(12)     invT ?s:SERVICE ?op:OPER;
(13)      invR !s !op;
(14)       terR ?s:SERVICE ?r:RESULT;
(15)        terT !s !r;
(16)         Loop[invT, terT, invR, terR]
(17)   endproc
(18) endproc
```

This `Choreography` defines that the first possible action is a component asking for its own registration (operation `register` defined by the second parameter in `invR !COSnaming !register`) with `COSnaming` (2). The request is passed to the `COSnaming` service (3) that performs a nonobservable action and returns the result (4) to the `Choreography`. Finally the result is sent back to the requester component (5). After the registration succeeds, a component looks for the service just registered (6). Then, following the same steps of the operation `register` (6–8), the component is allowed to make as many requests as it desires (9). This is specified by a loop that defines how two components interact to invoke an already-registered service (11–17).

6.5.3 CORBA as a Distributed Composite Connector

The view of CORBA as a distributed composite connector is the most detailed one if compared to the views presented in Sections 6.4.1 and 6.4.3. Figure 6.8 presents the elements of the CORBA software architecture. For simplicity, the services provided by the middleware have been grouped in the dashed box named `Service`. Additionally, the connectors `Network` and `ORBCore` (on the client and server sides) have been simplified.

The top-level specification of the CORBA distributed composite connector is shown in the following:

```
specification CORBAasDistributedConnector[invT, terT, invR,
                                           terR, invS, terS,
                                           invTN, terTN, invRN,
                                           terRN]: noexit
behaviour
(Client [invT, terT]
    |[invT, terT]|
 CORBAClient[invT, terT, invTN, terTN, invS,terS])
       |[invTN, terTN]|
   Network [invTN, terTN, invRN, terRN]
       |[invRN, terRN]|
(CORBAServer [invRN, terRN, invR, terR, invS, terS]
   |[invRN, terRN]|
 Server [invRN, terRN])
where
  (* behaviour specification *)
endspec
```

Three parts compose this specification: the client side (the processes `Client` and `CORBAClient`), the network (the process `Network`) and the server side (the processes `Server` and `CORBAServer`). A client request gets to CORBA through `CORBAClient`. The process `CORBAClient` receives the request, passes it to the proper set of services, and then sends it to `Network`. The process `Network` transports the request to `CORBAServer`, which forward the request to `Server`. The reply to `Client` takes a similar path in the reverse order.

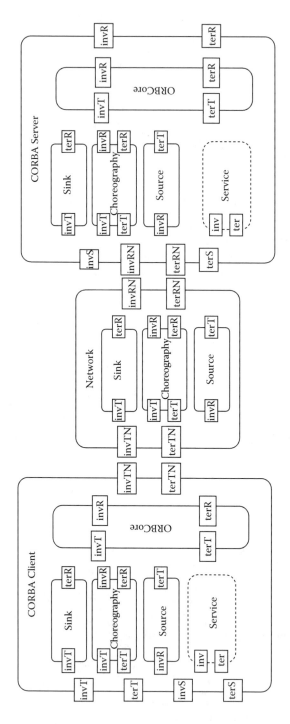

FIGURE 6.8:   CORBA as a distributed composite connector.

6.5.4 Temporal Properties of CORBA Software Architectures

A temporal property is expressed as a logic formula that is evaluated by a tool (the evaluator of the CADP Toolbox). The evaluator performs an on-the-fly verification of a property on a given labeled transition system (LTS) generated from the LOTOS specification. The temporal logic used to express the properties is called regular alternation-free mu-calculus and it is an extension of the alternation-free fragment of the modal mu-calculus with action predicates and regular expressions over action sequences.

The logic is built from three types of formulas: action formulas (A), regular formulas (R) and state formulas (F). An action formula is a logical formula built from basic action predicates and Boolean connectives. A regular formula is a logical formula built from action formulas and traditional expression operators. A regular formula R denotes a sequence of (consecutive) LTS transitions such that the word obtained by concatenating their labels belongs to the regular language defined by R. Finally, a state formula is a logical formula built from Boolean, modal and fixed-point operators. The axiom of the grammar is the F formula. These formulas enable us to define some interesting temporal properties of LOTOS specifications such as safety, liveness, and fairness.

As mentioned before, the adoption of LOTOS as an ADL enables us to check temporal properties and to compare CORBA software architecture specifications. The checking is performed by defining properties in the aforementioned temporal logic and checking them against specifications. In relation to the second task, the behavioral equivalence of the CORBA software architectures presented in Sections 6.4.1 through 6.4.3 have been verified. In particular, these specifications have been compared (in pairs) using a bisimulator (available in the CADP Toolbox) and defining a relation of observational equivalence. In practical terms, those specifications are observationally equivalent, i.e., from the point of view of an external observer, the CORBA specifications are equivalent. In this particular case, observational equivalence is rigorous enough given that application developers and standards bodies are not interested in internal details of how the middleware is actually implemented.

The temporal properties that have been defined for the CORBA software architectures are presented in the following:

Property 1 (deadlock freedom): It states that every state has at least one successor.

```
[ true* ] <true> true
```

Property 2 (safety property): A middleware service terminates before any other service starts.

```
[true*.'INV1\(!.Service1.\)\(!.*\)'.(not 'TER1\(!.Service1.\)
   \(!.*\)')*.'INV2\(!.Service2.\)\(!.*\)'] false
```

This states that every time `Service1` starts action `'INV1 \(!.Service1.\)
\(!.*\)'`, `Service2` cannot start action `'INV2 \(!.Service2.\) \(!.*\)'`
before `Service1` terminates action `'TER1 \(!.Service1.\) \(!.*\)'`.

Property 3 (safety property): The composition of the services in the composite connector defines that `Service1` is performed before `Service2`. The state formula in the following expresses this property:

```
[(not 'INV1 \(!.Service1.\) \(!.*\)' and true*).
 'INV2 \(!.Service2.\)\(!.*\)'] false
```

This states that it is not possible that `Service1` starts action `'INV1 \(!.Service1.\) \(!.*\)'` after `Service2` action `'INV2 \(!.Service2.\) \(!.*\)'`.

Property 4 (fairness)

```
[true* . 'INV1 \(!.Service1.\) \(!.*\)'.
(not 'TER1 \(!.Service1.\)\(!.*\)')*]
<(not 'TER1 \(!.Service1.\) \(!.*\)')*.
'TER1 \(!.Service1.\)\(!.*\)' > true
```

This formula expresses that after every invocation of `Service1` action `'INV1 \(!.Service1.\) \(!.*\)'`, all fair execution sequences will lead to its termination action `'TER1 \(!.Service1.\) \(!.*\)'`.

The CADP Toolbox took less than 100 ms to verify the aforementioned properties. This fact is understandable because the software architecture is the most abstract view of a software, which means that its behavioral specification does not include too many low level details of the software behavior. In fact, software architecture behavioral specifications are usually expected to be short.

6.6   Middleware Behavior Framework

While the first approach concentrated on defining templates (components, connectors, and configurations) used to formally describe middleware elements, this approach focuses on building a framework, i.e., a semispecification in the sense that one interested in specifying a middleware system's behavior only needs to reuse or extend the proposed templates [23].

The framework helps to formally describe middleware behavior by providing basic abstractions, which are generic and may be combined or extended in different ways in order to specify different middleware systems. Key in this framework is the fact that the abstractions are defined and organized according to their role in relation to the message request. A message request is any message that an application (e.g., client, server, sender, transmitter) sends to another application.

Hence, instead of adopting the traditional approach of organizing middleware systems in layers, the proposed abstractions are defined according to their role in the message request. In particular, the abstractions are grouped into classes related to storage, communication, dispatching, and mapping of message requests.

6.6.1 General Overview

As mentioned above, the framework includes abstractions that address a number of basic functionalities of middleware systems. The framework also defines how these abstractions work together to formalize different middleware models. For instance, abstractions may be combined to produce the specification of MOM, whilst they also may be joined together to define procedural middleware or a tuple space-based middleware.

The framework is message centric in the sense that basic elements are grouped according to how they act on the message. Figure 6.9 shows a general overview of the proposed approach in which the message is intercepted by middleware elements on the transmitter and receiver sides. It is worth noting that the message may be either a request in which a client (transmitter) asks for the execution of a task on the server (receiver), or simple information transferred between loosely coupled applications.

In the LOTOS specifications shown earlier, it was clear that synchronization and communication took place on same-named ports, e.g., `Component` and `Connector` communicated on ports `inv` and `ter`. But in the message-centric approach of this section, different-named ports are, in effect, wired together. This is accomplished via another process called `Interceptor` that is described in Section 6.6.3.

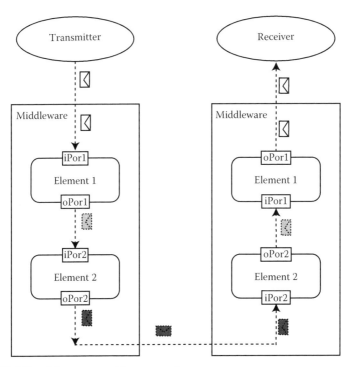

FIGURE 6.9: Message-centric approach.

6.6.2   Framework Abstractions

The abstractions of the framework are categorized into four classes, namely mappers (e.g., stub and skeletons), multiplexors (e.g., dispatcher), communication (e.g., communication channel), and storage (e.g., queue).

Whatever the class of the middleware element, it intercepts the message, processes it, and forward the message to the next element (local or remote). Only communication elements may forward the message to a remote one, i.e., an element is only accessible through the network. A noncommunication element may need to communicate with a remote element to carry out its task, but it does not send the message itself to a remote element. For instance, a transaction service may need to obtain a remote lock before passing the request to the next element of the middleware. The four classes of elements are described:

6.6.2.1   Mapper Elements

Mapper elements typically represent remote objects, serving as input points of the middleware. Their basic function is to (un)marshal application data (arguments and results) into a common packet level (e.g., GIIOP request), and are usually found in middleware systems that support client/server applications built in different programming languages (e.g., the client in Java and the server in C++). Furthermore, unconventional mappers may also compress data.

Figure 6.10 depicts a simple view of a typical mapper, named Stub. The specification is defined as follows:

```
(1) process Stub [iStub, oStub] : noexit :=
(2)  iStub ?m:Message;
(3)   oStub !marshalling (m);
(4)    iStub ?m:Message;
(5)     oStub !unmarshalling (m);
(6)      Stub [iStub, oStub]
(7) endproc
```

In this specification, the Stub receives a message sent by the transmitter and intercepted by the middleware (2), marshals it and passes it to the next element (3), and then waits for the reply from the receiver (4). The reply is also intercepted by the

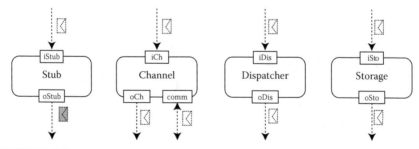

FIGURE 6.10:   Framework abstractions.

middleware and passed to the Stub (4) that takes responsibility of unmarshalling the reply (5).

6.6.2.2   Communication Elements

Communication elements act as an interface between the middleware and the operating system. They get a message and communicate it to a remote element (e.g., server, receiver). The structure of a communication element, named Channel, is shown in Figure 6.10.

The LOTOS specification of a communication element is defined as follows:

```
(1)  process Channel [iCh, oCh, comm] : noexit :=
(2)    Send [iCh, oCh, comm] ||| Receive [iCh, oCh, comm]
(3)  where
(4)    process Send [iCh, oCh, comm] : noexit :=
(5)      iCh ?m:Message;
(6)        comm !m;
(7)          oCh;
(8)            Send [iCh, oCh, comm]
(9)    endproc
(10)   process Receive [iCh, oCh, comm] : noexit :=
(11)     iCh;
(12)       comm ?m : Message;
(13)         oCh !m;
(14)           Receive [iCh, oCh, comm]
(15)   endproc
(16) endproc
```

In a similar way to Stub, the input (iCh), and output (oCh) ports serve as interception points of the element. However, communication elements have an additional port, named comm, used to communicate the message to or from a remote element. Additionally, the Channel is composed of Send and Receive processes that are responsible for send and receive messages, respectively. In the process Send, the port comm serves as an output port (6), whilst it is used as input port in the process Receive (12). In the first case, the Channel receives the message intercepted by the middleware (5), communicates it to a remote element (6), and then terminates the interception of Channel (7). In the second case, the Channel waits for a message (11), receives it (12) and then forwards the message to another element (13).

6.6.2.3   Dispatcher Elements

Dispatchers get the request and forward it to a particular object (service). The target object is defined by inspecting the request, which includes the object key. In fact, the dispatcher acts as a multiplexer inside the middleware. The general structure of a Dispatcher is depicted in Figure 6.10.

```
(1)  process Dispatcher [iDis, oDis] : noexit :=
(2)    iDis ?m:Message;
```

```
(3)    oDis !m !multiplexer(m);
(4)      Dispatcher [iDis, oDis]
(5) endproc
```

The dispatcher receives a message (2) and inspects it, through the function `multiplexer`, to obtain the destination object (3).

6.6.2.4 Storage Elements

Storage elements typically model buffers to store messages prior to their being sent, e.g., for asynchronous communication or to keep a copy of the message for recovery reasons (see Figure 6.10). The general structure of a `Storage` element is specified as follows:

```
(1) process Storage [iSto, oSto] (q: Queue): noexit :=
(2)   hide enq, fst, empt, deq in
(3)    Manager [iSto, oSto, enq, fst, empt, deq]
(4)       |[enq, fst, empt, deq]|
(5)    Queue [enq, fst, empt, deq] (q)
(6) where
(7)     (* behaviour specification *)
(8) endproc
```

In this particular specification, the storage element is modeled as a `Queue` that is administered by the `Manager`.

```
process Queue [enq, fst, empt, deq] (q:Queue) : noexit :=
  enq ?n:Nat;
     Queue [enq, fst, empt, deq] (enqueue (q, n))
     [] fst !first(q);
     Queue [enq, fst, empt, deq] (q)
     [] deq;
     Queue [enq, fst, empt, deq] (dequeue (q))
endproc
```

It is worth observing that, with minor changes to this storage element, it may be defined as a file.

6.6.3 Defining a Middleware

By using the basic abstractions defined in Section 6.6.2, middleware systems may be specified by combining them according to a particular middleware model. The general structure of any middleware specified using the framework is defined as follows:

```
specification TemplateMiddleware [invC,terC,invS,terS,comm] : noexit
  (* Abstract Data Types *)
behaviour
  (Transmitter[invC,terC] |[invC,terC]| LocalMiddleware[invC,terC,comm])
       |[comm]|
```

```
(RemoteMiddleware[invS,terS,comm] |[invS,terS]| Receiver[invS,terS])
where
   (* behaviour specification *)
endspec
```

The `Transmitter` sends a message to the `Receiver` by means of the middleware, which is made up of a local (`LocalMiddleware`) and a remote middleware
(`RemoteMiddleware`) that communicates with each other through the port `comm`
(e.g., it abstracts the whole network). Whatever the middleware model, its internal
structure is defined as follows (except for the number of components):

```
process Middleware [invC, terC, comm] : noexit :=
  hide iC1, oC1, iC2, oC2 in
     ((C1 [iC1,oC1] ||| C2 [iC2,oC2,comm])
         |[iC1, oC1, iC2, oC2]|
       Interceptor [invC,terC,iC1,oC1,iC2,oC2])
  where
     ...
  endproc
```

In this case, the middleware is composed of two components (`C1` and `C2`) and the
way they interact is defined in the process `Interceptor`. Its role is to listen on
all ports, and upon receiving a message on one, resend it on the logically connecting
port. Thus, the specification of `Interceptor` defines the intercomponent wiring
of the middleware, and allows the individual components to communicate without
having to match port names. From the standpoint of the components, the role of
`Interceptor` is transparent, so it is not shown in the middleware diagrams.

In this message-centric approach, each component intercepts the request in the
port `iCN` (input port of component N) that represents the point where the request
enters the component, processes the request and then passes it to the next component
through the port `oCN` (output port of component N) that represents the point where
the request exits the component.

6.6.4   Using the Framework

In order to illustrate how the elements introduced in Section 6.6.3 may be used to
specify middleware behavior, a simple CORBA-based middleware is initially
presented.

Figure 6.11 presents a client–server middleware where the local middleware (`LocalMiddleware`) is a composition of `stub` and `channel` elements. On
the server side (`RemoteMiddleware`), the middleware is more complex, as it is
composed by a communication element (`Channel`), a dispatcher (`Dispatcher`)
that forward the request to the proper skeleton, and some skeletons (`Skeletons`).
It is worth noting that additional middleware elements are easily added to the mid-
dleware just by including them in the parallel composition (|||) and changing the
`Interceptor` element.

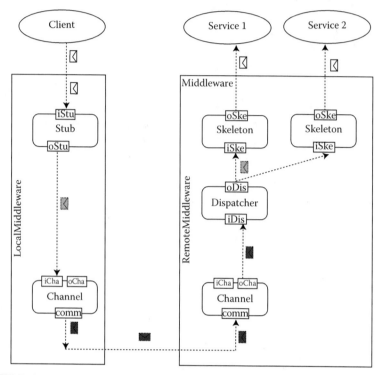

FIGURE 6.11:   Client–server middleware.

```
process RemoteMiddleware [invS, terS, comm] : noexit :=
  hide iSke, oSke, iCha, oCha, iDis, oDis in
    ((Skeleton [iSke, oSke] (1)
        ||| Skeleton [iSke, oSke] (2)
        ||| Channel [iCha, oCha, comm]
        ||| Dispatcher [iDis, oDis] )
   |[iSke, oSke, iCha, oCha, iDis, oDis]|
      Interceptor[invS,terS,iSke,oSke,iCha,oCha,iDis,oDis])
  where
      (* behaviour specification *)
  endproc
```

A MOM is characterized by the use of a queue for asynchronous communication. It is widely adopted to support loosely coupled applications. Figure 6.12 shows a simple MOM specified by using the basic abstractions defined in Section 6.6.2.

```
(1) process LocalMiddleware [send, receive, comm]: noexit :=
(2)   hide iSto, oSto, iCha, oCha in
(3)     (( Storage [iSto,oSto] ||| Channel [iCha, oCha, comm])
(4)           |[iSto, oSto, iCha, oCha]|
```

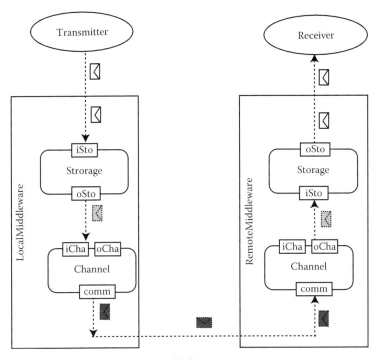

FIGURE 6.12: Message-oriented middleware.

```
(5)          Interceptor [send, receive, iSto, oSto, iCha, oCha])
(6) where
(7)     (* behaviour specification *)
(8) endproc
```

This MOM includes a Channel and Storage elements. The Channel is similar to one in Figure 6.11, whilst Storage is defined as presented in Section 6.6.2. The MOM executing on the transmitter side (LocalMiddleware) is usually similar to the one on the receiver side (RemoteMiddleware).

6.7 Modeling WSN Middleware in LOTOS

Advances in both wireless communications and electronics have enabled the development of low-cost, low-power, multifunctional sensor nodes that are small in size and communicate in short distances. These tiny sensor nodes, which consist of sensing, data processing, and communicating components, leverage the idea of sensor networks. A WSN is composed of a large number of sensor nodes that are densely deployed either inside a physical phenomenon or very close to it [1]. WSNs have been used in a multitude of applications that include industrial process control, fabrication plants, structural health monitoring, fluid pipelining monitoring, medical monitoring, environment and habitat monitoring, surveillance systems, etc.

WSN applications are usually very complex. In addition to formal approaches for specifying middleware behavior of desktop middleware systems, we present an approach to treat the complexity of specifying middleware for WSN through adopting LOTOS for formalizing their behavior [24]. The formalization has three main benefits: better understanding of application behavior, checking of desired properties of the application, and the possibility of simulating applications prior to build them. This approach is currently targeted for WSN systems based on TinyOS and nesC.

TinyOS is an operating system specially designed for running on nodes of WSNs. Applications in TinyOS are event driven in such a way that when an event occurs a handler may post a task that is scheduled by the TinyOS Kernel some time later. TinyOS applications are written in nesC [16].

Network embedded system C (nesC) [10] is a structured component-based dialect of C specially designed for building embedded systems like WSN applications. An application in nesC consists of a set of components linked together with bidirectional interfaces that serve as access points to the component. Two kinds of components may be defined in nesC, namely *configuration* and *module*. A configuration simply defines how components are put together, whilst a module represents the implementation itself. A component *provides* and *uses* interfaces that declare a set of functions (*commands*) the component must implement, and functions (*events*) that the user of the interface must implement. Every nesC application is described by a top-level configuration that wires components. Additionally, two basic elements of TinyOS applications include: the *Main* component that must be implemented in every application; and the standard interface named *StdControl* used to initialize/ start/stop TinyOS components, which includes three operations: *init*, *start*, and *stop*.

The proposed approach for specifying WSN middleware in LOTOS takes into account some basic principles. First, as a formal description technique, LOTOS is used to formally describe the behavior of nesC applications. Hence, typical constructs of nesC such as loop, decision, and assignment commands are not explicitly present in the specification. Second, the specification concentrates mainly on identifying and modeling interactions between nesC components. Third, the LOTOS state-oriented specification style [29] is extensively adopted. In this style, the whole system (or part of the system) is viewed as a single resource whose internal space is explicitly defined through a set of states and alternative sequence of interactions.

Following these principles, the LOTOS modeling process is carried out in four main steps [24]:

1. To identify the nesC components (modules and configurations) that make up the middleware. These components become LOTOS processes at the end of this step.

2. To identify how components (modules and configurations) found in Step 1 are wired and which interfaces are used to connect them. The way the components are wired defines how the LOTOS processes are composed, whilst the interfaces become synchronization ports in LOTOS.

3. To identify operations (nesC commands and events) of each interface identified in Step 2. For each interface, a LOTOS ADT is defined whose operations represent the nesC commands and events.

4. To identify interactions inside nesC module implementations. After identifying the modules in Step 1, their composition in Step 2, and the operations available to be invoked by each module in Step 3, it is needed to precisely identify the order in which the operations are invoked inside the modules. When a nesC operation is invoked, this indicates that an interaction takes place between the nesC modules. In LOTOS, these interactions represent synchronization events between two processes.

The following sections present how nesC first-class elements identified in the aforementioned steps are modeled in LOTOS.

6.7.1 Interface

nesC applications are built out of components with well-defined interfaces, which represent points of access to the components. The interaction between components occurs by invoking the functions of the interface. In LOTOS, processes interact through their synchronization ports. In this way, the nesC interface is defined as a LOTOS synchronization port, e.g., if a component provides three interfaces, it has three ports. Furthermore, as LOTOS ports have no direction (the notion of output/input ports does not exist), both the provided and used interfaces are simply modeled like ports. Actions occurring in the port give an idea whether the interface is a used or provided interface.

As mentioned above, each nesC interface has a set of functions (commands and events). As we are only interested in the module interactions or temporal ordering of events (not in their functionality), these functions are defined as LOTOS operations of an ADT defined for each interface [24]. In this way, the nesC interface *interface-name* leads to the definition of the LOTOS ADT I*interface-name*. The command *command-name* and event *event-name* are defined as type operations c_*command-name* and e_*event-name*, respectively.

In order to illustrate this approach, the interface SendMsg is presented in Figure 6.13. In this example, the nesC interface SendMsg is modeled as the LOTOS ADT ISendMsg, whilst its command send and event sendDone are defined as the operations c_send and c_event of type ISendMsg, respectively.

6.7.2 Module

Modules are first-class elements in the nesC language and have two basic parts: the set of interfaces provided and used by the module, and the implementation. Interfaces are modeled as defined in Section 6.7.1. Since the basic abstraction in LOTOS is the process, the nesC module is specified as a LOTOS process (see Figure 6.14). The implementation of each function (command or event) is defined in the behavior part of the process using the LOTOS choice operator ([ ]), and each command/event defines an option in the choice.

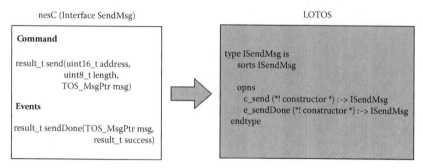

FIGURE 6.13:   nesC interface in LOTOS.

Modules implementing the TinyOS `StdControl` interface are defined using the state-oriented specification style [24]. Four states are then defined: the module starts in an initial state (state 0) prior to executing the nesC command `StdControl.init`; the second state (state 1) is reached when the module is initialized but not started yet (after the execution of the nesC command `StdControl.start`); the state when the module is already initialized and ready to execute the functions defined in the interface (state 2); and finally, the last state (state 9), when the module has been stopped after the execution of the nesC command `StdControl.stop`. On reaching state 9, the module is not able to execute any further action.

In the nesC example shown in Figure 6.14, the module (`MOD`) interacts with other modules (defined in the configuration where it is placed) through the interfaces

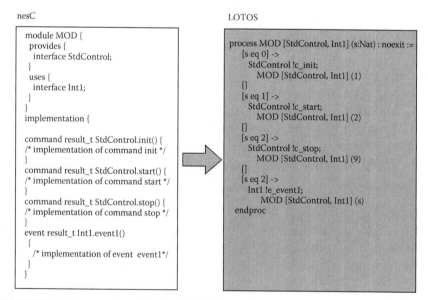

FIGURE 6.14:   nesC module in LOTOS.

StdControl (provides) and Int1 (uses). These interfaces becomes LOTOS synchronization ports. The interface StdControl has three operations, namely init, start, and stop (renamed to c_init, c_start and c_stop, respectively), defined as choices in the LOTOS choice operator ([]). The interface Int1 is used by this module, which means that the module must implement the events defined in the interface. In this case, only the event event1 is defined.

As this module implements the StdControl interface, the state-oriented LOTOS style is used. The statement below expresses that if process MOD is in state 0 ([s eq 0]) then a synchronization action may occur in the port StdControl if the value c_init is offered (StdControl !c_init). Then the module MOD is reinstantiated with the new state set to 1.

```
[s eq 0] ->
              StdControl !c_init;
              MOD [StdControl, Int1]  (1)
```

6.7.3   Configuration

In nesC, a configuration is a first-class element that defines the used and provided interfaces, the components that make up it and the way they are wired (implementation). A configuration is present in a nesC program in two situations: the initial configuration of an application; and part of another configuration. In the first case, the configuration is modeled as the LOTOS top-level process in the process hierarchy, which is defined as specification (see Section 6.2). In the second case, it is simply modeled as a LOTOS process. In both cases, the module implementation defines how the modules that make up the configuration are wired and how the interfaces are used to connect them [24].

If two modules within the configuration are wired through an interface, it means that they interact with each other. Hence, these two modules are placed in a LOTOS parallel composition with synchronization (each interface becomes a synchronization port according to Step 2). If two components are not wired, the parallel interleaving LOTOS operator (|||) is used.

In order to illustrate these elements, the specification bellow presents a nesC configuration:

```
configuration CONF {
}
implementation {
  components Mod1, Mod2, Mod3, Mod4;

  Mod1.Int1 -> Mod3.Int1;
  Mod1.Int1 -> Mod2.Int1;
  Mod2.Int2 -> Mod3.Int2;
  Mod2.Int3 -> Mod4.Int3;
}
```

Its respective LOTOS specification is as follows:

```
(1) specification CONF [Int1, Int2, Int3] : noexit
(2) behaviour
(3)   Mod1 [Int1]
(4)    |[Int1]|
(5)   Mod2 [Int1, Int2, Int3]
(6)    |[Int1, Int2, Int3]|
(7)   ( Mod4 [Int1] ||| Mod3 [Int1, Int3] )
(8) where
(9)   process Mod1 [Int1]: noexit
(10) :=   (* behaviour specification *) endproc
(11) process Mod2 [Int1, Int2, Int3]: noexit
(12) :=   (* behaviour specification *) endproc
(13) process Mod3 [Int1,Int3]: noexit
(14) :=   (* behaviour specification *) endproc
(15) process Mod4 [Int1]: noexit
(16) :=   (* behaviour specification *) endproc
(17)endspec
```

The nesC configuration CONF is defined as the LOTOS top-level specification (1) and consists of four components: Mod1 (9), Mod2 (11), Mod3 (13), and Mod4 (15). According to this approach, each nesC component is defined as a LOTOS process. These components are wired according to the nesC implementation clause: Mod1 is wired to Mod3/Mod2 through the interface Int1; Mod2 is wired to Mod3 through Int2; Mod2 is wired to Mod4 through Int3. In LOTOS, it means that Mod1 synchronizes simultaneously with Mod2 and Mod3 in port Int1 (4); Mod2 synchronizes with Mod1 in the port Int1 (4), with Mod3 in the port Int2 (6) and with Mod4 in the port Int3 (6); and Mod4 and Mod3 are not connected to each other (7).

6.8   Concluding Remarks

This chapter has presented three approaches for specifying middleware behavior in LOTOS: an approach based on software architecture principles, a framework for specifying middleware systems whatever their models, and an approach for specifying middleware for WSN.

In the initial approach, the middleware itself was defined as a connector and its structure is defined through software architecture elements (component, connectors, and configuration). Then, the middleware software architecture behavior was described in LOTOS. Due to its complexity, the middleware software architecture was presented at three abstractions levels that represent traditional views of the middleware: middleware as a simple connector (application developer's view), middleware as a composite connector (standards developer view) and middleware as a distributed composite connector (middleware developer's view). An approach based on software architecture concepts enables us to explicitly define the middleware internal structure and to separate computation and communication elements of middleware systems.

In the second approach, a framework was presented. The framework consists of a set of common elements usually found in the development of middleware systems. The framework is centered on the message request instead of the middleware layer. This facilitates the treatment of middleware complexity: simple abstractions are highly reusable, and it is easier to find specification errors and verify desired behavior properties. The way of composing middleware abstractions considers the order in which they intercept message requests, enormously facilitating composition of middleware abstractions.

The last approach focuses on formalizing the behavior middleware for WSN. The approach consists of modeling nesC and TinyOS elements in LOTOS. In order to carry out this task, we defined a set of specification steps in an approach to specify the main concepts of nesC in LOTOS, namely interfaces, modules, and configurations. The approach of formalizing nesC applications has some interesting benefits for those who build WSN applications in nesC. First, the specification process improves the understandability of the nesC application itself. Second, the emphasis on the interaction of components (instead of their functionality) in the LOTOS specification provides a better understanding of the way the components collaborate. Finally, the strategy adopted to specify nesC applications favours the high reuse of module specifications.

In closing, it is possible to note that software architecture principles are widely adopted to build distributed applications, but are rarely applied to the middleware systems. Our approach illustrates the benefits of doing so.

References

[1] I. F. Akyildiz, W. Su, Y. Sankarasubramaniam, and E. Cayirci. Wireless sensor networks: A survey. *Computer Networks*, 38:393–422, March 2002.

[2] R. J. Allen. *A Formal Approach to Software Architecture*. PhD thesis, School of Computer Science, Carnegie Mellon University, Pittsburgh, PA, 1997.

[3] D. Basin, F. Rittinger, and L. Viganò. A formal data-model of the CORBA security service. *SIGSOFT Software Engineering Notes*, 26(5):303–304, 2001.

[4] R. Bastide, P. Palanque, O. Sy, and D. Navarre. Formal specification of CORBA services: Experience and lessons learned. *ACM SIGPLAN Notices*, 35(10):105–117, 2000.

[5] R. Bastide, O. Sy, D. Navarre, and P. Palanque. A formal specification of the CORBA event service. In *Fourth International Conference on Formal methods for Open Object-based Distributed Systems (FMOODS)*, Norwell, MA, Kluwer Academic Publishers, 2000, pp. 371–395.

[6] T. Bolognesi and E. Brinksma. Introduction to the ISO specification language LOTOS. *Computer Networks and ISDN Systems*, 14:25–59, 1987.

[7] M. Chattopadhyay, S. Paul, S. Sanyal, and D. Das. An approach to formal specification and formal validation of facilities of a mobile middleware architecture. In *Proceedings of the First Mobile Computing and Wireless Communication International Conference MCWC 2006*, Aman, Jordan, Sept. 17–20, 2006, pp. 165–170.

[8] A. Coen-Porisini, M. Pradella, M. Rossi, and D. Mandrioli. A formal approach for designing CORBA-based applications. *ACM Transactions Software Engineering Methodology*, 12(2):107–151, 2003.

[9] H. Garavel, F. Lang, R. Mateescu, and W. Serwe. CADP 2006: A toolbox for the construction and analysis of distributed processes. In *19th International Conference on Computer Aided Verification (CAV)*, 2007, Berlin, Germany, pp. 1–5.

[10] D. Gay, P. Levis, R. von Behren, M. Welsh, E. Brewer, and D. Culler. The nesC language: A holistic approach to networked embedded systems. In *PLDI'03: Proceedings of the ACM SIGPLAN 2003 Conference on Programming Language Design and Implementation*, ACM, New York, NY, 2003, pp. 1–11.

[11] C. Ghezzi, D. Mandrioli, and A. Morzenti. TRIO, A logic language for executable specifications of real-time systems. *Journal of System and Software*, 12:107–123, 1990.

[12] J. Hugues, L. Pautet, and F. Kordon. Refining middleware functions for verification purpose. In *Monterey Workshop on Software Engineering for Embedded Systems: From Requirements to Implementation (MONTEREY'03)*, Chicago, IL, Sept. 2003, pp. 79–87.

[13] J. Hugues, T. Vergnaud, L. Pautet, Y. Thierry-Mieg, S. Baarir, and F. Kordon. On the formal verification of middleware behavioral properties. *Electronic Notes in Theoretical Computer Science*, 133:139–157, 2005.

[14] ISO/IEC. *Information Technology—Enhancements to LOTOS (E-LOTOS) ISO/IEC 15437*. International Organization for Standardization, 2001.

[15] D. Kreuz. Formal specification of CORBA services using object-Z. In *Second International Conference on Formal Engineering Methods*, Brisbane, Queensland, Australia, 1998, pp. 180–189.

[16] P. Levis. *TinyOS Programming*, June 2006. Available at http://csl.stanford.edu/pal/pubs/tinyos-programming.pdf.

[17] G. F. Lucero and J. Quemada. Specifying the ODP trader: An introduction to E-LOTOS. In *FORTE X/PSTV XVII '97: Proceedings of the IFIP TC6 WG6.1 Joint International Conference on Formal Description Techniques for Distributed Systems and Communication Protocols (FORTE X) and Protocol Specification, Testing and Verification (PSTV XVII)*, London, U.K., Chapman & Hall, 1998, pp. 127–142.

[18] N. Medvidovic, E. Dashofy, and R. Taylor. On the role of middleware in architecture-based software development. *International Journal of Software Knowledge Engineering*, 13(4):367–393, 2003.

[19] N. Medvidovic and R. N. Taylor. A classification and comparison framework for software architecture description languages. *IEEE Transactions on Software Engineering*, 26:70–93, August 2000.

[20] C. F. Ngolah, C. F. Ngolah, and Y. Wang. Formal specification of CORBA-based distributed objects and behaviors. In Yingxu Wang, (Ed.), *Fourth IEEE Conference on Cognitive Informatics (ICCI 2005)*, Irvine, CA, 2005, pp. 331–339.

[21] OMG. *Common Object Request Broker Architecture: Core Specification*. Object Management Group, March 2004.

[22] N. S. Rosa and P. R. F. Cunha. A software architecture-based approach for formalising middleware behaviour. *Electronic Notes in Theoretical Computer Science*, 108:39–51, 2004.

[23] N. S. Rosa and P. R. F. Cunha. A LOTOS framework for middleware specification. In *26th IFIP WG 6.1 International Conference on Formal Methods for Networked and Distributed Systems (FORTE'06)*, Paris, France. *Lecture Notes in Computer Science*, 4229:136–142, 2006.

[24] N. S. Rosa and P. R. F. Cunha. Using LOTOS for formalising wireless sensor network applications. *Sensors (Basel)*, 7:1447–1461, 2007.

[25] M. Shaw and D. Garlan. *Software Architecture: Perspectives on an Emerging Discipline*, Englewood Cliffs, NJ, Prentice-Hall, 1996.

[26] V. Subramonian, C. Gill, C. Sánchez, and H. B. Sipma. Reusable models for timing and liveness analysis of middleware for distributed real-time and embedded systems. In *6th ACM and IEEE International Conference on Embedded Software*, New York, ACM, 2006, pp. 252–261.

[27] Sun. *Java Transaction Service*. Sun Microsystems, December 1999. Available at: http://java.sun.com/javaee/technologics/jts/index.jsp.

[28] Sun. *Java Message Service*. Sun Microsystems, April 2002. Available at: http://java.sun.com/products/jms/docs.html.

[29] K. J. Turner and M. van Sinderen. *LOTOS Specification Style for OSI*. Technical Report PB93-173276, Technische University Twente, the Netherlands, 1992.

[30] V. Nalini, C. Talcott, and G. A. Agha. A formal model for reasoning about adaptive QoS-enabled middleware. *ACM Transactions on Software Engineering and Methodology*, 13(1):86–147, 2004.

[31] C. A. Vissers, G. Scollo, and M. van Sinderen. Architecture and specification style in formal descriptions of distributed systems. In Sudhir Aggarwal and Krishan K. Sabnani, (Eds.), *Eighth International Symposium on Protocol Specification, Testing and Verification VIII*, Atlantic City, NJ, 1988, pp. 189–204.

[32] C. A. Vissers, G. Scollo, and M. van Sinderen. Architecture and specification style in formal descriptions of distributed systems. *Theoretical Computer Science*, 89:179–206, 1991.

[33] C. A. Vissers, G. Scollo, M. van Sinderen, and E. Brinksma. *On the use of specification styles in the Design of Distributed Systems*, *Theoretical Computer Science*, 89(1): 179–206, 1991.

Chapter 7

Abstract Machine for Service-Oriented Mobility

Hervé Paulino

Contents

7.1   Introduction

The modeling of distributed computations in terms of software mobility (either code or state) has the ability to almost completely abstract the programmer from the underlying network. However, the security issues that this paradigm arises are too important to be overlooked [2,14], and cast a shadow over its design qualities. There is the need to protect the host from incoming agents and vice versa, and also to protect a network from continuous roaming and cloning. These well-founded security concerns have limited the use of the mobility paradigm, and, in our opinion, the immediate future lies on its application to local area networks and middleware layers. Restricting mobility to trusted administrative domains, or having complete knowledge of the code to migrate are key aspects for the softening of the security measures.

One other technology that has increased its role in middleware applications is service-oriented computing. Composition of loosely bound service-oriented components has proved to be a good paradigm for the modeling of distributed applications, mostly in heterogeneous environments, such as the ones that compose grids [10,11].

Combining software mobility and service-oriented computing seems to be a good approach for future middleware development. The modularity provided by the loosely bound service-oriented components can be complemented with advantages of mobility, such as the use of a more intuitive programming model for distributed computing, and the focus on local interaction. The fact of moving the client code toward the resources and interacting locally, unlike the usual communication paradigms (e.g., client–server), eliminates the need to maintain costly remote sessions.

It is thus our opinion that the seamless incorporation of these two paradigms in a single language can provide a powerful tool for the implementation of distributed applications, namely middleware. However, when looking at these two paradigms we see that one models network interaction in terms of services, regardless of their location, while the other is location-aware. Current mobility models, such as the distributed process calculi [1,3,8,12,24,29,31,32] use abstractions for network nodes, commonly called sites or nodes. In this work we propose a novel mobility model that is service oriented. The migration is done toward services, regardless of their location, rather than toward sites.

We present our model as the service-oriented mobility abstract machine (SOMAM), defining rules for service creation and client–server interaction. The machine is encoded onto a calculus that extends the (lexically scoped, distributed) LSD π-calculus [23] with basic objects, process definitions and expressions. The LSD π-calculus is, in turn, a form of the π-calculus [13,18] extended with support for distributed execution and mobility of resources, and with a well-studied semantics. The encoding provides the framework for the proof of a soundness result relative to the base calculus [19]. This is particularly important as it provides a form of language security, in the sense of being correct by design.

The outline of this chapter is as follows: Section 7.2 introduces the abstract machine, its syntax and semantics; Section 7.3 describes the encoding onto the process calculus; Section 7.4 draws the main lines for the implementation; and finally, in Sections 7.5 and 7.6, respectively, we compare our approach to existing work in the area and present some conclusions.

7.2 Service-Oriented Mobility Abstract Machine

A network in the SOMAM is described as a set of computing units (agents) plus a resolver for services. Agents are described as collections of threads running concurrently in the boundaries of a network node abstraction (host), sharing resources, namely code and address space. We distinguish between two types of agents: service providers and programs. The former are daemons that provide the implementation of a given service. They may also pose as clients to other providers; at this level there is no distinction between clients and servers whatsoever. The latter do not provide any service. They are simple clients that define the means for a user to interact with the network.

In the SOMAM, interagent action is service-driven. This ensures agent anonymity and provides the kind of loose bindings desirable to build resilient applications in highly dynamic networks. All service-related information, namely type and currently available implementations, are stored in the resolver.

We will begin by giving a general overview of the SOMAM and of the design adopted. The machine's definition is presented in two parts: the basic model, for which we provide a formal encoding and, a set of extensions that ease the programming and increase the model's capabilities. These extensions, although easily encodable in the base calculus, would introduce an extra layer of complexity that would be counterproductive.

7.2.1 General Overview and Design Choices

The focus of this work is service-oriented mobility. Therefore, we abstract ourselves from every aspect concerning the evolution of local computations, for example, object and explicit thread creation, interprocess communication, or concurrency management. Our intention is not to define a Turing-complete computing model, but rather to lay the foundations for service-driven mobility. In this context, we restrict our study to the primitives necessary for the creation of new agents and for interagent interaction (service-driven code mobility).

The service is the main abstraction in the SOMAM. It specifies the interface with which each of its implementations available in the network must comply. The machine assumes that all type checking must be performed at compile- or load-time, by matching the types for the services used or implemented by the agents, with the information kept in the resolver.

This procedure is borrowed from the MOB run-time system [21], where if a service known to the network is required or implemented by a program, then its locally inferred type must match the interface for that service kept in the resolver. If the service is introduced for the first time by the program (an interface for it does not yet exist in the resolver) then the type inferred for the service will become the adopted interface for the service in the network.

Service providers are abstracted by classes that define a concrete implementation of the service's interface. Without loss of generality, since the general case can be easily inferred, we decided to restrict the number of services implemented by a given provider to one. We also chose to restrict the set methods defined in the class to the ones defined in the service's interface. Private methods can be gathered in an special attribute that shares the remainder of the provider's state.

Since all service information is managed at compile- or load-time, no service definitions are required in an abstract machine program. Moreover, we assume that the loading operation places all the required code in the agent's code repository, and thus from the machine's point of view there are no class definitions in a program. They are all in the code repositories.

The handling of multiple requests concurrently is achieved by giving providers multithreading capability. Threads share the same heap space whilst having independent control data-structures. As stated before, the abstract machine does not include any support for concurrency management. We refer the reader to many widely studied models, such as the ones based on the π-calculus [13,18].

All interagent communication is driven by code mobility, more precisely by service-driven session uploading. In this initial approach, both code mobility and method invocations are synchronous. In the mobility case, this implies suspending the current thread until the result of the uploaded session arrives. To have a unique procedure for both local invocations and code mobility, we have chosen to simulate call stacks by suspending the current thread and launching a new one to execute the code of the invoked method.

Providers are daemons, and consequently do not have a termination point coded in their behavior. The SOMAM allows for their assisted termination by defining the **halt** event which must be explicitly posted in the network. It is only natural that, at the time of the reception of a halting event, a provider is running uploaded sessions. In order to avoid the abrupt termination of these sessions, without ever returning the result to the client, the definition of a deferred termination mechanism is required.

7.2.2   Base Model

Notation: Given a set X and one of its elements x, we write $X \cdot x$ to denote the set $(X - \{x\}) \cup \{x\}$.

7.2.2.1   Syntax

The abstract machine requires a set of identifiers and syntactic categories presented in Table 7.1. S denotes a service identifier, a and b represent network-wide unique identifiers for agents, which refer to as keys, l denotes a method label, and X stands for classes of providers and local objects, such as datatypes. A process is a sequence of instructions ranged over by P.

The set of methods in a class is given by a map of the form Method = MethodId $\mapsto$ Var* $\times$ Instruction*, ranged over by M. The code repository contains all the code

TABLE 7.1:   Syntactic categories.

S	∈	Service	Service identifier
a, b	∈	AgentKey	Agent key
h	∈	Host	Host name
x, y	∈	Var	Variable
v	∈	Value = Var	Value
X, Y	∈	Class	Class identifier
l	∈	Label	Method label
P	∈	Instruction*	Sequence of instructions, a program
M	∈	Method	Methods of a class
C	∈	Code	Code repository
r	∈	HeapRef	Heap reference
B	∈	Bindings	Set of bindings
H	∈	Heap	Address space for the agent
T	∈	Pool(running thread)	Thread
A	∈	Pool(agent)	Pool of agents
R	∈	NameService	Name resolver
N	∈	Network	Network

required by an agent and is defined by the map Code = Class ↦ Var* × Method × Class* × Service, ranged over by C. An heap reference stands for abstraction for an address in an address space and is denoted by r. To be more precise, heap references are qualified with the identifier of their hosting agent (e.g., reference r in the heap of agent a should be interpreted as r@a) and thus are unique in the network. To ease the reading of the rules we omit the qualifier of a heap reference when it is accessed from within its hosting agent. Special references null and self ∈ HeapRef denote, respectively, an undefined reference, and the reference where the closure of the agent can be found in the heap.

There are no primitive values in the machine; every value lives in the heap. Thus, language values only comprise variables* and are ranged over by v. A set of bindings is a map defined as Bindings = Var ↦ HeapRef, ranged over by B, and represents a map from identifiers in the code to references in the heap. A heap is a map defined as Heap = HeapRef ↦ ((Bindings × Class) ∪ (Var × RunningThread)) and represents the address space of an agent. It holds the heap representation of local objects and provider, and threads suspended, waiting for a result. These are stored along with the variable that will bound to the incoming result.

A thread is an element of the set RunningThread = Bindings × Instruction* × HeapRef ranged over by T. In the definition of a thread (B, P, r), r is a heap reference where the thread may place the result of its execution. Initial threads of a program do not return results, and thus the reference is null.

We denote a pool of a given category γ by Pool(γ) and the parallel composition of its elements by |.

An agent is an element of the set Agent = AgentKey × Host × Code × Heap × Pool(RunningThread), ranged over by A, and represents a multithreaded autonomous

---

* We distinguish between variables and values only to ease the reading of the rules.

TABLE 7.2: Syntax of a SOMAM network.

M	::=	$\{l_1 : (\tilde{x}_1, P_1), \ldots, l_n : (\tilde{x}_n, P_n)\}$
C	::=	$\{X_1 : (\tilde{x}_1, M_1, \tilde{X}_1, S_1), \ldots, X_n : (\tilde{x}_n, M_n, \tilde{X}_n, S_n)\}$
B	::=	$\{x_1 : r_1, \ldots, x_n : r_n\}$
K	::=	$(B, X) \mid (x, (B, P, r))$
H	::=	$\{r_1 : K_1, \ldots, r_n : K_n\}$
T	::=	$(B, P, r) \mid T \mid \mathbf{0}_T$
A	::=	$a(h, C, H, T)^{\circ} \mid A \mid a(h, C, H, T)^{\bullet} \mid A \mid \mathbf{halt}(a) \mid A \mid \mathbf{0}_A$
I	::=	$\{r_1, \ldots, r_n\}$
R	::=	$\{S_1 : (\alpha_1, I_1), \ldots, S_n : (\alpha_n, I_n)\}$
N	::=	A, R

TABLE 7.3: Syntax of a SOMAM program.

P	::=	$x = \mathbf{new} \ X \ (\tilde{v}) \ P$	Spawn of a new provider
	|	$x = S \ P \ \mathbf{in} \ P' \ \mathbf{error} \ P''$	Service-oriented process mobility
	|	$x = l \ (\tilde{v}) \ P$	Method invocation
	|	$\mathbf{return} \ v$	Return a result
	|	ϵ	Terminated process

computation. We write an agent (a, h, C, H, T) as $a(h, C, H, T)$ thus exposing the agent's key. Providers that received an halting event have dedicated rules and therefore must be distinguishable. We tag them with $\bullet$ $(a(h, C, H, T)^{\bullet})$, which indicates that the agent will terminate its execution as soon as it runs out of work. This behavior (and tag) can also be applied to programs, since their execution terminates as they reach the end of their code. When we want to refer explicitly to an agent that is not tagged for conclusion, we use $\circ$ as in $a(h, C, H, T)^{\circ}$. The absence of either tag indicates that the agent's state is not relevant in the given context.

The name resolver defined by map NameService $=$ Service $\mapsto$ Type $\times\ 2^{\text{HeapRef}}$, ranged over by R, represents a network-wide name service for obtaining the type of a service, ranged over by α and the set of agents that implement it, ranged over by I. Finally, a network is an element of the set Network $=$ Pool(Agent) $\times$ NameService, ranged over by N, and represents a network computation.

Based on the above definitions, we may write the syntax for a network as defined in the grammar presented in Table 7.2. The syntax for programs is given by the grammar in Table 7.3.

7.2.2.2 Syntactic Assumptions

We make some assumptions on the syntax of a program that must be checked at compile-time:

- All services referenced in a code repository and program are registered in the network

- The body of a method or a process P in $x = S \ P \ \mathbf{in} \ P' \ \mathbf{error} \ P''$ must terminate with the **return** instruction, and this instruction can only be applied in these contexts

- Method identifiers are pairwise distinct in a class definition, as well as parameters in class and method definitions

- A variable x is bound in P' in $x = $ **new** X $(\tilde{v})$ P', $x = S$ P **in** P' **error** P'' and $x = 1(\tilde{v})$ P'

- A variable x is bound in M in a class definition $(\tilde{x}, M, \tilde{X}, S)$ if it is one of the $\tilde{x}$, and in P in a method definition $(\tilde{x}, P)$ if it is one of the $\tilde{x}$

- A class identifier X is bound in an class definition $(\tilde{x}, M, \tilde{X}, S)$ if it is one of the $\tilde{X}$

The sets of free variables, free classes, and free services are defined accordingly. Well-formed programs are closed for variables, class identifiers, and service identifiers.

7.2.2.3 Initial and Final States

We describe the abstract machine from the point of view of the execution of one agent. Thus, when we start running an agent, the network may already have a pool of agents A running concurrently and distributed among the network nodes, together with the resolver R:

$$A, R$$

As stated in the initial discussion of this chapter, the code repository for the program is collected at load-time. Thus, when the agent is launched into the network it already contains all the code it requires. We have that the initial state of the execution of a program in host h is thus:

$$a(h, C, \emptyset, (\emptyset, P, null))^{\bullet} \mid A, R$$

where

  a is a fresh agent key
  P is the program itself executed by the agent's initial thread
  C is the repository resulting from the collection of the code required by P

Note that the agent's heap does not feature a `self` reference. This results from the fact that a program does not provide any service, nor has a closure in the heap. Moreover, the resolver remains unaltered, since no new service implementation is given. In consequence, a is not visible to the remainder of the network.

A program terminates its execution when it reaches the end of its code, and consequently runs out of threads to execute. As previously discussed, service providers are daemons by default and must be explicitly terminated by an **halt** event. This event does not instantly terminate the agent, since remote sessions may be in progress. The **halt** event will cause the provider to be ● tagged and to become unavailable to the remainder of the network. Thus, at the end of the execution of a program or a provider (in the conclusion mode) running in agent a, the configuration of the network will be of the form:

$$a(h, C, H, \mathbf{0})^{\bullet} \mid A', R'$$

This configuration will reduce into the terminated agent, which can then be garbage collected and produce the state:

$$A', R'$$

7.2.2.4 Congruence Rules

The computation in the abstract machine is driven by a set of reduction rules that operate over the left-most thread of the left-most agent in the respective pools. Thus, in order to be able to commute, associate and garbage-collect threads and agents in their pools, we need a set of congruence rules. These will allow for the rewriting of both pools into semantically equivalent ones, where the configuration is in conformity with the reduction rule to be applied. The congruence rules for SOMAM networks are

[AGENTSWAP]	$A \mid A', R \equiv A' \mid A, R$
[AGENTASSOC]	$A \mid (A' \mid A''), R \equiv (A \mid A') \mid A'', R$
[AGENTGC]	$\mathbf{0}_A \mid A, R \equiv A, R$
[THREADSWAP]	$a(h, C, H, T \mid T') \mid A, R \equiv a(h, C, H, T' \mid T) \mid A, R$
[THREADASSOC]	$a(h, C, H, T \mid (T' \mid T'')) \mid A, R \equiv a(h, C, H, (T \mid T') \mid T'') \mid A, R$
[THREADGC]	$a(h, C, H, \mathbf{0}_T \mid T) \mid A, R \equiv a(h, C, H, T) \mid A, R$

7.2.2.5 Reduction Rules

Each SOMAM instruction requires at least one machine transition to be processed. We begin by presenting [CONG] that allows reduction to occur under structural congruence, and proceed to the rules for the language constructs.

$$[\text{CONG}] \quad \frac{A, R \equiv A'', R \quad A'', R \to A''', R' \quad A'', R' \equiv A', R'}{A, R \to A', R'}$$

7.2.2.5.1 Launching a New Service Provider

Providers in the SOMAM are similar to objects in object-oriented languages, in the sense that they are instances of classes. Their instantiation, however, generates a new execution unit (agent) that is placed in the network.

An agent is self-contained, hosting in its repository all the code it requires. The repository of a freshly launched provider must contain the code closure for the provider's class, plus the one for the values passed to its constructor. The former is obtained by the *code* function that extracts the code closure for a set of classes from a given repository.

$$code : \text{Class} \times \text{Code} \times \text{Code} \mapsto \text{Code}$$
$$code(X \, \tilde{X}, C \cdot (X :(\tilde{x}, M, \tilde{X}', \_)), C') = code(\tilde{X} \, \tilde{X}', C, C' + \{X :(\tilde{x}, M, \tilde{X}', \_)\})$$
$$\text{if } X \notin dom(C')$$
$$code(X \, \tilde{X}, C, C') = code(\tilde{X}, C, C') \quad \text{if } X \in dom(C')$$
$$code(\epsilon, C, C') = C'$$

The latter is collected by function $copySeq_{ab}$, which is also used to build part of the agent's initial heap. The function creates a copy, at the new agent (b), of all the heap and code required by heap values mapped by the given references. The arguments are the code repository and heap of the original agent (a), and the sequence of references holding the values to be copied. The copy of each heap value is computed by function *copy*.

$$copySeq_{ab} : \text{Code} \times \text{Heap} \times \text{HeapRef}^* \mapsto \text{Code} \times \text{Heap} \times \text{HeapRef}^*$$
$$copySeq_{ab}(C, H, r \, \tilde{r}) = (C' + C'', H' + H'', r'\tilde{r}')$$
$$\text{where } copy_{ab}(C, H, r) = (C', H', r')$$
$$\text{and } copySeq_{ab}(C, H, \tilde{r}) = (C'', H'', \tilde{r}')$$
$$copySeq_{ab}(C, H, \epsilon) = (\emptyset, \emptyset, \epsilon)$$

$$copy_{ab} : \text{Code} \times \text{Heap} \times \text{HeapRef} \mapsto \text{Code} \times \text{Heap} \times \text{HeapRef}$$
$$copy_{ab}(C, H, \texttt{self@a}) = (\emptyset, \emptyset, \texttt{self@a})$$
$$copy_{ab}(C, H, r@a) = (\{X : C(X)\} + C', H' + \{r'@b : (\{\tilde{x} : \tilde{r}'\}, X)\}, r'@b)$$
$$\text{if } r@a \neq \texttt{self@a and } H(r@a) = (\{\tilde{x} : \tilde{r}\}, X)$$
$$\text{where } copySeq_{ab}(C, H, \tilde{r}) = (C', H', \tilde{r}')$$
$$\text{and } r'@b \notin H'$$

The provider's initial heap contains all the entries collected by the *copySeq* function plus the self entry (`self`) containing the provider's closure. The parent agent keeps a binding to new provider in x. The entry for service S in the resolver is updated with the reference holding the new provider.

[NEWPROVIDER]

$$\frac{C(X) = (\tilde{x}, M, \tilde{X}, S) \quad copySeq_{ab}(C, H, B(\tilde{v})) = (C', H', \tilde{r}) \quad R(S) = (\alpha, I) \quad \text{b fresh}}{\begin{array}{l} \texttt{a}(h, C, H, (B, x = \textbf{new } X(\tilde{v}) \, P, r) \mid T) \mid A, R \rightarrow \\ \quad \texttt{a}(h, C, H, (B + \{x : \texttt{self@b}\}, P, r) \mid T) \mid \\ \quad \texttt{b}(C' + code(X, C, \emptyset), H' + \{\texttt{self} : (\{\tilde{x} : \tilde{r}\}, X)\}, \mathbf{0}_T) \mid A, \\ \quad R + \{S : (\alpha, I + \{\texttt{self@b}\})\} \end{array}}$$

Note that no initial thread is provided. The new agent is a simple service provider.

7.2.2.5.2 Method Invocation Method invocations can only be performed locally, and thus, invoking a method in an object located in the address space of some other agent is not possible. The rule simulates a call stack by suspending the current thread, and by creating a new one to execute the body of the method. The environment of the new thread is obtained from the target object's environment modified with the values assigned to the method's parameters. The current thread is then suspended on a new reference (r'), waiting for the result. The variable (x) to hold the result once the thread resumes its execution is also stored.

[INV]

$$\frac{H(\mathsf{self}) = (B', X) \quad B(\tilde{v}) = \tilde{r} \quad C(X) = (\_, M, \_, \_) \quad M(1) = (\tilde{x}, P') \quad r' \notin H}{\begin{array}{c} a(h, C, H, (B, x = 1(\tilde{v})\, P, r) \mid T) \mid A, R \rightarrow \\ a(h, C, H + \{r' : (x, (B, P, r))\}, (B' + \{\tilde{x} : \tilde{r}\}, P', r') \mid T) \mid A, R \end{array}}$$

The conclusion of a method, and the consequent sending of its result, causes the suspended thread to resume its execution in the local pool. The result is bound to the variable defined for the purpose, and the reference on which the thread was waiting is discarded.

[RETURN]
$$\frac{B'(v) = r''}{\begin{array}{c} a(C, H \cdot r' : (x, (B, P, r)), (B', \mathbf{return}\ v, r') \mid T) \mid A, R \rightarrow \\ a(C, H, (B + \{x : r''\}, P, r) \mid T) \mid A, R \end{array}}$$

7.2.2.5.3 Service-Oriented Process Mobility All communication in the SOMAM is driven by code mobility. The code composing the session aimed at a particular service is uploaded to an on-the-fly chosen provider. The procedure is completely transparent to the user.

As in local method invocations, session uploading always launches a new thread in the target provider to execute the corresponding code. The difference lies in the fact that the result slot of the thread $(r''@a)$ is now a reference from the heap of the calling agent.

The process to migrate is abstracted in the set of variables that occur free in its code. At the time of the migration, these variables are bound to values from the current agent. By design we intend to omit all remote references other than services. For that purpose, we create a copy of all these values in the target provider, and bind them to the given set of free variables. The procedure is similar to the one used in remote method invocations with call-by-value semantics. The set of free variables is given by the *freeVars* function (defined as usual).

The target provider is picked from the currently available implementations of the service. We do not define any picking criteria, since we consider it to be an implementation detail.

[MOB]

$$\frac{\begin{array}{c} R(S) = (\alpha, I) \quad \exists\, \mathsf{self}@b \in I : b \neq a \quad \mathit{freeVars}(P) = \tilde{x} \\ \mathit{copySeq}_{ab}(C, H, B(\tilde{x})) = (C'', H'', \tilde{r}) \quad r' \notin H \end{array}}{\begin{array}{l} a(h, C, H, (B, x = S\, P\ \mathbf{in}\ P'\mathbf{error}\ P'', r) \mid T) \mid b(h', C', H', T') \mid A, R \rightarrow \\ \quad a(h, C, H + \{r' : (x, (B, P', r))\}, T) \mid \\ \quad b(h', C' + C'', H' + H'', T' \mid (\{\tilde{x} : \tilde{r}\}, P, r'@a)) \mid A, R \end{array}}$$

If no service implementation can be found, the **error** clause is triggered in the calling agent, and no result is obtained.

$$[\text{MobError}] \quad \frac{R(S) = (\alpha, I) \quad \forall\, b\, \texttt{self@b} \in R(S) \implies b = a}{a(h, C, H, (B, x = S\, P\ \textbf{in}\ P'\ \textbf{error}\ P'', r) \mid T) \mid A, R \rightarrow a(h, C, H, (B, P'', r) \mid T) \mid A, R}$$

The return value from a remote session is an object that must be cloned into the heap of the original agent. Once again to eliminate remote references, the cloning must include the heap and code closure of the value. To allow for the use of rule [RETURN], a new thread is spawned to handle the local placing of the result.

[REMOTERETURN]

$$\frac{copy_{ab}(C, H, B(v)) = (C'', H'', r')}{a(h, C, H, (B, \textbf{return}\ v, r@b) \mid T) \mid b(h', C', H', T') \mid A, R \rightarrow a(h, C, H, T) \mid b(h', C' + C'', H' + H'', (\{x : r'\}, \textbf{return}\ x, r) \mid T') \mid A, R}$$

7.2.2.5.4 Halt the Execution of a Provider The rule simply tags the agent with the • tag and removes the agent from the resolver, to avoid the arrival of new sessions. This causes the provider to become unavailable for the remainder of the network, whist continuing its execution until all remote sessions have concluded.

$$[\text{Halt}] \quad \frac{H(\texttt{self}) = (\_, X) \quad C(X) = (\_, \_, \_, \_, S) \quad R(S) = (\alpha, I)}{\textbf{halt}(a) \mid a(h, C, H, T) \mid A, R \rightarrow a(h, C, H, T)^{\bullet} \mid A, R + \{S : (\alpha, I|_{dom(I) - \{\texttt{self@a}\}})\}}$$

If the event aims at an agent already tagged, it is simply discarded.

$$[\text{HaltDiscard}] \quad \textbf{halt}(a) \mid a(h, C, H, T)^{\bullet} \mid A, R \rightarrow a(h, C, H, T)^{\bullet} \mid A, R$$

A tagged agent terminates its execution when its pool of threads is empty.

$$[\text{EndAgent}] \quad a(h, C, H, \mathbf{0}, \emptyset)^{\bullet} \mid A, R \rightarrow A, R$$

7.2.2.5.5 Program Thread Termination Most of the threads in the SOMAM terminate with the **return** instruction. Exception are the ones that define a program's initial behavior, since they do not return a result. Their life cycle ends when they run out of code to execute.

$$[\text{EndThread}] \quad a(h, C, H, (B, \epsilon, r) \mid T) \mid A, R \rightarrow a(h, C, H, T) \mid A, R$$

7.2.3 Refining the Model

The basic model has some restrictions that prevent the implementation of some classes of problems, namely the ones that require the singling out of a specific service implementation. To overcome this limitation, we associate a set of properties to every service provider. Examples of properties are the version of the implemented service, a logical name for its network location or the company that publishes the service.

This extra information is kept in the resolver with the remainder of the information. An entry must now include the properties for the service,[*] that we will denote by $\tilde{F}$, and the values associated to each provider, denoted by $\tilde{f}$. These must be constant primitive values to be defined by the language and include host identifiers. The definition for the resolver is given by

$$
\begin{aligned}
I &::= \{(\tilde{f}_1, r_1), \ldots, (\tilde{f}_n, r_n)\} \\
R &::= \{S_1 : (\alpha_1, \tilde{F}_1, I_1), \ldots, S_n : (\alpha_n, \tilde{F}_n, I_n)\}
\end{aligned}
$$

The properties to be associated to a provider are declared in its creation, as defined in the updated version of the [NEWPROVIDER] rule below.[†] The host property is mandatory and its instantiation is transparent to the user.

[NEWPROVIDER]

$$
\frac{C(X) = (\tilde{x}, M, \tilde{X}, S) \quad copySeq_{ab}(C, H, B(\tilde{v})) = (C', H', \tilde{r}) \quad R(S) = (\alpha, \text{host } \tilde{F}, I)}{|\tilde{F}| = |\tilde{f}| \quad b \in \text{AgentKey fresh}}
$$

$a(h, C, H, (B, x = \textbf{new } X < \tilde{f} > (\tilde{v}) P, r) \mid T) \mid A, R \rightarrow$
$\quad a(h, C, H, (B + \{x : \texttt{self@b}\}, P, r) \mid T) \mid$
$\quad b(C' + code(X, C, \emptyset), H' + \{\texttt{self} : (\{\tilde{x} : \tilde{r}\}, X)\}, \mathbf{0}_T) \mid A,$
$\quad R + \{S : (\alpha, \text{host } \tilde{F}, I + \{(h \tilde{f}, \texttt{self@b})\})\})$

Rule [ASYNCMOB] defines asynchronous session uploading, and illustrates the use of properties in service discovery. The choice of the target service is restricted by annotating the desired properties $<\tilde{F}' : \tilde{f}'>$. These must be a subset of the ones defined for the service in the resolver. In other words, it is not necessary to assign values for every property. The ones left undefined are interpreted as wildcards.

[ASYNCMOB]

$$
\frac{R(S) = (\alpha, \tilde{F}, I) \quad \exists (\tilde{f}, \texttt{self@b}) \in I : \{\tilde{F} : \tilde{f}\} \supset \{\tilde{F}' : \tilde{f}'\} \wedge b \neq a}{freeVars(P) = \tilde{x} \quad copySeq_{ab}(C, H, B(\tilde{x})) = (C'', H'', \tilde{r})}
$$

$a(h, C, H, (B, S < \tilde{F}' : \tilde{f}' > P \textbf{ in } < x > P'\textbf{error } P'', r) \mid T) \mid b(h', C', H', T') \mid A, R \rightarrow$
$\quad a(h, C, H + \{r' : \{\tilde{F} : \tilde{f}\}\}, (B + \{x : r'\}, P', r) \mid T) \mid$
$\quad b(h', C' + C'', H' + H'', T' \mid (B + \{\tilde{x} : \tilde{v}\}, P, \texttt{null})) \mid A, R$

In order for the continuation (P') to be aware of the properties of the selected provider, it may define a variable (x in the rule) that will be bound to the map holding this information in the heap. The access syntax is $x < F >$, e.g., props $< version >$ for instances props of x and version of F.

---

[*] Whether the properties are an integral part of the service's interface is a topic that is still under discussion. We have chosen to couple both these things, obliging each provider to supply a value for every property defined for the service.

[†] In this rule we adapt to sequences the set-related concept and notation of cardinality.

Another behavior easy to define is negative selection, e.g., obtain a provider whose property value is *not* a given value, *nor* is in a given set of values. For this purpose we define, respectively, the **not** and **notin** keywords.

7.2.4 Programming Examples

We illustrate the model with two small examples. To have a complete computing model, we assume the existence of the usual language constructs, such as branch selection and loops; primitive constants; some datatypes, such as lists; and input/output operations.

The first example illustrates a simple client–server relationship. Code in listing 7.1 exemplifies the interaction with service WebEngine that defines a method query with signature: List query(String).

The migrated session performs the desired query and returns the result. The operation is synchronous, causing the calling thread to wait for the result. Once it is available in lst, the print method of some I/O dedicated object sends it to the standard output.

Listing 7.1: Client code

```
List lst = WebEngine { return query("PDPA book"); } in
    io.print(lst);
error
    io.print("No engine found");
```

The second example shows how service-oriented mobility can be used to perform a task on a particular network, without knowing its composition in advance.

The task at hand is to install a given software package in all the hosts of a network. We begin by defining a service InstallHomeBase that must be running in every host willing to receive software updates. Its purpose is to provide the means for remote code to locally install packages by featuring the following three methods: isSoftwareInstalled to check if the software is already installed; install to install the software and obtain the status of the operation; and errorReport to report an error to the network administrator.

Listing 7.2: Remote software installation

```
List hostList = new List();
String software = "someSoftwarePackage";
while (true) {
  InstallHomeBase<host notin hostList> {
    if (!isSoftwareInstalled(software)) {
      byte[] soft = Repository { return getSoftware(software); }
      in {
        if (!install(soft)) errorReport("could not install service");
      }
      error errorReport("could not access repository");
    }
  }
  in <props> hostList.add(props<host>);
  error break;
}
io.print(hostList + " visited");
```

The code in listing 7.2 begins by finding an implementation of service Install-HomeBase, whose location is not in list hostList. It then uploads a session to perform the installation. If the package is not already installed, a new session is created to fetch the package from a repository, here represented by service Repository with method getSoftware. The package comes in form of a byte array to be handled by the install method. If the installation or the quest for a repository fails, an error is generated by using the errorReport tool.

Note that the software variable occurs free in both the InstallHomeBase and Repository sessions, and thus its instantiation is not bound to either of them.

The location of every InstallHomeBase provider retrieved is saved in the hostList list. When no new provider is found, all of the currently available implementations of InstallHomeBase in the network have been visited, and the work is done.

By means of syntactic sugar we can give a lighter, more expressive and intuitive clothing to this code, as presented in listing 7.3.

Listing 7.3: Remote software installation (v2)

```
String software = "someSoftwarePackage";
forall (InstallHomeBase) {
  if (!isSoftwareInstalled(software)) {
    byte[] soft = Repository { return getSoftware(software); }
    in {
       if (!install(soft)) errorReport("could not install service");
    }
    error errorReport("could not access repository");
  }
}
io.print(hostList + " visited");
```

7.3 Encoding in a Process Calculus

In this section, we present the encoding of the base model into an extension of the LSD π-calculus [23] that incorporates some of the features of the Typed Concurrent Objects (TyCO) process calculus [28]. We propose an encoding that will give us the framework upon which we aim to prove an operational correspondence between the abstract machine and the calculus [19]. This is particularly important as it provides a form of language security often overlooked when defining high-level constructs.

We have chosen this calculus because of its object-based syntax, which provides a good framework to encode SOMAM agents, and its simple model of distribution first proposed in Ref. [29] and further refined in the LSD π-calculus [23]. Another important aspect is the implementation support provided by the compiler and the run-time system featured in the TyCO language release.* This implementation will give us the platform we need for a future implementation. Finally, we considered our own familiarity with the calculus, its associated language and implementation.

---

* http://www.dcc.fc.up.pt/tyco/.

7.3.1 Distributed Typed Concurrent Objects

In this section, we present an extension of the LSD π-calculus that incorporates some of the features of the TyCO calculus. The TyCO calculus is a form of asynchronous π-calculus, featuring objects, asynchronous method invocations, and process definitions as fundamental abstractions. Objects are sets of methods that are placed in channels. A method is selected by sending an asynchronous message targeted at the channel that holds the object. Thus, these messages are in fact method invocations. Process definitions allow for the abstraction of a process on a set of parameters, and enable recursion.

Distributed TyCO (DiTyCO) was introduced in Ref. [29], presenting the guidelines for incorporating distribution and mobility in TyCO. The features introduced were lexical scoping for identifiers, located computations or sites, and code mobility (driven by lexical scope).

A subcalculus of DiTyCO was later selected for a more in-depth study of its semantics, LSDπ. This calculus differs from DiTyCO in that it does not have objects as inputs nor process definitions. The calculus that we use as the target of the encoding includes these constructs from DiTyCO; it extends LSDπ with DiTyCO-like objects; DiTyCO-like process definitions to replace replicated inputs $(a\ ?\ *\ (\tilde{x}) = P)$; variables; and expressions over primitive types. Adding objects to LSDπ does not raise any concerns, since the semantics of the calculus require only minor changes. The same happens with the addition of expressions and primitive values. The substitution of replicated input with process definitions is not that easy. It requires the addition of a new set of rules to handle the migration of definitions, such as the one defined in Ref. [16]. Although the calculus presented here is not exactly the same proposed in Refs. [16,29], since it uses some of the features particular to the LSD π-calculus we will still refer to it as DiTyCO.

7.3.1.1 Syntax

The grammar in Table 7.4 defines the syntax of the DiTyCO networks. The required syntactic categories are presented in Table 7.5. Some syntactic restrictions are assumed for processes, namely the parameters $\tilde{x}$ of a process definition $U(\tilde{x}) = P$, or in a method definition $l(\tilde{x}) = P$, are pairwise distinct; the labels l_i are pairwise distinct in a method collection $\{l_1(\tilde{x}_1) = P_1 \cdots l_n(\tilde{x}_n) = P_n\}$; and the definitions U_i in a process $U_1(\tilde{x}_1) = P_1 \cdots U_n(\tilde{x}_n) = P_n$ are pairwise distinct.

7.3.1.2 Semantics

The operational semantics of our calculus is based on the one for LSDπ described in Ref. [23]. The only exceptions are the rules regarding process definitions which are inherited from DiTyCO (thus inferred from the TyCO calculus). Here we only give a general overview of the calculus, presenting the rules more relevant for its comprehension. The base for a more in-depth study can be found in the bibliography. We begin with a general description of the structural congruence rules:

TABLE 7.4: Syntax of DiTyCO networks.

N, O	$::=$	$\mathbf{0}$	Terminated network
	$\mid$	$N \mid O$	Concurrent composition
	$\mid$	$(\nu\, s)\ N$	New located site
	$\mid$	$(\nu\, g)\ N$	New located channel
	$\mid$	$\mathbf{def}\ X_1@s(\tilde{x}_1) = P_1 \cdots$	
		$\quad X_n@s(\tilde{x}_n) = P_n\ \mathbf{in}\ N$	Located recursive definition
	$\mid$	$s[P]$	Site running a process
P, Q	$::=$	$\mathbf{0}$	Terminated process
	$\mid$	$P \mid Q$	Concurrent composition
	$\mid$	$(\nu\, s)\ P$	New located site
	$\mid$	$(\nu\, n)\ P$	New channel
	$\mid$	$u\, !\, l(\tilde{v})$	Asynchronous message
	$\mid$	$u\, ?\, \{l_1(\tilde{x}_1) = P_1 \cdots l_n(\tilde{x}_n) = P_n\}$	Object
	$\mid$	$\mathbf{def}\ U_1(\tilde{x}_1) = P_1 \cdots U_n(\tilde{x}_n) = P_n\ \mathbf{in}\ P$	Recursive definition
	$\mid$	$U(\tilde{v})$	Instantiation

TABLE 7.5: Syntactic categories of DiTyCO networks.

r, s	$\in$	TySite	Sites
a, b	$\in$	TySChannel	Simple channels
$a@s$	$\in$	TySChannel@TySite	Located channels
u, v	$\in$	TyValue = TySChannel $\cup$ TySChannel@TySite	Values
n, m	$\in$	TyName = TyValue $\cup$ TySite	Names
g	$\in$	TyGlobal = TySChannel@TySite $\cup$ TySite	Global names
x, y, x	$\in$	TyVar	Variables
X, Y	$\in$	TySDef	Simple definitions
$X@s$	$\in$	TySDef@TySite	Located definitions
U, V	$\in$	TyDef = TySDef $\cup$ TySDef@TySite	Definitions
l	$\in$	TyLabel	Labels
M	$\in$	TyMeth = TyLabel $\mapsto$ TySChannel $\times$ TyProc	Method collections
P, Q	$\in$	TyProc	Processes
N, O	$\in$	TyNet	Network

[S-Scos1]	$(\nu\, a@s)\, s[P] \equiv s[(\nu\, a@s)\, P)]$	if $a \notin fn(P)$
[S-Scos2]	$(\nu\, a@s)\, s[P] \equiv s[(\nu\, a)\, P)]$	if $a@s \notin fn(P)$
[S-Migo]	$s[a@s\, !\, l(\tilde{v})] \equiv s[a\, !\, l(\tilde{v})]$	
[S-Migi]	$s[a@s\, ?\, M] \equiv s[a\, ?\, M]$	
[S-Ncomm]	$\mathbf{def}\ U(\tilde{x}) = P\ \mathbf{in}$	
	$\qquad (\nu\, n)\, E \equiv (\nu\, n)\, \mathbf{def}\ U(\tilde{x}) = P\ \mathbf{in}\ E$	if $n \notin fn(P)$
[S-Dcos]	$\mathbf{def}\ X@s(\tilde{x}) = P\ \mathbf{in}$	
	$\qquad s[Q] \equiv s[\mathbf{def}\ X(\tilde{x}) = P\ \mathbf{in}\ Q]$	
[S-Inst]	$s[X@s(\tilde{v})] \equiv s[X(\tilde{v})]$	

Rules [S-Scos1] and [S-Scos2] allow for the scope of names to be local to a process. Function $fn()$ returns the set of names that occur free in a process. Rules

[S-MIGO] and [S-MIGI] indicate that, regarding communication, channels bound to a site can be viewed as simple channels within that site.

Rules regarding process definitions are inherited from DiTyCO (thus inferred from the TyCO calculus). Here we present [S-NCOMM] for the commutation of definitions with name creation, and rules [S-DCOS] and [S-INST] that apply the same lexical scoping concepts over definitions that we find in rules [S-SCOS] and [S-MIGO].

In rules [S-MIGO], [S-MIGI], and [S-INST] simple names are implicitly located at the site where the process is executing. When these names are sent over the network their lexical scope must be preserved. For that purpose we define a function σ to handle name translation over intersite communication:

$$\sigma_{rs}(a) \stackrel{\text{def}}{=} a@r \qquad\qquad \sigma_{rs}(X) \stackrel{\text{def}}{=} X@r$$
$$\sigma_{rs}(a@s) \stackrel{\text{def}}{=} a \qquad\qquad \sigma_{rs}(X@s) \stackrel{\text{def}}{=} X$$
$$\sigma_{rs}(a@s') \stackrel{\text{def}}{=} a@s' \quad s \neq s' \qquad \sigma_{rs}(X@s') \stackrel{\text{def}}{=} X@s' \quad s \neq s'$$

We may now define the reduction rules over networks and processes. Reduction occurs by communication between objects and messages and by instantiation of definitions, and it is always local. $M\sigma_{rs}$, $P\sigma_{rs}$, and $\tilde{v}\sigma_{rs}$ denote the application of the σ function to all the free channels and definitions in M, P, and $\tilde{v}$.

[R-COMM] $\qquad a\,?\,\{\ldots l(\tilde{x}) = P \ldots\} \mid a\,!\,l(\tilde{v}) \to P\{\tilde{v}/\tilde{x}\}$

[R-MIGO] $\qquad\qquad\qquad\qquad r[a@s\,!\,l(\tilde{v})] \to s[a\,!\,l(\tilde{v}\sigma_{rs})] \qquad r \neq s$

[R-MIGI] $\qquad\qquad\qquad\qquad r[a@s\,?\,M] \to s[a\,?\,M\sigma_{rs}] \qquad r \neq s$

[R-SITE] $\qquad\qquad\qquad\qquad \dfrac{P \to Q}{s[P] \to s[Q]}$

[R-INST] $\quad$ **def** $X_1@s(\tilde{x}_1) = P_1 \,\cdots\, X_n@s(\tilde{x}_n) = P_n$ **in** $s[X_i(\tilde{v})] \to$
$\qquad\qquad$ **def** $X_1@s(\tilde{x}_1) = P_1 \,\cdots\, X_n@s(\tilde{x}_n) = P_n$ **in** $s[P_i\{\tilde{v}/\tilde{x}_i\}]$

[R-FETCH] $\quad$ **def** $X_1@r(\tilde{x}_1) = P_1 \,\cdots\, X_n@r(\tilde{x}_n) = P_n$ **in** $s[X_i@r(\tilde{v})] \to$
$\qquad\qquad$ **def** $X_1@r(\tilde{x}_1) = P_1 \,\cdots\, X_n@r(\tilde{x}_n) = P_n$ **in**
$\qquad\qquad s[\textbf{def } X_1(\tilde{x}_1) = P_1\sigma_{rs} \,\cdots\, X_n(\tilde{x}_n) = P_n\sigma_{rs} \textbf{ in } X_i(\tilde{v})] \quad r \neq s$

Rule [R-COMM] defines reduction through communication by selecting the method with label l of the object placed at a. The resulting process is the body of the method whose parameters are replaced by the given values. Rule [R-INST] defines another type of reduction, instantiation, that creates an instance of the X process definition. Similarly with the previous rules, the resulting process is the body of the process definition whose parameters are replaced by the arguments of the instantiation.

The migration over networks is defined based on the principle that all reduction occurs locally within sites. Rule [R-SITE] forces local reduction only. It states that sites reduce only if the process running within their boundaries reduces. Rules [R-MIGO] and [R-MIGI] allow the shipment of messages and objects targeted at a channel to the site where it is located. The behavior for process definitions is different. Rule [R-FETCH] extends the rule for the download of definitions found in Ref. [17], rather than uploading the instantiation to the site of origin. The extension widens the

TABLE 7.6: Syntax of expressions.

e	::=	$e \ bop \ e$ | $uop \ e$ | (e) | c	Expressions
bop	::=	**and** | **or** | $+$ | $-$ | $*$	Binary operations
	|	$/$ | $\%$ | $\char94$ | $==$ | $/=$	
	|	$<$ | $>$ | $<=$ | $>=$	
uop	::=	$-$ | **not**	Unary operations

scope of the rule to download all the code in the original definition, not just X_i, to the site performing the instantiation. We chose to download the entire group of definitions to minimize code references to the site from where the code was downloaded.

To enable us to write a simpler and cleaner encoding, we add expressions (e) and primitive values (c) to this calculus, as well as conditional execution (the **if** instruction). Expressions are given by the grammar in Table 7.6, and the syntax of the **if** instruction is **if** e **then** P **else** Q.

The static semantics of the calculus can be derived from the type-system for LSDπ [23], with the rules for objects, definitions, and expressions taken from the type-system of TyCO [28]. Both these type-systems have been formally studied, and provide type-safety, such as the absence of protocol errors of well-typed programs, and the formally proved substitution and subject-reduction lemmas. No study has been made regarding definitions in a distributed context.

7.3.2 Encoding

In this section, we present the formal encoding of the SOMAM in the form of a map that transforms a state of the abstract machine into a DiTyCO network. Our encoding uses a top-down approach from the point of view of the structure of the abstract machine. Therefore, we begin by presenting the encoding for the whole network, and move down to its components.

The encoding of the network consists of encoding the pool of providers currently in execution and the resolver. We define a map $[\![\ ]\!]_\varphi$ to encode a state (A, R) into a DiTyCO network. The map is defined for a set of agents with keys $a_1, \ldots, a_m$ in A and

$$R = \{S_1 : \{a_i \mid i \in [1, j] \ j < m\}, \ldots, S_n : \{a_i \mid i \in [1, k] \ k < m\}\}$$

Notice that the main encoding map will resort to other maps of appropriate domains, to provide the encoding for the components of the state. For simplicity we use the same notation $[\![\ ]\!]_\varphi$ for all these maps.

φ is the set of environment variables required to pass top-level information to the encoding of the lower-level components of the abstract machine. The environment variables are

- REF is a map that, given a heap reference or a service, returns the channel holding its encoding. To ease the reading we denote $\varphi(\text{REF}(r))$ by $\text{ch}(r)$ and $\varphi(\text{REF}(S))$ by $\text{ch}(S)$.

- RET is the channel to handle the reception of the result of the thread currently being encoded.

- KEY is the string representation of the key of the provider currently being encoded.

Notation: The DiTyCO variable that encodes a SOMAM variable x is denoted by x^x. The DiTyCO definition variable that encodes a SOMAM class X is denoted by X^X. The DiTyCO label that encodes a SOMAM method label l is denoted by l^1. These notations are extended to sequences of both variables and classes.

7.3.2.1 Network

To encode the network we need to define a set of DiTyCO sites plus a set of channels and definitions at the network level, since their scope encompasses several sites. The ServiceRegistry definition abstracts an object target of the encoding of an entry of the resolver. Each entry must be encoded in a different channel to solve type conflicts that may appear between different service registries. Maps are implemented in DiTyCO as lists of pairs (or equivalent). Since different services have distinct interfaces, type conflicts may arise from the fact that a list is only typeable in DiTyCO if each of its elements has the same type. This requirement makes the encoding of R as a map implementation in DiTyCO impossible, since, in general, services have different types.

The object abstracted in ServiceRegistry is parameterized by the type and the list of providers of the service, and features methods register, remove, and find. These allow, respectively, for the registry, removing, and retrieving of a provider, given its key (key). The key is a constant used to enable provider comparison, since DiTyCO does not allow direct comparison of channels. This mechanism is used to ensure that a process does not migrate to its own site, and to identify the provider to remove from the set in the remove method. Unique keys are generated from agent identifiers by an abstract function *agentKey*(), that can be easily implemented by resorting to strings.

When encoding a SOMAM network the scope of some channels have to encompass several of the generated sites, and thus must be created at network level. These are the channels holding service implementations, that must be visible in their site of origin and in the site hosting the resolver, and the ones used to encode process mobility. The execution of a remotely uploaded process produces a result that must be sent to the original agent. This behavior is implemented by passing a channel whose scope must encompass both the sites in the operation (the provider and the requirer).

On the other hand, channels used to encode local heap values only encompass most of the agent's components. We could create them at the agent's level, but, for the sake of simplicity, we opt to also create them here, at the network level.* This

---

* The scope of only locally visible channels can be narrowed to the respective site by the means of DiTyCO structural congruence.

concentrates the creation of all heap encoding channels in function *netChs* that, given the network, inspects the heap of each agent and the provider to generate all the necessary channels:

$$netChs : \text{Network} \times \text{TyNet} \mapsto \text{TyNet}$$

$$netChs(\texttt{a}(\texttt{h, C, H} \cdot (\texttt{r} : \_), \texttt{T}) \mid \texttt{A, R, } N) \ = \ (\nu \ \texttt{ch}(\texttt{r})@s^{\texttt{a}})$$
$$(netChs(\texttt{a}(\texttt{h, C, H, T}) \mid \texttt{A, R, } N))$$
$$netChs(\texttt{a}(\texttt{h, C, } \emptyset, \texttt{T}) \mid \texttt{A, R, } N) \ = \ netChs(\texttt{A, R, } N)$$
$$netChs(\mathbf{0}_{\texttt{A}}, \texttt{R} \cdot \texttt{S} : (\alpha, \texttt{I}), N) \ = \ (\nu \ \texttt{ch}(\texttt{S})@s^{\texttt{R}}) \ (netChs(\mathbf{0}_{\texttt{A}}, \texttt{R, } N))$$
$$netChs(\mathbf{0}_{\texttt{A}}, \emptyset, N) \ = \ N$$

The heap representation of a given object or provider mapped by r in agent $\texttt{a}$ will be placed in channel $\texttt{ch}(\texttt{r})@s^{\texttt{a}}$. *netChs* also creates the channels required to encode the resolver. For each registered service $\texttt{S}$, it creates channel $\texttt{ch}(\texttt{S})@s^{\texttt{R}}$. We thus have the encoding of a network given by

$$[\![\texttt{A, R}]\!]_{\varphi} \ \overset{\text{def}}{=} \ (\nu \ s^{\texttt{R}}) \ (\nu \ s^{\texttt{a}_1}) \ \dots \ (\nu \ s^{\texttt{a}_n})$$
$$\textbf{def } \mathsf{ServiceRegistry}@s^{\texttt{R}}(\mathsf{self \ type \ providerList}) = \mathsf{self} \ ? \ \{$$
$$\mathsf{register(key \ provider)} = \dots$$
$$\mathsf{remove(key)} = \dots$$
$$\mathsf{find(key \ replyTo)} = \dots$$
$$\}$$
$$\mathsf{EmptyStringMap}@s^{\texttt{R}}(\mathsf{self}) = \mathsf{self} \ ? \ \{\dots\}$$
$$\mathsf{StringMap}@s^{\texttt{R}}(\mathsf{self}) = \mathsf{self} \ ? \ \{\dots\}$$
$$\textbf{in } (netChs(\texttt{A}, s^{\texttt{R}}[[\![\texttt{R}]\!]_{\varphi}] \mid [\![\texttt{A}]\!]_{\varphi}))$$

where φ contains in REF the channels associated to all the heap references and services collected in function *netChs*. The target calculus is not location-aware and thus the location of a site is not expressible.

7.3.2.2 Resolver

As explained above, each entry of the resolver map is encoded in a different channel. This channel $(\texttt{ch}(\texttt{S})@s^{\texttt{R}})$ holds the list of providers of a given service $\texttt{S}$. In other words, it holds the channel's target of the encoding of each service.

Each element of the map is thus registered by creating an instance of the ServiceRegistry definition placed at $\texttt{ch}(\texttt{S})$. This instance keeps a map that, for each provider, associates the agent's key to the service provided. Definitions EmptyStringMap and StringMap implement this map from strings to channels holding providers of a given service. The *toString* function returns the string representation of the argument. We assume that types are stored in some string representation.

$$\llbracket R \cdot S : (\alpha, \{\texttt{self@}a_1, \ldots, \texttt{self@}a_n\}) \rrbracket_\varphi \overset{\text{def}}{=}$$

$$(\nu\, a_0)\, (\nu\, a_1)\, \ldots (\nu\, a_n)\, ($$

  $\mathsf{EmptyStringMap}(a_0)\ |$

  $\mathsf{StringMap}(a_1\ \mathit{agentKey}(\texttt{a}_1)\ \mathsf{ch}(\texttt{self})@s^{\texttt{a}_1}\ a_0)\ |\ \cdots\ |$

  $\mathsf{StringMap}(a_n\ \mathit{agentKey}(\texttt{a}_n)\ \mathsf{ch}(\texttt{self})@s^{\texttt{a}_n}\ a_{n-1})\ |$

  $\mathsf{ServiceRegistry}(\mathsf{ch}(\texttt{S})\ \mathit{toString}(\alpha)\ a_n)$

  $)\ |\ \llbracket R \rrbracket_\varphi$

$$\llbracket \emptyset \rrbracket_\varphi \overset{\text{def}}{=} \mathbf{0}$$

7.3.2.3 Agents

An agent with key $\texttt{a}$ is encoded into a site $s^{\texttt{a}}$. Within its boundaries runs a process composed of a set of built-in definitions and channels defined at top level (gathered in function $\mathsf{localChs}$) and, in their scope, the encoding of the agent's components:

$$\llbracket \texttt{a}(\texttt{h}, \texttt{C}, \texttt{H}, \texttt{T})^\circ\ |\ \texttt{A} \rrbracket_\varphi \overset{\text{def}}{=} s^{\texttt{a}}[\mathsf{localChs}(\textbf{false}, \textbf{def}\ \llbracket \texttt{C} \rrbracket_{\varphi'}\ \textbf{in}\ (\llbracket \texttt{H} \rrbracket_{\varphi'}\ |\ \llbracket \texttt{T} \rrbracket_{\varphi'}))]\ |\ \llbracket \texttt{A} \rrbracket_\varphi$$

$$\llbracket \texttt{a}(\texttt{h}, \texttt{C}, \texttt{H}, \texttt{T})^\bullet\ |\ \texttt{A} \rrbracket_\varphi \overset{\text{def}}{=} s^{\texttt{a}}[\mathsf{localChs}(\textbf{true}, \textbf{def}\ \llbracket \texttt{C} \rrbracket_{\varphi'}\ \textbf{in}\ (\llbracket \texttt{H} \rrbracket_{\varphi'}\ |\ \llbracket \texttt{T} \rrbracket_{\varphi'}))]\ |\ \llbracket \texttt{A} \rrbracket_\varphi$$

$$\llbracket \textbf{halt}(\texttt{a})\ |\ \texttt{A} \rrbracket_\varphi \overset{\text{def}}{=} \mathsf{ch}(\texttt{self})@s^{\texttt{a}}\ !\ \mathsf{halt}()\ |\ \llbracket \texttt{A} \rrbracket_\varphi$$

$$\llbracket \mathbf{0}_{\texttt{A}} \rrbracket_\varphi \overset{\text{def}}{=} \mathbf{0}$$

where $\varphi' = \varphi + \{\textsc{key} : \mathit{agentKey}(\texttt{a})\}$ adds the key of the agent to the environment of the encoding of its components. Halting events are translated into invocations of the halt method on the target provider.

Function $\mathsf{localChs}$, defined below, receives a boolean value that indicates whether the agent is tagged for conclusion or not (ht), and the DiTyCO process with the encoding of the components (P). The function resorts to the $\mathsf{AgentManager}$ definition, and its instance amg, to keep track of the contents in the agent's heap, the number of threads currently in execution, and whether the agent has been tagged for conclusion. The parameters of the definition are the number of currently executing remote sessions (n); a channel holding a list of processes that perform the discarding of each element in the heap (objs); and a boolean value indicating whether the agent is tagged for conclusion (halt). The objs parameter stores, for each heap element, a process that removes the element itself from the heap. This requires some sort of list implementation provided by definitions $\mathsf{EmptyList}$ and List. Besides the usual list operations, these allow for the triggering of the stored processes, by means of the $\mathsf{trigger}$ method.

$localChs(ht, P) \overset{\text{def}}{=}$ **def** EmptyList(self) = self ? {...}

                List(self) = self ? {...}

                AgentManager(self n objs halt) =

                self ? {

                  addObj(obj) =

                     (ν replyTo) (

                     objs ! put(obj replyTo) |

                     replyTo ? (list) = AgentManager(self n list halt))

                  addThread(replyTo) =

                     AgentManager(self n+1 objs halt) | replyTo ! ()

                  return(replyTo) =

                     if n == 1 **and** halt == **true then** objs ! trigger()

                     else (AgentManager(self n-1 objs halt) | replyTo ! ())

                  halt() =

                     AgentManager(self n objs **true**)

                }

              in (ν amg) (ν objs)

              (EmptyList(objs) | AgentManager(amg 0 objs ht) | P)

The set of methods featured in amg are addObj that adds a process to discard an heap element; addThread, invoked whenever a new remote session is launched, that increases the number of accounted threads; halt that tags the agent by setting the halt flag to **true**; and return, invoked whenever a remote session terminates, that tests the two conditions necessary to terminate the agent's execution: *no threads in execution* and *the agent is tagged*. If these conditions are satisfied, the agent's heap is discarded by triggering the processes stored in objs. With no heap and no threads, all the agent's code reduces to **0**, terminating its execution.

7.3.2.4 Code Repository

Classes for both providers and local objects are encoded in dedicated definitions parameterized by the attributes of the given class, and the built-in self parameter, the channel where the object is placed. These are the only names that occur free in the body of the definition for local objects. In order to be able to move code from one site to another without leaving remote references, we want the process definitions associated to SOMAM classes to be self-contained. In other words, not to have any free names.[*]

---

[*] When a definition is downloaded, the free variables it contains are transformed into network references, bound to their site of origin. See the definition of the σ translation function in DiTyCO.

$$[\![C \cdot X : (\tilde{x}, M, \tilde{X}, \_)]\!]_\varphi \overset{\text{def}}{=} X^X(\text{self } \widetilde{x^X}) = \text{self} \ ! \ \{$$

$$[\![M, X^X, \text{self } \widetilde{x^X}]\!]$$

$$\text{discard}() = \mathbf{0}$$

$$\text{clone}(\text{buddy replyTo}) =$$

$$X^X(\text{self } \widetilde{x^X}) \ |$$

$$\text{buddy} \ ! \ () \ |$$

$$\text{buddy} \ ? \ () = (v \ a_1) \ \ldots (v \ a_n) \ ($$

$$x^{X_1} \ ! \ \text{clone}(\text{buddy } a_1) \ |$$

$$a_1 \ ? \ (y_1) = (x^{X_2} \ ! \ \text{clone}(\text{buddy } a_2) \ |$$

$$\ldots$$

$$a_{n-1} \ ? \ (y_{n-1}) = (x^{X_n} \ ! \ \text{clone}(\text{buddy } a_n) \ |$$

$$a_n \ ? \ (y_n) = (v \ \text{ref}) \ ($$

$$X^X(\text{ref } y_1 \ \ldots \ y_n) \ |$$

$$\text{replyTo} \ ! \ (\text{ref}) \ )$$

$$) \ \ldots)$$

$$)$$

$$\}$$

$$[\![C]\!]_\varphi$$

The self object contains the encoding of the methods of the class (M), plus methods discard and clone. The purpose of the first is to remove the object from the self channel, thus eliminating it from the heap. The second returns a clone of the object, located at a remote site. The cloning is done by placing the creation of a new instance of the definition in a channel from the target site (buddy) that is received as argument. The semantics of remote instantiations is given by rule [R-FETCH] of the calculus, which allows the downloading of sets of definitions, and thus, can be used to download all the code required by the X^X definition. This provides the means for the target site to automatically hold all the code required by the new instance.

To instantiate the definition we need to provide values for its self parameter and for the ones associated with the attributes of the class. All these are created under the scope of buddy, thus bound to the remote site. The value for self is a new channel, while the values for the attributes of the class are obtained by recursively cloning the attributes of the current instance. This process creates a closure of all the code required by the class.

Classes related to providers require two extra parameters: key, to hold the agent's key, three extra methods, halt, getRemoteChannels, and getKey, and the modification of the implementation of method clone, since providers are not cloned. The result of the operation is the provider itself.

$$[\![C \cdot X : (\tilde{x}, M, \tilde{X}, S)]\!]_\varphi \overset{\text{def}}{=} X^X(\text{self key } \widetilde{x^x}) = \text{self ? } \{$$

$$\cdots$$

$$\text{clone}(\text{buddy replyTo}) =$$
$$X^X(\text{self key } \widetilde{x^x}) \mid \text{replyTo ! (self)}$$
$$\text{halt}() =$$
$$\text{ch}(S)@s^R \text{ ! remove(key)} \mid \text{amg ! halt}()$$
$$\text{getRemoteChannels}(\text{replyTo}) =$$
$$X^X(\text{self key } \widetilde{x^x}) \mid (\nu \text{ buddy}) \text{ replyTo ! (buddy amg)}$$
$$\text{getKey}(\text{replyTo}) =$$
$$X^X(\text{self key } \widetilde{x^x}) \mid \text{replyTo ! (key)}$$
$$\}$$
$$[\![C]\!]_\varphi$$

Method halt removes the agent from the resolver and invokes the method with the same name in amg to place the tag. The getRemoteChannels method provides the remote (buddy) channel required by the clone method, plus the local manager needed to update the number of running threads on session uploading. getKey returns the agent's key.

The encoding of the methods of the class requires the identifier and the parameters of the DiTyCO definition associated to the class currently being encoded. The last two are required to generate the recursive call to the definition that simulates the behavior of a persistent object. Besides the recursive call, the resulting code contains the encoding of the body of the method. The environment is updated with the variable to handle the method's return value.

$$[\![M \cdot 1 : (\tilde{x}, P), X, \tilde{y}]\!]_\varphi \overset{\text{def}}{=} [\![\{1 : (\tilde{x}, P)\}, X, \tilde{y}]\!]_\varphi \, [\![M, X, \tilde{y}]\!]_\varphi$$

$$[\![\{1 : (\tilde{x}, P)\}, X, \tilde{y}]\!]_\varphi \overset{\text{def}}{=} l^1(\widetilde{x^x} \text{ replyTo}) = X(\tilde{y}) \mid [\![(\emptyset, P, \texttt{null})]\!]_{\varphi = \varphi + \{\text{RET:replyTo}\}}$$

7.3.2.5 Heap

The heap contains two kinds of elements: references for objects and providers (of the form $(r : (B, X))$), and references holding a thread waiting for the computation of a result (of the form $(r : (x : (B, P, r')))$). The first are encoded in a simple instantiation in $\text{ch}(r)$ of the definition associated to the given class, plus the registry in the local agent manager of the process to discard the value from the heap. The values for the instantiation of the definition are the channels holding the attributes of the instance. Providers require a value for the extra key parameter, retrieved from the environment.

References waiting for a result are encoded in an instance of dedicated ResultHandler definition. The object abstracted contains method getRemote Channels (as the encoding for classes) plus method handleResult to restart the execution of the thread suspended on the result. In contrast to the other heap elements, we define a ResultHandler definition for each reference holding a suspended thread. This is required due to the fact that the type of the object abstracted in the definition is bound to the type of the result to handle. Thus, a single ResultHandler definition would force every result to be of the same type.

$$[\![\mathtt{H} \cdot \mathtt{self} : (\{\tilde{\mathtt{x}} : \tilde{\mathtt{r}}\}, \mathtt{X})]\!]_\varphi \overset{\text{def}}{=} X^X(\mathsf{ch}(\mathtt{r})\ \varphi(\text{KEY})\ \mathsf{ch}(\tilde{\mathtt{r}}))\ |$$
$$(\nu\, a)(a\,?()=\mathsf{ch}(\mathtt{r})\,!\,\mathsf{discard}()|\mathsf{amg}\,!\,\mathsf{addObj}(a))\ |$$
$$[\![\mathtt{H}]\!]_\varphi$$

$$[\![\mathtt{H} \cdot \mathtt{r} : (\{\tilde{\mathtt{x}} : \tilde{\mathtt{r}}\}, \mathtt{X})]\!]_\varphi \overset{\text{def}}{=} X^X(\mathsf{ch}(\mathtt{r})\ \mathsf{ch}(\tilde{\mathtt{r}}))\ |$$
$$(\nu\, a)(a\,?()=\mathsf{ch}(\mathtt{r})\,!\,\mathsf{discard}()|\mathsf{amg}\,!\,\mathsf{addObj}(a))\ |$$
$$[\![\mathtt{H}]\!]_\varphi$$

$$[\![\mathtt{H} \cdot \mathtt{r} : (\mathtt{x}, (\mathtt{B}, \mathtt{P}, \mathtt{r}'))]\!]_\varphi \overset{\text{def}}{=} \textbf{def } \mathsf{ResultHandler}(\mathsf{self}) = \mathsf{self}\,?\,\{$$
$$\mathsf{getRemoteChannels}(\mathsf{replyTo}) =$$
$$\mathsf{ResultHandler}(\mathsf{self})\ |$$
$$(\nu\, \mathsf{buddy})\,\mathsf{replyTo}\,!\,(\mathsf{buddy}\ \mathsf{amg})$$
$$\mathsf{handleResult}(x^X) =$$
$$[\![(\mathtt{B}, \mathtt{P}, \mathtt{r}')]\!]_\varphi$$
$$\}\ \textbf{in } \mathsf{ResultHandler}(\mathsf{ch}(\mathtt{r}))\ |\ [\![\mathtt{H}]\!]_\varphi$$

$$[\![\varnothing]\!]_\varphi \overset{\text{def}}{=} 0$$

7.3.2.6 Pool of Threads

The translation of a pool of running threads consists of translating each thread individually, and composing in parallel the resulting DiTyCO processes. We define a rule for threads that place results in the local heap, extracting the channel hosting the target reference, and a rule for threads that output a result to a remote reference. These require the result to be placed in the heap of the original agent, which involves the creation of a clone of the value at the target heap. In the environment of the thread running the session, the channel to handle the result (a) is in charge of cloning and forwarding the result to the calling agent. The r reference holding the suspended thread is only known to the thread that must locally place the result.

$$[\![(\mathtt{B}, \mathtt{P}, \mathtt{r}) \mid \mathtt{T}]\!]_\varphi \overset{\mathrm{def}}{=} [\![(\mathtt{B}, \mathtt{P}, \mathtt{null})]\!]_{\varphi=\varphi+\{\textsc{ret}:\mathsf{ch}(\mathtt{r})\}} \mid [\![\mathtt{T}]\!]_\varphi$$

$$[\![(\mathtt{B}, \mathtt{P}, \mathtt{r@b}) \mid \mathtt{T}]\!]_\varphi \overset{\mathrm{def}}{=} [\![(\mathtt{B}, \mathtt{P}, \mathtt{null})]\!]_{\varphi=\varphi+\{\textsc{ret}:\mathsf{ch}(\mathtt{r@s^b})\}} \mid [\![\mathtt{T}]\!]_\varphi$$

$$[\![(\mathtt{B}, \mathtt{P}, \mathtt{null})]\!]_{\varphi \cdot \textsc{ret}:\mathsf{ch}(\mathtt{r})@s^b} \overset{\mathrm{def}}{=} (\nu\, a)\, ($$

$$\qquad (\nu\, b)\, ($$

$$\qquad\qquad \mathsf{amg}\,!\,\mathsf{addThread}(b) \mid$$

$$\qquad\qquad b\,?\,() = [\![(\mathtt{B}, \mathtt{P}, \mathtt{null})]\!]_{\varphi=\varphi+\{\textsc{ret}:a\}}$$

$$\qquad) \mid$$

$$\qquad a\,?\,(x) = (\nu\, b)\, ($$

$$\qquad\qquad \mathsf{amg}\,!\,\mathsf{return}(b) \mid$$

$$\qquad\qquad b\,?\,() = (\nu\, a)\, ($$

$$\qquad\qquad\qquad \mathsf{ch}(\mathtt{r})@s^b\,!\,\mathsf{getRemoteChannels}(a) \mid$$

$$\qquad\qquad\qquad a\,?\,(\mathsf{buddy}\ \mathsf{amg}) =$$

$$\qquad\qquad\qquad\qquad \mathsf{buddy}\,!\,() \mid$$

$$\qquad\qquad\qquad\qquad \mathsf{buddy}\,?\,() =$$

$$\qquad\qquad\qquad\qquad\qquad (\nu\, b)\, ($$

$$\qquad\qquad\qquad\qquad\qquad\qquad x\,!\,\mathsf{clone}(\mathsf{buddy}\ b) \mid$$

$$\qquad\qquad\qquad\qquad\qquad\qquad b\,?\,(x^x) =$$

$$\qquad\qquad\qquad\qquad\qquad\qquad\qquad [\![(\emptyset, \mathbf{return}\ x, \mathtt{null})]\!]_{\varphi=\varphi+\{\textsc{ret}:\mathsf{ch}(\mathtt{r})\}})$$

$$\qquad\qquad)))$$

$$[\![\mathbf{0}]\!]_\varphi \overset{\mathrm{def}}{=} \mathbf{0}$$

Session uploading is the sole operation that modifies the number of threads in an agent, since local invocations simply replace one thread by another. Hence, the execution must be delimited by the updating of the local thread accounting, i.e., by local agent manager interaction.

The third rule is a simple transformation that we need to operate in order to apply the previous rule in the encoding of session uploading.

To allow for SOMAM threads to be constructed from DiTyCO processes, as is required by method bodies, we define the following rule:

$$[\![(\emptyset, \mathtt{P}, \mathtt{null})]\!]_\varphi \{\widetilde{x^x}/\mathsf{ch}(\tilde{r})\} \overset{\mathrm{def}}{=} [\![(\{\tilde{x} : \tilde{r}\}, \mathtt{P}, \mathtt{null})]\!]_\varphi$$

7.3.2.7 Program Instructions

The next step* is to present the encoding for each of the SOMAM instructions. Here we assume that all references (or associated channels in the environment variable RET) defined to hold the result of the thread are local to the agent (site).

7.3.2.7.1 Creation of a New Provider The encoding of the creation of a new provider (b) to be placed in a reference self, requires (1) a new site for the provider identified by a fresh site identifier s^b, which must be located at the same host of the site for the current agent; (2) a new channel $(ch(self)@s^b)$ where the heap representation of the provider will be placed; (3) a process to perform the registry of the new service implementation in the resolver; (4) replacing all the occurrences of x^x in the continuation of the current agent by the channel holding the new provider (the agent that creates the new provider keeps a binding for it); and (5) a process to launch the execution of the new agent. This process creates the new instance of the provider's class. For that it must also create clones of the values given to its constructor. The code required by these attributes is automatically downloaded by the instantiation of the definition (see rule [R-FETCH] of the DiTyCO calculus).

The encoding of the creation of a new provider b instance of class X with entry $C(X) = (\tilde{x}, M, \tilde{X}, S)$ in the repository and $B(\tilde{v}) = \tilde{r}$ is

$$
\begin{aligned}
&[\![(B, x = \textbf{new } X(v_1 \ldots v_n) \, P, \text{null})]\!]_{\varphi \cdot \text{RET:ch}(r)} \overset{\text{def}}{=} \\
&(\nu \, s^b) \, (\nu \, ch(self)@s^b) \, (\\
&\quad [\![(B + \{x : self@b\}, P, \text{null})]\!]_{\varphi \cdot \text{RET:ch}(r)} \mid \\
&\quad ch(S)@s^R \, ! \, \text{register}(ch(self)@s^b \, agentKey(b)) \mid \\
&\quad (\nu \, a@s^b) \, (\\
&\qquad a@s^b \, ! \, () \mid \\
&\qquad a@s^b \, ? \, () = \\
&\qquad\quad \text{localChs}(\textbf{false}, \\
&\qquad\qquad \textbf{def } [\![code(X, C, \emptyset)]\!]_{\varphi + \{\text{KEY}:agentKey(b)\}} \, \textbf{in} \, (\nu \, a_1) \ldots (\nu \, a_n) \, (\\
&\qquad\qquad\quad ch(r_1) \, ! \, \text{clone}(a@s^b \, a_1) \mid \\
&\qquad\qquad\quad a_1 \, ? \, (y_1) = (ch(r_2) \, ! \, \text{clone}(a@s^b \, a_2) \mid \\
&\qquad\qquad\quad \ldots \\
&\qquad\qquad\quad a_{n-1} \, ? \, (y_{n-1}) = (ch(r_n) \, ! \, \text{clone}(a@s^b \, a_n) \mid \\
&\qquad\qquad\quad a_n \, ? \, (y_n) = (\\
&\qquad\qquad\qquad X^X(ch(self)@s^b \, agentKey(b) \, y_1 \ldots y_n) \mid \\
&\qquad\qquad\qquad (\nu \, a) \, (a \, ? \, () = ch(self)@s^b \, ! \, \text{discard}() \mid \text{amg} \, ! \, \text{addObj}(a)) \\
&\qquad\qquad\quad) \ldots))) \\
&\quad)))
\end{aligned}
$$

We begin by creating site s^b and channel $ch(self)@s^b$. In their scope we place the continuation of the current agent, the registry of the new provider in the resolver, and, in the scope of a new channel $a@s^b$, the provider's initial state. A reduction in $a@s^b$ causes the later process to migrate to s^b and execute there.

The registry in the resolver map requires a **register** method call targeted at the channel that is dedicated to that service. The arguments of the call are the channel where the provider is answering $(ch(self)@s^b)$ and the provider's key.

The process that defines the provider's initial state uses macro localChs to create the required built-in definitions and channels. The instantiation of definition X^X receives locally located clones of the values given to the constructor, along with the agent's key. All the code required by X^X (given $code(X, C, \emptyset)$) is uploaded, in contrast with the code required by the arguments, downloaded during the cloning operations. The uploaded code terminates with the process to remove the agent from the local heap.

7.3.2.7.2 Service-Oriented Process Mobility   The first step is to find a provider of the required service. This is done by invoking method find in the object responsible for the registry of the service in the network $(\mathsf{ch}(S)@s^R)$. The encoding, for $B(\tilde{x}) = \tilde{r}$, is thus given by

$$
[\![(B, x = S \; P \; \textbf{in} \; P' \; \textbf{error} \; P'', \texttt{null})]\!]_{\varphi \cdot \mathrm{RET:ch}(r)} \overset{\text{def}}{=}
$$
$$
(v \; b) \; (
$$
$$
\quad \mathsf{ch}(S)@s^R \; ! \; \mathsf{find}(\varphi(\mathrm{KEY}) \; b) \; |
$$
$$
\quad b \; ? \; \{
$$
$$
\quad \mathsf{found}(x) = (v \; \mathsf{ch}(r')) \; (
$$
$$
\qquad [\![r' : (x, (B, P', r))]\!]_{\varphi} \; |
$$
$$
\qquad (v \; a) \; (
$$
$$
\qquad\quad x \; ! \; \mathsf{getRemoteChannels}(a) \; |
$$
$$
\qquad\quad a \; ? \; (\mathsf{buddy} \; \mathsf{amg}) =
$$
$$
\qquad\quad \mathsf{buddy} \; ! \; () \; |
$$
$$
\qquad\quad \mathsf{buddy} \; ? \; () = (v \; b) \; (
$$
$$
\qquad\qquad x \; ! \; \mathsf{getKey}(b) \; |
$$
$$
\qquad\qquad b \; ? \; (\mathsf{agentKey}) = (v \; a_1) \; \ldots (v \; a_n) \; (
$$
$$
\qquad\qquad\quad \mathsf{ch}(r_1) \; ! \; \mathsf{clone}(\mathsf{buddy} \; a_1) \; |
$$
$$
\qquad\qquad\quad a_1 \; ? \; (x^{x_1}) = (\mathsf{ch}(r_2) \; ! \; \mathsf{clone}(\mathsf{buddy} \; a_2) \; | \; \ldots |
$$
$$
\qquad\qquad\qquad a_{n-1} \; ? \; (x^{x_{n-1}}) = (\mathsf{ch}(r_n) \; ! \; \mathsf{clone}(\mathsf{buddy} \; a_n) \; |
$$
$$
\qquad\qquad\qquad\quad a_n \; ? \; (x^{x_n}) = [\![(\emptyset, P, \texttt{null})]\!]_{\varphi = \varphi + \{\mathrm{KEY:agentKey}, \mathrm{RET:ch}(r')\}}
$$
$$
\qquad\qquad\quad) \ldots)
$$
$$
\qquad\quad))))
$$
$$
\quad \mathsf{notFound}() = [\![(B, P'', \texttt{null})]\!]_{\varphi} \varphi \cdot \mathrm{RET} : \mathsf{ch}(r)
$$
$$
\quad \})
$$

The outcome of the invocation of method find triggers two different local behaviors. If a provider is found, a reference to hold the suspended thread and handle the result must be created, and the code of the session uploaded. This code is abstracted on a set of free names that must be gathered. The *call by value* semantics approach requires the creation of clones at the target site.[*] Note that, in order to relate to its new location, the environment of the migrated process is updated with the key of the target agent, obtained by resorting to the getKey method. Also note that the set of bindings of the uploaded session is built from the generated clones.

---

[*] The procedure is the same as the one used in the creation of a new provider.

If no provider is found, the **error** handling code is selected. Note that the amg channel obtained from the agent hosting the service is required by the encoding of $(\emptyset, P, \text{null})$.

7.3.2.7.3 Method Invocation The encoding creates a heap reference to hold the result (r'), suspends the current thread upon it, notifies the local manager that a new thread is going to be launched, and finally invokes the method with arguments $\text{ch}(\tilde{r})$ (for $B(\tilde{v}) = \tilde{r}$) and with $\text{ch}(r')$ (the channel to where the result must be sent). Once more for $B(\tilde{v}) = \tilde{r}$ we have

$$
[\![(B, x = l(\tilde{v})\ P, \text{null})]\!]_{\varphi \cdot \text{RET}:\text{ch}(r)} \overset{\text{def}}{=} (\nu\ \text{ch}(r'))\ (
$$
$$
[\![\{r' : (x, (B, P, r))\}]\!]_{\varphi}\ |
$$
$$
\text{ch}(\texttt{self})\ !\ l^1(\text{ch}(\tilde{r})\ \text{ch}(r'))
$$
$$
)
$$

7.3.2.7.4 Returning a Result The execution of the **return** instruction sends the value to the channel that will handle its placing (or cloning in remote sessions). Note that the rule is only applicable if no reference to hold the result is defined. This forces the RET environment variable to be defined.

$$
[\![(B \cdot \{v : r\}, \textbf{return}\ (v), \text{null})]\!]_{\varphi} \overset{\text{def}}{=} \varphi(\text{RET})\ !\ \text{handleResult}(\text{ch}(r))
$$

7.4 Implementation

In this section, we discuss how the SOMAM can be implemented. A common approach would be to build a dedicated virtual machine. However, the choice of a TyCO-based calculus as the target of our encoding allows us to use the existing run-time infrastructure for the TyCO language [6,16]. This infrastructure supports the distribution and mobility constructs defined for DiTyCO in Ref. [16], and so, for the sake of uniformity, we shall refer to it as DiTyCO. The syntax of the language is slightly different from the one presented for the calculus, since it does not support network-level operations, such as channel creation. These operations must be performed within a site.

With this framework, we only need to implement a SOMAM to DiTyCO language compiler, since we can use the already available DiTyCO compiler to produce the final code. In this context, the encoding of the previous section can now be used as the base for our compilation scheme.

7.4.1 Compiler

As illustrated in Figure 7.1, the compilation process is divided into two steps. The first is the SOMAM to DiTyCO compiler that produces an intermediate file to be supplied as input to the second stage, the DiTyCO language compiler. This compiler

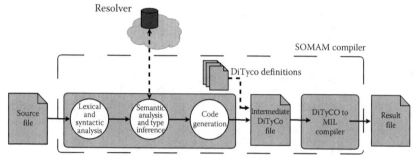

FIGURE 7.1:   SOMAM compiler.

generates the final program, written in an intermediate language called Multithreaded Intermediate Language (MIL) [22,30].

Only well-typed source programs can be successfully compiled into DiTyCO. This implies a network connection to check if the types locally inferred for the required and/or provided services match the ones already defined in the network, as is explained in Section 7.2.1. This means that the first step of the compiler is also responsible for inferring the types of the services used in the program, and checking them against the types registered in the resolver. The resolver, as with all SOMAM agents, is compiled into a DiTyCO site. Thus, the network interaction required to perform the type-checking must be implemented in DiTyCO. To enable disconnected compilation, the type-checking stage can be pushed to the beginning of the run-time execution. This, however, requires a mechanism to store this extra information in the MIL file, in order to be available at run-time.

The output generated by the first stage of the compiler is mostly the adaptation of the encoding to the concrete syntax of the language. Exceptions are the cases that require operations that are simply not available in the language, such as site and remote channel creation. These operations are compiled into methods of a primitive object that resorts to a run-time library of native code.* This procedure is common in the DiTyCO framework, e.g., to include string operations. We resort to it a second time, in order to allow for the resolver to access the typing-system at run-time. In the original platform, the type-system is only used by the compiler.

7.4.2   Run-Time System

A SOMAM network is now mapped into several DiTyCO sites running SOMAM agents plus a site running the resolver (Figure 7.2). The run-time system for each of these is composed of the original DiTyCO run-time plus two libraries: the one to handle the extra primitive object, and the one to access the type-system.

---

* The DiTyCO run-time is implemented in Java, and thus, the mentioned run-time libraries are composed of sets of Java classes.

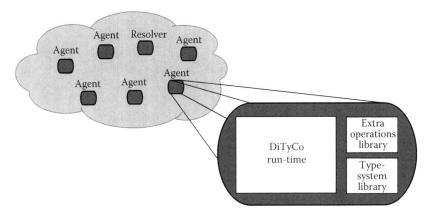

FIGURE 7.2: Run-time system for a SOMAM network.

7.5 Related Work

Software mobility has been a research topic for several years and many different approaches to the matter have been proposed. In this section, we overview the work that more closely relates to the SOMAM, beginning by identifying some aspects that cut across different approaches.

Mobility can be classified as strong or weak, according to the transfer (or not) of the computation's execution state. Strong migration provides a more intuitive programming model, since the migration of the execution state allows for the transparent resumption of the computation at the exact same point where it was interrupted. Weak migration relies solely on the migration of code and data, which, when moving computations, requires the implementation of some reception code to trigger the execution.

A common ground is the fact that all systems, until now, model mobility in terms of network node abstractions. In other words, computation moves toward network nodes (either physical or logical). In this context, the approach followed by the SOMAM is unprecedented.

Many languages that cope with mobility have some formal framework that allows for the proof of correctness properties. These languages are mostly different forms or extensions of the π-calculus [13,18]. They define process algebras that, in most cases, were initially aimed to model concurrent processes and later where extended to cope with distribution and mobility.

The Distributed π-calculus [12] is an extension of the π-calculus to distributed settings. Hosts are abstracted in a new syntactic category, named location, that defines the boundaries for process execution. Mobility is weak and explicit. Movement is triggered by the use of a primitive (**go**) with the indication of which code to migrate and the target location, e.g., **go** $l.P$ migrates process P to location l.

The LSD π-calculus [23] is also a distributed extension of the π-calculus. The approach, however, is different. Remote channels are qualified with their site (location) of origin, and an input or output aimed at such a channel causes the value or process to migrate to the given site. This means that the lexical scope of the channel transparently triggers the mobility, causing reduction to be always local to the channel's site of origin, e.g., $a@s\ ?\ () = P$ causes P to migrate to s, the site hosting channel a. The DiTyCO language [29] extends these concepts to handle TyCO objects and process definitions.

Mobile ambients [3] is a calculus for distributed computations where processes run in the boundaries of nested localities, named ambients. Mobility is expressed as entering (**in**) or exiting (**out**) an ambient, e.g., $n[(\textbf{in}\ m.P)\ |\ Q]$ causes ambient n running processes P and Q to be executed within the boundaries of ambient m. Note that the whole ambient moves, indicating we are in the presence of strong mobility. In the example, **in** does not only move P but also the remaining processes running in the ambient (Q).

Jocaml [4] is a programming language that applies the concepts of the Join calculus [7] to the Objective Caml language. Computations are structured in trees, i.e., an agent and its tree of subagents. Moving an agent involves moving its whole tree. The syntax is close to the one found in Distributed π.

The X-KLAIM [1] programming language is an implementation of the KLAIM model [5] with ad hoc extensions to incorporate higher order constructs, asynchronous reading of tuple-spaces and hierarchical structured networks. Mobility is strong and is expressed in terms of writing a process in a remote tuple-space. This is done by resorting to a modified version of the Linda **eval** operation.

Nomadic Pict [32] grows from the Nomadic π-calculus [27], also an extension of the π-calculus. It features two levels of language primitives. The lower level is location-dependent, while the higher provides some location independence. Mobility is location-dependent and is expressed much in the same way as in Distributed π, using a **migrate to** primitive. The difference, however, is that **migrate to** causes the whole site to move, and thus, mobility is strong.

MOB [21] is a high-level language that has been in the genesis of this work. It allows for the programming of mobile agents that interact by providing and requiring services. A formal semantics has been defined in Ref. [20], its main difference to the SOMAM being the fact that, although resource bindings are expressed in terms of services, mobility is still done toward hosts.

Not all languages or systems that support mobility have a formal background. Most of the existing systems rely on the Java language and its run-time system. Systems, such as Aglets [15], Mole [26], and Voyager [9], define a set of base classes that the programmer must extend, in order to define its mobile agents. Mobility is weak, since the running thread's run-time state cannot be accessed from the language, nor can it be completely serialized, due to the use of native code. Thus, mobility is done by sending a serialized object down a stream and indicating which is the method to be executed on arrival. These details are usually hidden by the API classes. Some higher-level abstractions, such as itinerary patterns, have also been proposed.

We conclude with the Acute [25] programming language, which is built on top of Objective Caml. The approach followed is similar to the one used in Java, with the difference that migration of multithreaded computations is possible. Thus, no migration primitive is featured. Moving computations is achieved by using the **thunkify** operation that is able to atomically serialize multithreaded computations and obtain a *thunk*, which can afterward be moved across the network.

7.6 Conclusions and Future Work

In this chapter, we described the SOMAM, an abstract machine that defines client–server interaction by resorting to service-driven session uploading. Unlike other models, the SOMAM completely abstracts the programmer from the network. Existing models, such as process calculi or Java based, always resort to some kind of node abstraction to incorporate mobility. This novel approach allows for the seamless combination of both services and mobility, since all network-related operations are modeled in terms of services.

It is our opinion that this combination provides an intuitive framework for the implementation of distributed applications. This is even more so when applied to today's highly dynamic and volatile networks. The use of services to model client–server relationships provides the means for dynamic service discovery and binding. No longer must the failure of a component have a disruptive action on the system. On the other hand, mobility has the ability to eliminate the need for costly remote sessions, which is a major feature in environments with frequent disconnection and intermittent bandwidth.

To validate our approach, we proposed an encoding map from abstract machine states into DiTyCO networks. This map provided us the framework upon which we can prove a soundness result relative to the calculus. The encoding can also be used as a base for a compilation scheme to DiTyCO programs. This will allow us to use the available infrastructure for DiTyCO to build a compiler and run-time system for SOMAM computations.

Future work will follow two major research topics: (1) articulation between the SOMAM model and the common mobile agent features, such as concurrency and resource access restrictions, and (2) porting of the model to common programming languages, such as Java, yielding a real-world framework that provides mobility as a tool for service interaction in service-oriented middleware architectures.

References

[1] L. Bettini, R. de Nicola, and R. Pugliese. X-Klaim and Klava: Programming mobile code. In *TOSCA 2001*, volume 62. Elsevier Science, Udine, Italy, 2001.

[2] R. R. Brooks. Mobile code paradigms and security issues. *IEEE Internet Computing*, 8(3):54–59, 2004.

[3] L. Cardelli and A. Gordon. Mobile ambients. In *Foundations of Software Science and Computation Structures (FoSSaCS'98). Lecture Notes in Computer Science*, Lisbon, Portugal, March 28–April 4, 1378:140–155, 1998.

[4] S. Conchon and F. Le Fessant.   Jocaml: Mobile agents for Objective-Caml. In *First International Symposium on Agent Systems and Applications (ASA'99)/Third International Symposium on Mobile Agents (MA'99)*, Palm Springs, CA, October 3–6, IEEE Computer Society, 1999, pp. 22–29.

[5] R. de Nicola, G. Luigi Ferrari, and R. Pugliese. Klaim: A kernel language for agents interaction and mobility. *IEEE Transactions on Software Engineering (Special Issue on Mobility and Network Aware Computing)*, 24(5):315–337, 1998.

[6] A. Figueira, H. Paulino, L. Lopes, and F. Silva. Distributed typed concurrent objects: A programming language for distributed computations with mobile resource. *Journal of Universal Computer Science*, 8(9):745–760, 2003.

[7] C. Fournet and G. Gonthier.   The reflexive chemical abstract machine and the join-calculus.   In *ACM SIGPLAN-SIGACT Symposium on Principles of Programming Languages (POPL'96)*, St. Petersburg Beach, FL, 21–24 January, ACM Press, 1996, pp. 372–385.

[8] C. Fournet, G. Gonthier, J.-J. Lévy, L. Maranget, and D. Rémy. A calculus of mobile agents. In U. Montanari and V. Sassone, (Eds.), *Proceedings of the 7th International Conference on Concurrency Theory (CONCUR'96)*, Pisa, Italy, August 26–29, *Lecture Notes in Computer Science*, 1110:406–421, 1996.

[9] G. Glass. Overview of voyager: ObjectSpace's product family for state-of-the-art distributed computing. Technical Report, CTO ObjectSpace, 1999.

[10] Wang GuiLing, Li YuShun, Yang ShengWen, Miao ChunYu, Xu Jun, and Shi MeiLin. Service-oriented grid architecture and middleware technologies for collaborative e-learning. *SCC*, 2:67–74, 2005.

[11] A. Harrison and I. Taylor.   Service-oriented middleware for hybrid environments.   In *ADPUC '06: Proceedings of the 1st International Workshop on Advanced Data Processing in Ubiquitous Computing (ADPUC 2006)*, p. 2, New York, ACM Press, 2006.

[12] M. Hennessy and J. Riely.   Resource access control in systems of mobile agents. *Information and Computation*, 173(1):82–120, 2002.

[13] K. Honda and M. Tokoro. An object calculus for asynchronous communication. In *European Conference on Object-Oriented Programming (ECOOP'91)*, Geneva, Switzerland. *Lecture Notes in Computer Science*, 512:141–162, July 15–19, 1991.

[14] W. Jansen and T. Karygiannis. NIST Special Publication 800-19—Mobile agent security. Special Publication 800-19, National Institute of Standards and Technology, 1999. Available from http://csrc.nist.gov/mobilesecurity/publications.html.

[15] D. B. Lange and M. Oshima. *Programming and Deploying Java Mobile Agents with Aglets*. Addison-Wesley, Reading, MA, 1998.

[16] L. Lopes, Á. Figueira, F. Silva, and V. T. Vasconcelos. A concurrent programming environment with support for distributed computations and code mobility. In *Proceedings of the IEEE International Conference on Cluster Computing (Cluster'2000)*, Saxony, Germany, November 28–December 1, IEEE Computer Society, 2000, pp. 297–306.

[17] L. Lopes, F. Silva, Á. Figueira, and V. Vasconcelos. DiTyCO: An experiment in code mobility from the realm of process calculi. Presented at the *5th Mobile Object Systems Workshop (MOS'99)*, Lisbon, Portugal, June 14–18, 1999.

[18] R. Milner, J. Parrow, and D. Walker. A calculus of mobile processes (parts I and II). *Information and Computation*, 100(1):1–77, 1992.

[19] H. Paulino. The SOMAM is sound relative to the DiTyCO calculus: Sketching the proof. Technical Report, CITI—Universidade Nova de Lisboa, Lisbon, Portugal, 2008. Available from http://www-asc.di.fct.unl.pt/~herve/papers/.

[20] H. Paulino and L. Lopes. Mob core language and virtual machine (rev 0.2). Technical Report, CITI—Universidade Nova de Lisboa & LIACC - Universidade do Porto, Lisbon, Portugal, 2006. Available from http://www-asc.di.fct.unl.pt/~herve/papers/.

[21] H. Paulino and L. Lopes. A programming language and a run-time system for service-oriented computing with mobile agents. *Software—Practice and Experience*, 38(6), 2008.

[22] H. Paulino, P. Marques, L. Lopes, V. Vasconcelos, and F. Silva. A multi-threaded asynchronous language. In *7th International Conference on Parallel Computing Technologies (PaCT'03)*, Novosibirsk, Russia. *Lecture Notes in Computer Science*, 2763:316–323, September 15–19, 2003.

[23] A. Ravara, A. Matos, V. Vasconcelos, and L. Lopes. Lexically scoped distribution: What you see is what you get. In *FGC: Foundations of Global Computing*, volume 85(1) of *Electronic Notes in Theoretical Computer Science*. Elsevier Science, 2003.

[24] A. Schmitt and J.-B. Stefani. The M-calculus: A higher-order distributed process calculus. In *POPL '03: Proceedings of the 30th ACM SIGPLAN-SIGACT Symposium on Principles of Programming Languages (POPL)*, New Orleans, LA, January 15–17, ACM Press, 2003, pp. 50–61.

[25] P. Sewell, J. J. Leifer, K. Wansbrough, F. Z. Nardelli, M. A.-Williams, P. Habouzit, and V. Vafeiadis. Acute: High-level programming language design for distributed computation. In *ICFP '05: Proceedings of The 10th ACM SIG-PLAN International Conference on Functional Programming*, Tallinn, Estonia, September 26-28, ACM Press, 2005, pp. 15–26.

[26] M. Straßer, J. Baumann, and F. Hohl. Mole—A Java based mobile agent system. In M. Muehlhaeuser, (Ed.), *Special Issues in Object Oriented Programming, Workshop Reader of the 10th European Conference on Object-Oriented Programming (ECOOP 96)*, Linz, Austria, July 8–12, Heidelberg, 1997, pp. 301–308. Available at dpunkt.verlag.

[27] A. Unyapoth. Nomadic Pi Calculi: Expressing and verifying infrastructure for mobile computation. PhD thesis, University of Cambridge, Cambridge, U.K., 2001.

[28] V. Vasconcelos. Typed concurrent objects. In *European Conference on Object-Oriented Programming (ECOOP'94)*. Bologna, Italy, July 4–8, *Lecture Notes in Computer Science*, 821:100–117, 1994.

[29] V. Vasconcelos, L. Lopes, and F. Silva. Distribution and mobility with lexical scoping in process calculi. In *Workshop on High Level Programming Languages (HLCL'98)*, Nice, France, September 12, volume 16(3) of *Electronic Notes in Theoretical Computer Science*, Elsevier Science, 1998, pp. 19–34.

[30] V. T. Vasconcelos and F. Martins. A Multithreaded typed assembly language. In *Proceedings of TV06 - Multithreading in Hardware and Software: Formal Approaches to Design and Verification*, Seattle, WA, August 21–22, 2006.

[31] J. Vitek and G. Castagna. Seal: A framework for secure mobile computations. In H. Bal, B. Belkhouche, and L. Cardelli, (Eds.), *Internet Programming Languages. Lecture Notes in Computer Science*, Chicago, IL, 1686:47–77, 1998.

[32] P. T. Wojciechowski and P. Sewell. Nomadic Pict: Language and infrastructure design for mobile agents. *IEEE Concurrency*, 8(2):42–52, 2000.

Chapter 8

Specifying and Implementing Secure Mobile Applications

Andrew Phillips

Contents

8.1   Introduction

The Internet has grown substantially in recent years, and an increasing number of applications are now being developed to exploit this distributed infrastructure. Mobility is an important paradigm for such applications, where mobile code is supplied on demand and mobile components interact freely within a given network. However, mobile applications are difficult to develop. Not only do they involve complex parallel interactions between multiple components, but they must also satisfy strict security requirements. This chapter describes how the problem of specifying and implementing secure mobile applications can be addressed using a process calculus.*

The Internet is used for a wide variety of distributed applications that can benefit from mobile software. Specific examples include search engines, mining of data repositories, scripting languages for web browser animations, peer-to-peer file sharing systems, financial trading software, consumer auction sites, and travel reservation systems. Unfortunately, these applications are often hindered by two phenomena that cannot be abstracted away in a distributed setting: network latency and network failure (see Ref. [29] for technical definitions and related discussion). Network latency refers to the interval of time between the departure of a message from one machine and its arrival on another machine. Two common factors of network latency are network congestion and the use of a slow network interface. Network failure, on the other hand, refers to a break in the connectivity between two machines on a network. This can occur for a variety of reasons. In some cases, network congestion can cause certain packets to be lost, resulting in a temporary disconnection between two machines. In other cases, a machine can be brought down by a direct attack from another machine and become isolated from the rest of the network. Alternatively, a machine can be physically unplugged from the network for a period of time.

In spite of the large increase in network bandwidth over the last few years [14,28],[†] network latency and failure are still very much an issue. This is due in part to the large increase in the size of files being transmitted over a network, such as audio and video content, particularly at peak times. Another reason is the increase in unsolicited network traffic. For example, certain viruses and worms can have a devastating effect on networks, albeit for a limited time period. Finally, with the growing popularity of mobile devices such as laptops, handhelds, and Internet-enabled mobile phones, the effects of network latency and failure are becoming increasingly apparent. This is because mobile devices can have comparatively slower connection speeds than fixed machines, and tend to connect and disconnect from networks more frequently.

---

* This chapter is an extended version of Ref. [33] and is based on the author's PhD thesis [31].
† Recent Internet growth statistics are available from http://www.dtc.umn.edu/mints/.

In order to minimize the effects of network latency and failure, distributed applications are relying increasingly on mobile software in the form of mobile code, and mobile agents. By definition, mobile code refers to program code that can be sent from one machine to another over a network. The code itself has no state, and can only begin executing after reaching its destination. A mobile agent, on the other hand, refers to an autonomous program that can stop executing, move through a network to a new machine, and continue executing at its new location. In the general case, a mobile agent can autonomously travel to an itinerary of multiple destinations, preserving its state after each move.

Mobile software, in the form of mobile code and mobile agents, can help to minimize the effects of network latency and failure in a number of distributed applications. Some of the benefits of mobile agents are described in Ref. [12]. It is somewhat ironic that many of the problems related to the use of mobile devices can in part be solved by mobile software.

Mobile applications can be roughly grouped into three main categories: mobile code, resource monitoring, and information retrieval. Applets, scripting languages, and data mining are examples of mobile code, where a single piece of code is sent to a remote machine. In the case of data mining, agents are dispatched to data warehouses, such as those containing consumer information or news archives, to look for general trends in the data. Online trading and electronic commerce are examples of resource monitoring, and travel agencies, search engines, and peer-to-peer applications are examples of information retrieval. A broader survey of the various categories of mobile applications can be found in Ref. [41]. It is worth noting that a large proportion of these applications involve mobile software agents traveling between hardware devices over wide area networks (WANs) such as the Internet. Even applications that only require mobile code can be expressed using the agent paradigm, giving programmers the flexibility to extend these applications and make full use of agents as appropriate. For example, in cases where a client uploads mobile code to a server, the client license may expire while the client is disconnected. If mobile agents are used, the agent responsible for uploading the code can move to a renewal site, negotiate the renewal of the license and then return to the server to continue executing the code. Perhaps one of the simplest categories of mobile applications is resource monitoring, and one such application is used as a running example for this chapter. Information retrieval applications are more complex, since they require sophisticated algorithms for allowing agents to communicate with each other as they move between data repositories. One such application is presented at the end of this chapter.

8.1.1 Process Calculi for Mobile Applications

Process calculi have recently been proposed as a promising formalism for specifying and implementing secure mobile applications. Calculi can be thought of as simple programming languages, which provide a concise description of computation that facilitates rigorous analysis. They have a precise syntax and computable operational semantics that are both formally defined, together with an execution state that is implicit in the terms of the calculus. This contrasts with many alternative models of computation, including automata models, where the execution state needs to be given explicitly as a separate component.

Calculi have been used successfully for many years to model various forms of computation. An important example is the λ-calculus [13], which captures the essence of functional computation in a small number of terms. This enables a concise description of functional computation that supports formal reasoning about the correctness of algorithms. Many functional programming languages including Standard ML [27] have been based on the λ-calculus, which provides a solid theoretical foundation.

More recently, the π-calculus [26] has been developed as a model for concurrent computation. Many argue that the π-calculus achieves for concurrent computation what the λ-calculus does for functional computation, capturing the essence of computation in a small number of terms and facilitating formal reasoning.

With the advent of mobile programming, there has been considerable research on calculi for mobile computation. In particular, the Nomadic π-calculus [37] demonstrated the feasibility of using process calculi to specify and implement applications involving location-independent communication between mobile agents. The Ambient calculus [10] was also introduced to model the hierarchical topology of modern networks, and many variants of ambients were subsequently proposed. One variant of ambients that seems well suited to modeling mobile applications is the Boxed Ambient calculus. In general, Boxed Ambient calculi have been used to model and reason about a variety of issues in mobile computing. The original paper on Boxed Ambients [5] shows how the Ambient calculus can be complemented with finer-grained and more effective mechanisms for ambient interaction. In Ref. [6] Boxed Ambients are used to reason about resource access control, and in Ref. [16] a sound type system for Boxed Ambients is defined, which provides static guarantees on information flow. Recently, several new variants of Boxed Ambients have been proposed, which seek to improve on the foundations of the original calculus. In particular, the Safe Boxed Ambient calculus [23] uses cocapabilities to express explicit permissions for accessing ambients, and the Noninterfering Boxed Ambients (NBA) calculus [7] seeks to limit communication and migration interferences in Boxed Ambients. Boxed Ambient calculi can also benefit from many of the analysis techniques of the Ambient calculus, most notably Ambient Logics [9]. Although research on calculi for mobile applications is still in its early stages, already the potential benefits of such calculi are beginning to be widely recognized. In spite of these theoretical advances, there has been little research on how Boxed Ambient calculi can be correctly implemented in a distributed environment. Furthermore, Ambient calculi in general have not yet been used to model real-world mobile applications. This chapter demonstrates that it is feasible to develop a distributed programming language for mobile applications, based on a variant of the Ambient calculus. Furthermore, it is feasible to derive a provably correct algorithm for executing these applications, and to refine this algorithm to executable program code.

The chapter is structured as follows. Section 8.2 presents a novel calculus for specifying mobile applications, known as the Channel Ambient (CA) calculus. The calculus is inspired by previous work on calculi for mobility, including the π-calculus, the Nomadic π-calculus and the Ambient calculus. Section 8.3 presents an abstract machine for the CA calculus, known as the Channel Ambient Machine (CA Machine). The abstract machine is a formal specification of a runtime for executing calculus processes, which bridges a gap between the specification and implementation of mobile applications. Section 8.4 presents a runtime for executing processes of the CA calculus, known as the CA Runtime. The runtime is implemented by defining a direct mapping from the CA Machine to functional program code. The Channel Ambient Language is also presented, together with an example mobile application, in which a mobile agent monitors resources on a remote server. Finally, Section 8.5 presents an agent tracker application, which keeps track of the location of registered client agents as they move between trusted sites in a network.

8.2   Channel Ambient Calculus

Process calculi were described in Section 8.1 as a promising formalism for specifying and implementing secure mobile applications. In particular, Boxed Ambient calculi [7] were identified as a foundation for specifying applications that execute on networks with a hierarchical structure. Boxed Ambient calculi benefit from a range of analysis techniques originally developed for the Ambient calculus, including type systems, equational theories, and Ambient Logics. They also enforce more rigid security measures by preventing ambient boundaries from being dissolved. This allows for new and richer security mechanisms to be developed, including various type systems and methods for control flow analysis. Many variants of Boxed Ambients have been defined, together with analysis techniques that can be used across multiple variants. Boxed Ambient calculi therefore seem a natural starting point when looking for a calculus to specify and implement secure mobile applications.

Although numerous variants of Boxed Ambients have been proposed, it can be argued that an essential feature of mobile applications is lacking from these variants, namely the existence of channels as first class entities. At the lowest level, channels are the building blocks of some of the most widely used protocols in mobile applications, including TCP/IP. Channels are also a fundamental programming abstraction, since they correspond to the notion of methods or services provided by an application component. In particular, channels allow a given component to provide multiple services, each of which is invoked using a separate service channel. From a security perspective, channels are also fundamental, since they correspond to the notion of keys that can be used to regulate communication and migration of components in a mobile application. For example, a private channel can be used by a single component to establish a private communication with another component or to gain exclusive access to a location. Conversely, a public channel can be used to provide a public communication service or to enable public access to a location.

This section presents a calculus for specifying mobile applications, known as the CA calculus. The calculus was first described in an extended abstract [33], which formed the basis of the author's PhD thesis [31]. The calculus is inspired by previous work on calculi for mobility, including the π-calculus, the Nomadic π-calculus, and the Ambient calculus. In many respects, the calculus can be considered a variant of Boxed Ambients in which channels are defined as first class entities, allowing ambients to communicate with each other and move in and out of each other over channels.

8.2.1 Syntax of Calculus Processes

The CA calculus uses the notion of an ambient, first presented in Ref. [10], to model the components of a mobile application. An ambient is an abstract entity that can be used to model a machine, a mobile agent, or a module. In this calculus, ambients are named, arranged in a hierarchy, and can interact by sending messages to each other and moving in and out of each other over channels.

The syntax of the CA calculus is presented in Definition 8.1 in terms of processes P, Q, R, actions α and values $a, b, ..., z$. It is assumed that a, b, c represent ambient names; x, y, z represent channel names; n, m represent ambient or channel names; and u, v represent arbitrary values. In an applied version of the calculus, these values can include names, constants, tuples, etc. A corresponding graphical syntax is presented in Definition 8.2. The processes P, Q, R of the calculus have the following meaning:

Null 0 does nothing and is used to represent the end of a process.

Parallel $P \mid Q$ executes process P in parallel with process Q.

Restriction $vn\ P$ executes process P with a private name n.

Ambient $a\boxed{P}$ executes process P inside an ambient a. The ambient a represents a component of a mobile application, such as a machine, a mobile agent, or a module.

Action $\alpha.P$ tries to perform the action α and then execute process P. The action can involve either a communication or a migration.

Replication $!\alpha.P$ repeatedly tries to perform the action α and then execute process P.

$P, Q, R ::=$	**0**	Null	$\alpha ::=$	$a \cdot x \langle v \rangle$	Sibling output
	$\mid P \mid Q$	Parallel	$\mid$	$x^\uparrow \langle v \rangle$	Parent output
	$\mid vn\ P$	Restriction	$\mid$	$x(u)$	Internal input
	$\mid a\boxed{P}$	Ambient	$\mid$	$x^\uparrow(u)$	External input
	$\mid \alpha.P$	Action	$\mid$	$\mathrm{in}\ a \cdot x$	Enter
	$\mid !\alpha.P$	Replication	$\mid$	$\mathrm{out}\ x$	Leave
			$\mid$	$\overline{\mathrm{in}}\ x$	Accept
			$\mid$	$\overline{\mathrm{out}}\ x$	Release

DEFINITION 8.1 Syntax of CA

DEFINITION 8.2 *Graphical Syntax of* CA

The graphical syntax is reminiscent of various informal representations for concurrent processes, in which each process or thread is represented as a vertical bar. The representation is particularly reminiscent of Message Sequence Charts [19], where each component is represented as a box above a process, labeled with the component name, and parallel composition is represented as a collection of adjacent processes. Unlike Message Sequence Charts, each vertical bar is also labeled with the current state of the process. In addition, restriction is represented as a dotted ring around a process, labeled with the restricted name. The actions α of the calculus have the following meaning:

Sibling Output $a \cdot x \langle v \rangle$ tries to send a value v on channel x to a sibling ambient a.

Parent Output $x^{\uparrow} \langle v \rangle$ tries to send a value v on channel x to the parent ambient.

External Input $x^{\uparrow}(u)$ tries to receive a value u on channel x from outside the current ambient.

Internal Input $x(u)$ tries to receive a value u on channel x from inside the current ambient.

Enter $\text{in}\, a \cdot x$ tries to enter a sibling ambient over channel x.

Leave $\text{out}\, x$ tries to leave the parent ambient over channel x.

Accept $\overline{\text{in}}\, x$ tries to accept a sibling ambient over channel x.

Release $\overline{\text{out}}\, x$ tries to release a child ambient over channel x.

8.2.2 Reduction Rules for Executing Calculus Processes

The reduction rules of the CA calculus describe how a calculus process can be executed. Each rule is of the form $P \longrightarrow P'$, which states that the process P can evolve to P' by performing an execution step. The reduction rules are presented in Definition 8.3, where $P_{\{v/u\}}$ assigns the value v to the value u in process P, and $P \equiv Q$ means that process P is equal to process Q. A corresponding graphical representation of the reduction rules is presented in Definition 8.4:

$$a\boxed{b \cdot x\langle v\rangle.P \mid P'} \mid b\boxed{x^\uparrow(u).Q \mid Q'} \longrightarrow a\boxed{P \mid P'} \mid b\boxed{Q_{\{v/u\}} \mid Q'} \quad (8.1)$$

$$a\boxed{x^\uparrow\langle v\rangle.P \mid P'} \mid x(u).Q \longrightarrow a\boxed{P \mid P'} \mid Q_{\{v/u\}} \quad (8.2)$$

$$a\boxed{\text{in}\, b \cdot x.P \mid P'} \mid b\boxed{\overline{\text{in}}\, x.Q \mid Q'} \longrightarrow b\boxed{Q \mid Q' \mid a\boxed{P \mid P'}} \quad (8.3)$$

$$b\boxed{a\boxed{\text{out}\, x.P \mid P'} \mid \overline{\text{out}}\, x.Q \mid Q'} \longrightarrow b\boxed{Q \mid Q'} \mid a\boxed{P \mid P'} \quad (8.4)$$

$$P \longrightarrow P' \Rightarrow P \mid Q \longrightarrow P' \mid Q \quad (8.5)$$

$$P \longrightarrow P' \Rightarrow \nu n\, P \longrightarrow \nu n\, P' \quad (8.6)$$

$$P \longrightarrow P' \Rightarrow a\boxed{P} \longrightarrow a\boxed{P'} \quad (8.7)$$

$$Q \equiv P \longrightarrow P' \equiv Q' \Rightarrow Q \longrightarrow Q' \quad (8.8)$$

DEFINITION 8.3   *Reduction in* CA

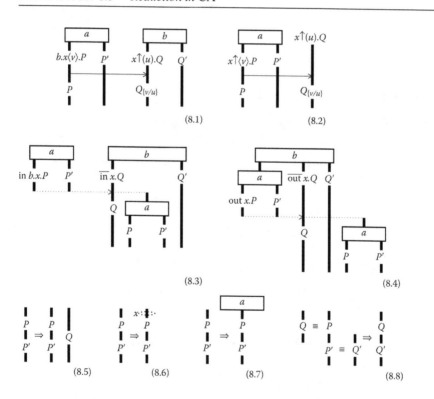

DEFINITION 8.4   *Graphical Reduction in* CA

(8.1) An ambient can send a value to a sibling over a channel. If an ambient a contains a sibling output $b \cdot x \langle v \rangle.P$, and there is a sibling ambient b with an external input $x^\uparrow(u).Q$, then the value v can be sent to ambient b along channel x, and assigned to the value u in process Q.

(8.2) An ambient can send a value to its parent over a channel. If an ambient a contains a parent output $x^\uparrow \langle v \rangle.P$, and there is an internal input $x(u).Q$ in parallel with a, then the value v can be sent along channel x, and assigned to the value u in process Q.

(8.3) An ambient can enter a sibling over a channel. If an ambient a contains an enter $\texttt{in}\, b \cdot x.P$, and there is a sibling ambient b with an accept $\overline{\texttt{in}}\, x.Q$, then a can enter b over channel x.

(8.4) An ambient can leave its parent over a channel. If an ambient a contains a leave $\texttt{out}\, x.P$, and there is a parent ambient with a release $\overline{\texttt{out}}\, x.Q$, then a can leave its parent over channel x.

(8.5) A reduction can occur inside a parallel composition. If a process P can reduce to P', then the reduction can also take place in parallel with a process Q.

(8.6) A reduction can occur inside a restriction. If a process P can reduce to P' then the reduction can also take place if P has a restricted name n.

(8.7) A reduction can occur inside an ambient. If a process P can reduce to P' then the reduction can also take place if P is inside an ambient a.

(8.8) Equal processes can perform the same reduction. If a process P can reduce to P', Q is equal to P and Q' is equal to P', then Q can reduce to Q'.

The graphical reduction rules are reminiscent of Message Sequence Charts, where time proceeds vertically downward, and communication between parallel components is represented by a solid arrow from a thread inside the sender to a thread inside the receiver. Unlike standard Message Sequence Charts, a given component can also move in or out of another component. This is represented as a dotted arrow from a thread inside the migrating component to a thread inside the destination component. In addition, each vertical bar is labeled with the current state of the thread, which can be modified as a result of an interaction.

8.2.3 Substitution of Free Values Inside Processes

Substitution in the CA calculus is used to replace one free value by another. The expression P_σ denotes the application of a substitution σ to a process P, where σ is a substitution that maps a given value to another (different) value, and v_σ applies the substitution σ to the value v. An example of a substitution is $P_{\{u',v'/u,v\}}$, which replaces u with u' and v with v' in process P. If v is not in the domain of σ then $v_\sigma = v$. By convention, it is assumed that there is no overlap between the set of bound names of a process and the set of substituted values given by the range of σ.

The expression $\mathrm{fn}(P)$ denotes the set of free values $\mathrm{fn}(P)$ of a process P in the CA calculus. The definition is standard, and relies on the definition of the set of bound values $\mathrm{bn}(P)$, where restriction $vn\ P$ binds the name n in process P, and internal input $x(u).P$ and external input $x^\uparrow(u).P$ bind the value u in process P. The set of bound values $\mathrm{bn}(\alpha)$ of an action α is defined as $\{u\}$ for $\alpha = x(u)$ and $\alpha = x^\uparrow(u)$, and $\emptyset$ otherwise. As usual, processes of the calculus are assumed to be equal up to renaming of bound values.

8.2.4 Structural Congruence Rules for Equating Calculus Processes

$$\mathbf{0} \mid P \equiv P \tag{8.9}$$

$$P \mid Q \equiv Q \mid P \tag{8.10}$$

$$P \mid (Q \mid R) \equiv (P \mid Q) \mid R \tag{8.11}$$

$$\mathrm{fn}(\alpha) \nsubseteq \mathrm{bn}(\alpha) \Rightarrow\ !\alpha.P \equiv \alpha.(P \mid\ !\alpha.P) \tag{8.12}$$

$$vn\ \mathbf{0} \equiv \mathbf{0} \tag{8.13}$$

$$vn\ vm\ P \equiv vm\ vn\ P \tag{8.14}$$

$$n \notin \mathrm{fn}(Q) \Rightarrow (vn\ P) \mid Q \equiv vn\ (P \mid Q) \tag{8.15}$$

$$a \neq n \Rightarrow a\boxed{vn\ P} \equiv vn\ a\boxed{P} \tag{8.16}$$

$$va\ a\boxed{\mathbf{0}} \equiv \mathbf{0} \tag{8.17}$$

DEFINITION 8.5   *Structural congruence in* CA

The structural congruence rules of the CA calculus describe what it means for two processes to be equal. The rules are presented in Definition 8.5 and are mostly standard, apart from the rule for replication (8.12). As usual, structural congruence is the least congruence that satisfies the rules in Definition 8.5

8.2.5 Syntax Abbreviations for Frequently Used Processes

$$z \notin \mathrm{fn}(x, v, P) \Rightarrow x\langle v\rangle.P \triangleq vz\,(z\boxed{x^\uparrow\langle v\rangle.z^\uparrow\langle\rangle} \mid z().P)$$

$$z \notin \mathrm{fn}(a, x, v, P) \Rightarrow a/x\langle v\rangle.P \triangleq vz\,(z\boxed{a\cdot x\langle v\rangle.z^\uparrow\langle\rangle} \mid z().P)$$

DEFINITION 8.6   *Syntax abbreviations in* CA

A number of convenient abbreviations can be defined for the CA calculus, in order to improve the readability of the calculus syntax. Standard syntactic conventions are used, including writing α as an abbreviation for $\alpha.\mathbf{0}$ and assigning the lowest precedence to the parallel composition operator. In addition, local output $x\langle v\rangle.P$ and child output $a/x\langle v\rangle.P$ can be encoded using parent output and sibling output, respectively, as described in Definition 8.6.

Note that the structural congruence rule $va\;a\boxed{\mathbf{0}} \equiv \mathbf{0}$ allows the empty ambient z to be garbage-collected after the value v has been sent. The encodings of local output and child output are straightforward enough to justify the use of syntactic abbreviations, rather than extending the syntax and reduction rules of the calculus itself.

8.2.6 Using the Calculus Syntax to Model TCP/IP Networks

The CA calculus can be used to model mobile applications that execute in a wide range of networks. Depending on the choice of network, the syntax of the calculus can be constrained to model the properties of the underlying network protocols. In this chapter, mobile applications are assumed to execute on networks that support the widely used TCP/IP version 4 protocol. Although TCP/IP networks are highly complex, it is possible to make a number of abstractions in order to obtain a high-level view of the main network properties. Based on this high-level view, the syntax of the calculus can be constrained to distinguish between two types of ambients: sites s and agents g. Sites represent hardware devices that are assumed to have a fixed network address, while agents represent software programs that can move in and out of sites and other agents. In practice, the name of a site corresponds to an IP address, while the name of an agent corresponds to a simple identifier.

The following constraints can be placed on the syntax of the CA calculus, in order to model the behavior of sites and agents:

Sites A process inside a site s is constrained so that it cannot contain a sibling output to an agent. This reflects the assumption that agents do not have a network address, and therefore cannot be reached directly by sites over a network. In addition, a process inside a site s is constrained so that it cannot contain an enter or a leave. This reflects the assumption that sites have a fixed network address. Note that this constraint does not prevent a site from physically moving around in the network. It merely ensures that a given site remains in the same logical location with respect to other sites.

Agents A process inside an agent g is constrained so that it cannot contain a site. This reflects the assumption that hardware sites cannot be contained inside software agents.

In a flat network topology, all sites are assumed to execute in parallel with each other and a given site can potentially communicate with any other site in the network. In a hierarchical topology, sites can be logically contained inside other sites to form local area networks (LANs). As a result, a given site is unable to communicate directly with another site that is not in the same LAN.

For such hierarchical networks it is useful to distinguish between two types of sites: ordinary sites and gateways. A gateway is a site that can logically contain other sites to form a LAN, while an ordinary site cannot contain any other sites. In practice, a gateway acts as a bridge between two networks: the local network it contains and the global network in which it is contained. As a result, a given gateway is usually assigned two network addresses, one for the local network and one for the global network. By definition, the CA calculus allows only a gateway to have a single name, which is used by other ambients to interact with the gateway. Based on this definition, the name of a gateway corresponds to its global address. The local address is not needed at the calculus level, since child ambients do not need to explicitly use the name of their parent in order to interact with it. Therefore, the local address is used merely as an implementation mechanism, to allow messages or agents from child ambients to be correctly routed to the parent gateway.

An example of how sites can be used to model the topology of LANs and WANs is illustrated in Figure 8.1. In this example, each site executes on a separate machine, with a given IP address. The sites 192.168.0.2–192.168.0.4 are part of a LAN inside a gateway with local address 192.168.0.1 and global address 82.35.60.43. The ambients inside the LAN will send messages or agents to a default parent, which will automatically be routed to the local address of the gateway. The ambients outside the LAN will send messages or agents to the global address of the gateway. Thus, the local and global addresses allow the gateway to distinguish between local and global interactions.

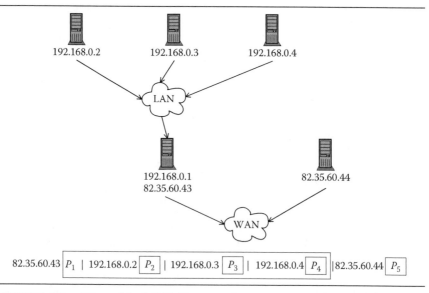

FIGURE 8.1: Hierarchical TCP/IP networks and their corresponding calculus representation.

8.2.7 Using Reduction to Model Network Execution

The CA calculus can be used to model the execution of mobile applications on networks that support the TCP/IP. Note that the details of setting up of a TCP/IP session are below the level of abstraction of the calculus, which only models individual interactions. A host with address IP_1 can be modeled as a site with name IP_1, a port number n can be modeled as a channel with name n and communication between hosts can be modeled using the communication primitives of the calculus. For example, a server with IP address 82.35.60.43 running an ftp service on port 21 and a telnet service on port 23 can be modeled as a site with name 82.35.60.43 containing replicated external inputs on channels 21 and 23. A corresponding client with IP address 82.35.60.44 that interacts with the server can be modeled as a site with name 82.35.60.44 containing a sibling output to the server on the corresponding channels:

$$82.35.60.43 \boxed{!21^{\uparrow}(args).Ftp \mid !23^{\uparrow}(args).Telnet \mid Server}$$
$$\mid 82.35.60.44 \boxed{82.35.60.43 \cdot 21 \langle v \rangle.P \mid Client} \mid Network$$

The migration of agents between hosts on the network can also be modeled using the primitives of the calculus. For example, a server with IP address 82.35.60.43 that accepts an agent on port 3001 can be modeled as a site with name 82.35.60.43 containing an accept on channel 3001. In general, when two ambients interact over a network the channel corresponds to a port number, and when two ambients interact locally the channel corresponds to a simple identifier.

A private communication channel established between two hosts can be modeled using channel restriction. For example, a private *ssh* channel that was established between a client and a server using the SSH protocol can be modeled as

$$\nu ssh \ (client \boxed{server \cdot ssh \langle n \rangle.P \mid Client} \mid server \boxed{ssh^{\uparrow}(m).Q \mid Server}) \mid Network$$

The use of restriction to limit the scope of the *ssh* channel between client and server guarantees, at an abstract level, that other entities in the network cannot interfere with communication on this channel. For a more detailed model of establishing private channels over a network, encryption and decryption mechanisms can be added as an extension to the calculus, in the style of Refs. [1,4].

Up to this point, TCP/IP networks have been modeled in the CA calculus by mapping IP addresses to sites and port numbers to channels. Implicitly, this approach assumes that each site corresponds to a separate device on the network, with its own IP address. An alternative, more flexible approach is to map each site to a socket address, where a socket address consists of an IP address and a port number. This allows multiple sites to run on the same device, where each site uses a separate port on the device. Arbitrary string names can then be used as channels and multiple channels can share the same port, resulting in a significantly more flexible interaction model. This can be implemented by adding a thin layer of multiplexing above the TCP/IP, in order to allow a given site to interact on an arbitrary number of named channels.

8.2.8   Resource Monitoring Application

This section describes how the CA calculus can be used to specify an example mobile application, in which a mobile agent monitors a resource on a remote server.

Figure 8.2 presents a formal specification of the resource monitoring application, expressed as a process of the CA calculus:

- The *client* tries to send its name and an acknowledgment channel *ack* to the server on the *register* channel. After sending the registration request, the client waits for a login channel *x* on the acknowledgment channel. After receiving the login channel, the client creates a new *monitor* agent, which tries to leave on the *logout* channel, enter the server on the login channel and then execute the process *P*. In parallel, the client tries to release an agent on the logout channel and then execute the process *Q*. The client can also execute other processes in parallel, represented by the process *C*.

- The *server* continually listens on the *register* channel for a client name *c* and an acknowledgment channel *k*. Each time a registration request is received, the server creates a new *login* channel. The server tries to send the login channel to the client on the acknowledgment channel, accept an agent on the login channel and then execute the process *R*. The server can handle multiple requests concurrently, represented by the process *S*.

- The network can contain arbitrary agents and machines, represented by the process *N*. These agents and machines can potentially try to interfere with the interactions between the client and the server.

Figure 8.3 uses the graphical CA calculus to formally describe an execution scenario for the resource monitoring application. The corresponding textual execution scenario is presented in Figure 8.4:

1. The client sends its name and an acknowledgment channel to the server on the register channel.

2. The server creates a new *login* channel and sends it to the client on the acknowledgment channel.

3. The client creates a new monitor agent, which leaves the client on the logout channel.

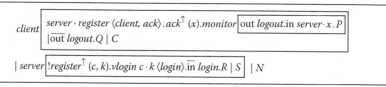

FIGURE 8.2:   Resource monitoring specification.

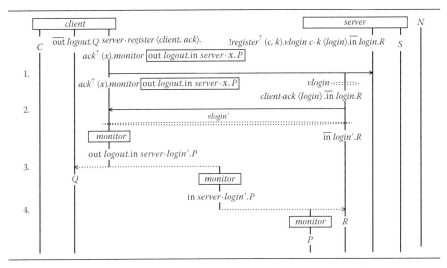

FIGURE 8.3:   Graphical calculus execution scenario, where $login' \notin \text{fn}(P, R)$.

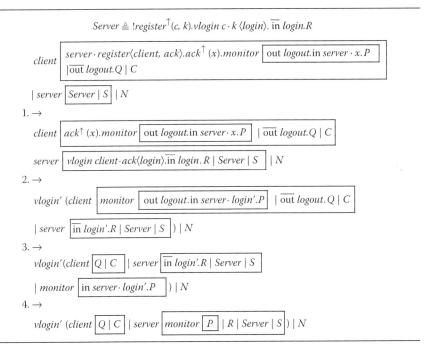

FIGURE 8.4:   Calculus execution scenario, where $login' \notin \text{fn}(P, Q, R, S, C)$.

4. The monitor agent enters the server on the login channel and executes the process P, which monitors the resource on the server. In parallel, the server executes the process R, which forward information about the resource to the monitor.

Note that the graphical representation offers greater flexibility in displaying the scope of restricted names than the corresponding textual representation. In particular, the scope of the ring for the *login'* channel can be graphically adjusted to represent the fact that *login'* $\notin$ fn(Q, S, C).

This resource monitoring example illustrates how the CA calculus can be used to specify various security mechanisms for mobile applications. On receiving a registration request, the server creates a fresh login channel and sends this to the client over an acknowledgment channel. The login channel acts as a key, which the client can use to send a monitor agent to the server. The server will only allow a single monitor to enter using this key, thereby ensuring strict access control to the server. Similarly, the monitor agent can only leave the client on the logout channel. This prevents other agents that do not know the name of the logout channel from leaving the client without permission. Furthermore, other processes inside the client that do not know the name of the acknowledgment channel cannot interfere with communication from the server, and therefore will be unable to acquire the login channel. More generally, a wide range of analysis techniques that have been developed for related calculi can also be applied to the CA calculus, in order to reason about the security properties of applications. Some of these techniques are discussed in Ref. [31], including proving safety to prevent runtime errors, using channel types to ensure reliable communication, and using syntactic constraints to prevent ambient impersonation.

8.3   Channel Ambient Machine

The CA calculus was designed as a high-level formalism for specifying mobile applications. A given application can first be specified as a process of the calculus, and a number of security properties can then be verified for the specification. Once an application has been formally specified in this way, the next stage is to develop a corresponding implementation. One way to achieve this is to use a runtime to execute calculus processes. This approach bridges a gap between specification and implementation by allowing the specification to be executed directly. In addition, the approach ensures that any security properties of the calculus specification are preserved during execution, provided the runtime is implemented correctly. Support for language interoperability can also be provided in the runtime, allowing the mobile and distributed aspects of an application to be specified in the calculus, and the remaining local and functional aspects to be written in a chosen target language. The interaction between these two aspects can be achieved using the communication primitives of the calculus, allowing a clean separation of concerns in the spirit of modern coordination languages such as provided in Ref. [3]. More importantly, the abstract machine provides an efficient way of executing a large number of parallel threads that constantly synchronize and move between machines. In order to ensure that the runtime is implemented correctly, an abstract machine can be defined as a formal specification of how the runtime should behave. The correctness of the abstract machine can then be verified with respect to the underlying calculus. This section presents an abstract machine for the CA calculus, known as the CA Machine (CAM). A proof of correctness is given in Section 8.3.8.

8.3.1 Syntax of Machine Terms

$$V ::= vn \; V \quad \text{Restriction} \quad (8.18)$$
$$\mid \quad A \quad \text{List} \quad (8.19)$$

$$z ::= \underline{z} \quad \text{Blocked} \quad (8.23)$$
$$\mid z \quad \text{Unblocked} \quad (8.24)$$

$$A, B, C ::= \quad [] \quad \text{Empty} \quad (8.20)$$
$$\mid \alpha.P :: C \quad \text{Action} \quad (8.21)$$
$$\mid a \boxed{A} :: C \quad \text{Ambient} \quad (8.22)$$

DEFINITION 8.7 *Syntax of* CA Machine, *where* (::) *denotes list composition. For convenience,* $!\alpha.P$ *is written as syntactic sugar for an expanded replicated action* $\alpha.(P \mid !\alpha.P)$.

The syntax of the CA Machine is defined using terms U, V, where each term represents a corresponding calculus process. In general, a machine term is a list of actions $\alpha.P, \ldots, \alpha'.P'$ and ambients $a\boxed{A}, \ldots, a'\boxed{A'}$, with a number of top-level restricted names $n, \ldots, n'$. The actions and ambients in the list can be either blocked or unblocked, and each ambient contains its own list. The notation $\alpha.P$ denotes either a blocked action $\underline{\alpha}.P$ or an unblocked action $\alpha.P$, while the notation $a\boxed{A}$ denotes either a blocked ambient $\underline{a}\boxed{A}$ or an unblocked ambient $a\boxed{A}$. The full syntax of the CA Machine is presented in Definition 8.7. Therefore, a machine term can be viewed as a tree of actions and ambients with a number of top-level restricted names, where the nodes of the tree are ambients, and the leaves are actions:

$$vn, \ldots, vn' \; \alpha.P :: \ldots :: \alpha'.P' :: a\boxed{A} \ldots :: a'\boxed{A'} :: []$$

The machine executes a given term by scheduling an unblocked action $\alpha.P$ somewhere in the tree. If there is a corresponding blocked coaction then the two actions can interact and a reduction can occur. If there is no corresponding blocked coaction, then the scheduled action is blocked. If all of the actions in a given ambient $a\boxed{A}$ are blocked then the ambient itself is blocked to $\underline{a}\boxed{A}$. The machine continues scheduling actions in this way until all the actions and ambients in the tree are blocked, and no more reductions can occur.

8.3.2   Using Construction to Encode a Process to a Machine Term

In order for a process to be executed by the CA Machine, it must first be converted to a corresponding machine term. This can be achieved by defining a suitable encoding function. The encoding function (P) encodes a given process P to a corresponding machine term using a construction operator, as described in Definition 8.8. The construction $P : []$ adds the process P to the empty list $[]$.

$$(P) \triangleq P : []$$

DEFINITION 8.8   *Encoding* CA *to* CA Machine

In general, construction is used to add a process to a machine term or to add an ambient and a term to a machine term. The construction $P : V$ adds the process P to the term V, and the construction $a \boxed{U} : V$ adds the ambient a and the term U to the term V. Note that the symbol (:) is overloaded, and carries a different type depending on the context in which it is used. In the first case, it takes a process and a term as arguments, and in the second case, it takes an ambient name and two terms as arguments. Also note the distinction between the function (:), which expands processes and terms so they can be added to a list, and the syntactic construct (::), which denotes an action or ambient at the head of a list. The full definition of construction is given in Definition 8.9, where $\text{fn}(V)$ is the set of free names in V.

A process is added to a term by extending the scope of any restricted names in the term to the top level, and then adding the process to the list inside the scope of the restricted names (8.25). A process is added to a list by discarding the null process (8.26), adding each process in a parallel composition separately (8.27), expanding replicated actions (8.29) and extending the scope of any restricted names in the process to the top level (8.28). Any processes inside an ambient are first encoded to corresponding terms, which can then be added to the list (8.30). It is worth emphasizing here that a parallel composition $P_1 \mid \ldots \mid P_N$ will ultimately be added to a term A as a list of processes $P_1 : \ldots : P_N : A$, in sequence. Individual processes will then be selected for execution using a suitable scheduling algorithm.

An ambient and a term are added to a term by extending the scope of any restricted names in both terms to the top level (8.33), and then placing the ambient at the head of the list inside the scope of the restricted names (8.34).

8.3.3   Using Selection to Schedule a Machine Term

In order to execute a term of the CA Machine, the term must be matched with the left-hand side of one of the execution rules. This can be achieved by defining

$$n \notin \mathrm{fn}(P) \Rightarrow P:(vn\ V) \triangleq vn\ (P:V) \tag{8.25}$$

$$\mathbf{0}:A \triangleq A \tag{8.26}$$

$$(P \mid Q):A \triangleq P:Q:A \tag{8.27}$$

$$n \notin \mathrm{fn}(P:A) \Rightarrow (vm\ P):A \triangleq vn\ (P_{\{n/m\}}:A) \tag{8.28}$$

$$\mathrm{fn}(\alpha) \not\subseteq \mathrm{bn}(\alpha) \Rightarrow {!}\alpha.P:A \triangleq \alpha.(P \mid {!}\alpha.P):A \tag{8.29}$$

$$a\boxed{P}:A \triangleq a\boxed{P:[]}:A \tag{8.30}$$

$$\alpha.P:A \triangleq \alpha.P::A \tag{8.31}$$

$$n \notin \mathrm{fn}(a,V) \Rightarrow a\boxed{vn\ U}:V \triangleq vn\ (a\boxed{U}:V) \tag{8.32}$$

$$n \notin \mathrm{fn}(a,A) \Rightarrow a\boxed{A}:vn\ V \triangleq vn\ (a\boxed{A}:V) \tag{8.33}$$

$$a\boxed{A}:C \triangleq a\boxed{A}::C \tag{8.34}$$

DEFINITION 8.9 *Construction in* CA *Machine*

a selection function, which rearranges a given term to match another term. The selection function is a simplification of the structural congruence rules of the calculus, which takes into account the additional constraints on the syntax of machine terms.

The full definition of selection is given in Definition 8.10, where the append function $A@A'$ is used to concatenate the list A with the list A', and $A \succ A'$ means that the term A can be rearranged to match the term A'. A list can match itself, an action or ambient inside a list can be brought to the front of the list (8.35)–(8.36), and a list can be rearranged inside an ambient (8.37). Unlike structural congruence, the selection function is neither symmetric nor transitive. This minimizes the amount of rearranging that is needed to match a term with the left-hand side of one of the execution rules, thereby increasing the efficiency of the machine.

$$A@\alpha.P::A' \succ \alpha.P::A@A' \tag{8.35}$$

$$A@b\boxed{B}::A' \succ b\boxed{B}::A@A' \tag{8.36}$$

$$A \succ A' \Rightarrow a\boxed{A}::C \succ a\boxed{A'}::C \tag{8.37}$$

DEFINITION 8.10 *Selection in* CA *Machine*

8.3.4   Using Unblocking to Prevent Deadlocks during Execution

The CA Machine distinguishes between blocked and unblocked actions, in order to efficiently schedule a sequence of actions to execute. In principle, the execution rules of the machine are similar to those of the calculus, except that a given interaction takes place between an action and a corresponding blocked coaction. As a result, the machine needs to prevent both an action and a corresponding coaction from being blocked simultaneously. If this happens, the blocked action and coaction may remain deadlocked, each waiting indefinitely for the other to unblock. Such deadlocks could arise when an ambient containing a blocked action moves to a new location containing a corresponding blocked coaction. In order to ensure that deadlocks do not occur during a migration, an unblocking function is used to unblock the contents of an ambient when it moves to a new location. This allows the ambient to rebind to its new environment, by giving any blocked actions in the ambient the chance to interact with their new location.

$$\lfloor [] \rfloor \triangleq [] \qquad (8.38)$$

$$\lfloor \alpha.P :: C \rfloor \triangleq \alpha.P :: \lfloor C \rfloor \qquad (8.39)$$

$$\lfloor a\boxed{A} :: C \rfloor \triangleq a\boxed{A} :: \lfloor C \rfloor \qquad (8.40)$$

DEFINITION 8.11   *Unblocking in* CA *Machine*

The unblocking function $\lfloor A \rfloor$ unblocks all of the top-level actions in a given list A. The full definition of unblocking is given in Definition 8.11. The function is used to unblock the contents of a given ambient $a\boxed{A}$ when the ambient moves to a new location. Since the execution rules only allow adjacent ambients to interact, nested ambients inside A cannot interact directly with the new environment. Therefore, only the top-level actions in A need to be unblocked. For example, suppose the ambient a contains a blocked parent output, and has just moved inside an ambient b, which contains a corresponding blocked internal input. Unblocking the contents of a allows it to check its new environment for potential interactions, such as communicating with its new parent:

$$b\boxed{x(m).Q :: a\boxed{\lfloor x^{\uparrow}\langle n\rangle.P :: A \rfloor} :: B} :: D$$

On the other hand, suppose ambient a contains a child c with a blocked parent output. In this case the contents of c do not need to be unblocked, since its immediate environment has not changed. In particular, there will be no blocked actions inside a that were not already present before the migration took place:

$$b\left[\,x(m).Q::a\ \left\lfloor c\left[\,x^{\uparrow}\langle n\rangle.P::C\,\right]::A\right\rfloor\,\right]::B\ ::D$$

8.3.5 Using Reduction to Execute a Machine Term

The reduction rules of the CA Machine describe how a machine term can be executed. By definition, the relation $V \longrightarrow V'$ is true if the machine can reduce the term V to the term V' in a single execution step. The reduction rules of the machine are derived from the reduction rules of the calculus. In the calculus reduction rules, an ambient can send a value to a sibling or to its parent over a channel, and can enter a sibling or leave its parent over a channel. Each of these rules is mapped to four corresponding reduction rules in the machine, to allow an action to interact with a corresponding blocked coaction and vice versa, and to block an action or coaction if no interaction can take place. As with the calculus, there is also a rule to allow reduction inside a restriction and inside an ambient, and to allow matching terms to perform the same reductions. However, unlike the calculus, there is no rule to allow reduction inside a list composition, since each rule is defined over the entire length of a list. Note that the entire list needs to be checked before an action can be blocked, since the machine needs to ensure that there is no corresponding blocked coaction. If the entire list is not checked, then a given action could be blocked even though a suitable blocked coaction may be present in another part of the list. This would result in a form of execution deadlock, in which both an action and a corresponding coaction are blocked simultaneously.

The full definition of reduction is given in Definition 8.12, where $A \succ A'$ means that the term A can be rearranged to match the term A', and $\lfloor A \rfloor$ unblocks any top-level blocked actions in A, as defined previously.

8.3.6 Modeling Execution on TCP/IP Networks

The reduction rules of the CA Machine can be used to model the execution of mobile applications on TCP/IP networks with broadcast. The main rules used are explained below:

(8.41) An ambient a can send a value v to a site b on channel x using TCP/IP. The blocked external input on channel x inside site b can be implemented by binding a socket to a port x inside a machine with IP address b, and waiting for a connection on port x. The sibling output to b can be implemented by connecting a socket to IP address b on port x, and then sending the value v over the network from a to b.

(8.45) A site a can send a value v to its parent over a channel x using TCP/IP. The blocked external input on channel x inside the parent can be implemented by binding a socket to a port x inside the parent machine. The output to the parent can be implemented by connecting a socket to the IP address

$$C \succ b\boxed{x^\uparrow(u).Q::B}::C' \Rightarrow a\boxed{b \cdot x\langle v\rangle.P::A}::C \longrightarrow a\boxed{P:A}:b\boxed{Q_{\{v/u\}}:B}:C' \quad (8.41)$$

$$C \nsucc b\boxed{x^\uparrow(u).Q::B}::C' \Rightarrow a\boxed{b \cdot x\langle v\rangle.P::A}::C \longrightarrow a\boxed{b \cdot x\langle v\rangle.P::A}::C \quad (8.42)$$

$$C \succ a\boxed{b \cdot x\langle v\rangle.P::A}::C' \Rightarrow b\boxed{x^\uparrow(u).Q::B}::C \longrightarrow a\boxed{P:A}:b\boxed{Q_{\{v/u\}}:B}:C' \quad (8.43)$$

$$C \nsucc a\boxed{b \cdot x\langle v\rangle.P::A}::C' \Rightarrow b\boxed{x^\uparrow(u).Q::B}::C \longrightarrow b\boxed{x^\uparrow(u).Q::B}::C \quad (8.44)$$

$$C \succ \underline{x(u).Q}::C' \Rightarrow a\boxed{x^\uparrow\langle v\rangle.P::A}::C \longrightarrow a\boxed{P:A}:Q_{\{v/u\}}:C' \quad (8.45)$$

$$C \nsucc \underline{x(u).Q}::C' \Rightarrow a\boxed{x^\uparrow\langle v\rangle.P::A}::C \longrightarrow a\boxed{x^\uparrow\langle v\rangle.P::A}::C \quad (8.46)$$

$$C \succ a\boxed{x^\uparrow\langle v\rangle.P::A}::C' \Rightarrow x(u).Q::C \longrightarrow a\boxed{P:A}:Q_{\{v/u\}}:C' \quad (8.47)$$

$$C \nsucc a\boxed{x^\uparrow\langle v\rangle.P::A}::C' \Rightarrow x(u).Q::C \longrightarrow \underline{x(u).Q}::C \quad (8.48)$$

$$C \succ b\boxed{\overline{in}\,x.Q::B}::C' \Rightarrow a\boxed{in\,b \cdot x.P::A}::C \longrightarrow b\boxed{Q:a\boxed{P:\lfloor A\rfloor}:B}:C' \quad (8.49)$$

$$C \nsucc b\boxed{\overline{in}\,x.Q::B}::C' \Rightarrow a\boxed{in\,b \cdot x.P::A}::C \longrightarrow a\boxed{in\,b \cdot x.P::A}::C \quad (8.50)$$

$$C \succ a\boxed{in\,b \cdot x.P::A}::C' \Rightarrow b\boxed{\overline{in}\,x.Q::B}::C \longrightarrow b\boxed{Q:a\boxed{P:\lfloor A\rfloor}:B}:C' \quad (8.51)$$

$$C \nsucc a\boxed{in\,b \cdot x.P::A}::C' \Rightarrow b\boxed{\overline{in}\,x.Q::B}::C \longrightarrow b\boxed{\overline{in}\,x.Q::B}::C \quad (8.52)$$

$$B \succ \overline{out}\,x.Q::B' \Rightarrow b\boxed{a\boxed{out\,x.P::A}::B}::C \longrightarrow b\boxed{Q:B'}:a\boxed{P:\lfloor A\rfloor}:C \quad (8.53)$$

$$B \nsucc \overline{out}\,x.Q::B' \Rightarrow b\boxed{a\boxed{out\,x.P::A}::B}::C \longrightarrow b\boxed{a\boxed{out\,x.P::A}::B}::C \quad (8.54)$$

$$B \succ a\boxed{out\,x.P::A}::B' \Rightarrow b\boxed{\overline{out}\,x.Q::B}::C \longrightarrow b\boxed{Q:B'}:a\boxed{P:\lfloor A\rfloor}:C \quad (8.55)$$

$$B \nsucc a\boxed{out\,x.P::A}::B' \Rightarrow b\boxed{\overline{out}\,x.Q::B}::C \longrightarrow b\boxed{\overline{out}\,x.Q::B}::C \quad (8.56)$$

$$V \longrightarrow V' \Rightarrow \nu n\,V \longrightarrow \nu n\,V' \quad (8.57)$$

$$B \succ A \wedge A \longrightarrow V' \Rightarrow B \longrightarrow V' \quad (8.58)$$

$$A \longrightarrow V' \Rightarrow a\boxed{A}::C \longrightarrow a\boxed{V'}:C \quad (8.59)$$

$$A \nsucc \alpha.P::A' \wedge A \nsucc b\boxed{B}::A' \Rightarrow a\boxed{A}::C \longrightarrow a\boxed{A}::C \quad (8.60)$$

DEFINITION 8.12   *Reduction in* CA Machine

of the parent on port x and sending the value v over the network from a to its parent.

(**8.49**) An agent a can enter a site b on channel x using TCP/IP. The blocked accept on channel x inside site b can be implemented by binding a socket to a port x inside a machine with IP address b, and waiting for a connection on port x. The enter to b can be implemented by connecting a socket to IP address b on port x, and then sending the agent a over the network in serialized form to b. The agent a can resume execution on arrival.

(**8.58**) The machine can nondeterministically select a given ambient or action to be executed. This rule reflects the inherent nondeterminism of distributed networks, in which parallel reductions can occur in any order.

(**8.59**) The machine can execute the contents of a site independently of other sites. This rule is fundamental for distribution, since it allows sites to be independently executed on different physical machines, by different runtimes. The different sites can then interact with each other using the protocols of the underlying network.

Although the above reduction rules can be readily applied to arbitrary TCP/IP networks, the rules (8.43), (8.47), and (8.51) require additional support for network broadcast. In particular, rule (8.43) allows an external input inside a given site b to interact with a blocked sibling output inside a remote site a. By definition, the rule requires site b to poll all the sites in the network until it finds a site a with a suitable blocked sibling output. This is because an external input does not specify a particular site with which to interact, and can therefore potentially interact with any of the sites in a network. Such interactions can be readily implemented in a LAN by broadcasting to all the sites in the network, but do not scale to WANs with potentially large numbers of sites. More precisely, assume S is the number of sites in a WAN, N is the number of external inputs in b, and R is the number of corresponding sibling outputs to b over a given time period. In cases where $S \times N \gg R$ it is significantly more efficient for the external inputs in site b to block and wait for corresponding sibling outputs to b from remote sites, rather than polling all the sites in the network. This is particularly apparent on the Internet, where S is of the order of millions and R is typically of the same order as N. One way to enforce this constraint is to prevent sibling outputs to remote sites from blocking. This ensures that a given external input inside a site will always block first, and wait for a corresponding sibling output. A similar argument can be applied to the rule that allows a value to be received from a child site (8.47) and the rule that allows an agent to be accepted from a remote site (8.51).

A number of constraints can be placed on the reduction rules of the CA Machine, in order to model the execution of mobile applications on TCP/IP networks without broadcast. The constraints avoid the use of network broadcast by ensuring that a sibling output to a site, an enter to a site and a parent output to a site are never blocked. Instead, these actions repeatedly try to interact with a corresponding blocked coaction in a remote site until a synchronization can occur. In some cases, a synchronization may never occur and the action will remain unblocked indefinitely. In these cases, the interval between synchronization attempts can be increased exponentially over time in order to avoid causing a denial of service attack.

8.3.7   Resource Monitoring Application

This section describes how the CA Machine can be used to execute an example application, in which a mobile agent monitors resources on a remote site.

Figure 8.5 shows how the calculus specification of Figure 8.2 can be encoded to a corresponding machine term, using the encoding function $[P] = P : []$. The unguarded *client* and *server* ambients in the calculus process are encoded to unblocked *client* and *server* ambients in the corresponding machine term. All unguarded parallel compositions are encoded to list compositions. Since there are no unguarded restrictions in the calculus process, the scope of restricted names remains unchanged in the corresponding machine term. The replicated external input on the register channel is expanded according to the corresponding construction rule. For convenience, the replicated action is represented in its unexpanded form.

Once the calculus process has been encoded to a corresponding machine term, it can then be executed by the machine. Figure 8.6 uses a graphical representation of the CA Machine to describe an execution scenario for the encoded term. The representation is based on the graphical syntax of the CA calculus presented in Section 8.2. In addition, the graphical syntax of the machine distinguishes between blocked and unblocked actions and ambients, by placing a gray box under any actions or ambients that are blocked. Each reduction step in the figure is numbered, and the corresponding explanation for each number is given below:

1. The server blocks a replicated external input on the *register* channel, waiting to receive a value on this channel.

2. The client blocks a release on the *logout* channel, waiting to release an agent on this channel.

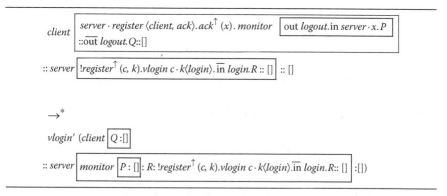

FIGURE 8.5:   Application encoding and execution.

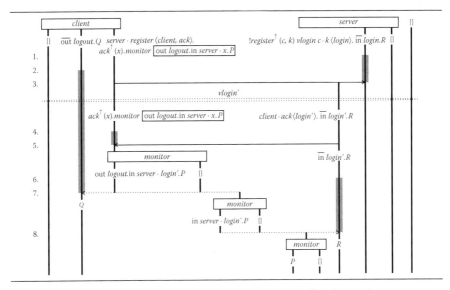

FIGURE 8.6: Graphical machine scenario, where $login' \notin \text{fn}(P, Q, R)$.

3. The client sends its name and an acknowledgment channel to the server on the register channel.

4. The client blocks an external input on the acknowledgment channel, waiting to receive a value on this channel.

5. After creating a globally unique *login'* channel, the server sends this channel to the client on the acknowledgment channel.

6. The server blocks an accept on the login channel, waiting to accept an agent on this channel.

7. After the client creates a *monitor* agent, the agent leaves the client on the logout channel, and the client executes the process Q.

8. The monitor agent enters the server on the login channel and then executes the process P, which monitors the resource on the server. In parallel, the server executes the process R, which forward information about the resource to the monitor.

8.3.8   Correctness of the Channel Ambient Machine

This section proves the correctness of the CA Machine with respect to the CA calculus. Additional proof details are given in Ref. [31].

8.3.8.1   Proving Safety to Prevent Runtime Errors

Safety ensures that the machine always produces a valid term after each execution step. This ensures that the machine does not produce any runtime errors when executing a given term. Theorem 8.1 states that if the machine reduces a term V to V' then V' is a valid machine term.

THEOREM 8.1   *(Reduction Safety)*
$$\forall V.V \in \text{CAM} \wedge V \longrightarrow V' \Rightarrow V' \in \text{CAM}$$

PROOF   By induction on Definition 8.12 of reduction in CA Machine.   ▯

$$[\![vn\ V]\!] \triangleq vn\ [\![V]\!] \tag{8.61}$$
$$[\![[]]\!] \triangleq \mathbf{0} \tag{8.62}$$
$$[\![\alpha.P :: C]\!] \triangleq \alpha.P \mid [\![C]\!] \tag{8.63}$$
$$[\![a\boxed{A} :: C]\!] \triangleq a\boxed{[\![A]\!]} \mid [\![C]\!] \tag{8.64}$$

DEFINITION 8.13   *Decoding* CA Machine *to* CA

8.3.8.2   Proving Soundness to Ensure Valid Execution Steps

Soundness ensures that each execution step in the machine corresponds to a valid execution step in the calculus. This ensures that the machine always performs valid execution steps when executing a given term. The correspondence between machine execution and calculus execution is defined using a decoding function $[\![V]\!]$, which maps a given machine term V to a corresponding calculus process (see Definition 8.13). In general, the decoding function maps the null term to the null process, list construction to parallel composition, blocked machine actions to calculus actions, and blocked machine ambients to calculus ambients. Restricted names, unblocked actions, and unblocked ambients are preserved by the mapping.

Once a decoding function from machine terms to calculus processes has been defined in this way, it is possible to prove the soundness of the machine. Theorem 8.2 states that if the machine reduces a term V to V' then the calculus can reduce the decoding of V to the decoding of V' in at most one step.

THEOREM 8.2   *(Reduction Soundness)*
$$\forall V.V \in \text{CAM} \wedge V \longrightarrow V' \Rightarrow [\![V]\!] \longrightarrow [\![V']\!] \vee [\![V]\!] \equiv [\![V']\!]$$

PROOF By induction on Definition 8.12 of reduction in CA Machine. The decoding function is applied to the left- and right-hand side of each reduction rule and the result is shown to be a valid reduction in CA. ☐

8.3.8.3 Proving Completeness to Ensure Accurate Execution

Completeness ensures that each execution step in the calculus can be matched by a corresponding sequence of execution steps in the machine, up to reordering of machine terms. This ensures that the machine can match all possible execution steps of the calculus when executing the encoding of a given process. The reordering of terms can be defined using a structural congruence relation $V \equiv U$, which allows a given term V to be reordered to match a term U. In general, the structural congruence relation allows segments of a list to be permuted, successive restricted names to be permuted, and unused restricted names to be discarded. Theorem 8.3 states that if the calculus can reduce a process P to P' then the machine can reduce the encoding of P to the encoding of P' in two steps, up to structural congruence.

THEOREM 8.3 *(Reduction Completeness)*
$$\forall P.P \in CA \wedge P \longrightarrow P' \Rightarrow (\![P]\!) \longrightarrow \longrightarrow \equiv (\![P']\!)$$

PROOF By induction on Definition 8.3 of reduction in CA. ☐

8.3.8.4 Proving Liveness to Prevent Deadlocks

Liveness ensures that the machine always produces a deadlock-free term after each execution step. This ensures that the machine does not deadlock when executing a given term. Intuitively, a term is deadlocked if it is unable to match a reduction of the corresponding calculus process. In practice, a term is deadlocked if it contains both a blocked action and a corresponding blocked coaction. For example, the following term contains an ambient a with a blocked sibling output to ambient b on channel x. It also contains a sibling ambient b with a blocked external input on channel x:

$$a \boxed{ \underline{b \cdot x \langle n \rangle} . P :: A } :: b \boxed{ \underline{x^\uparrow (m)} . Q :: B } :: C$$

Since the sibling output and external input are both blocked they cannot interact, because there is no rule that allows an interaction between two blocked actions. In contrast, the interaction is possible in the corresponding calculus process:

$$a \boxed{ \underline{b \cdot x \langle n \rangle} . P \mid [\![A]\!] } \mid b \boxed{ \underline{x^\uparrow (m)} . Q \mid [\![B]\!] } \mid [\![C]\!] \longrightarrow a \boxed{ P \mid [\![A]\!] } \mid b \boxed{ Q_{\{n/m\}} \mid [\![B]\!] } \mid [\![C]\!]$$

A term is also deadlocked if it contains an unblocked action or ambient inside a blocked ambient. For example, the following term contains a blocked ambient a

with an internal input on channel x. Ambient a also contains an ambient b with a parent output on channel x:

$$\boxed{a \,\middle|\, b \,\middle|\, \boxed{x^\uparrow\langle n\rangle.P :: B} \,\middle|\, :: x(m).Q :: A} :: C$$

Since ambient a is blocked, the parent output and external input cannot interact, because there is no rule that allows an interaction inside a blocked ambient. In contrast, the interaction is possible in the corresponding calculus process:

$$\boxed{a \,\middle|\, b \,\middle|\, \boxed{x^\uparrow\langle n\rangle.P \mid [\![B]\!]} \mid x(m).Q \mid [\![A]\!]} \mid [\![C]\!] \longrightarrow \boxed{a \,\middle|\, b \,\middle|\, \boxed{P \mid [\![B]\!]} \mid Q_{\{n/m\}} \mid [\![A]\!]} \mid [\![C]\!]$$

Conversely, a term is deadlock-free if it is able to match all reductions of the corresponding calculus process. In general, a term is deadlock-free if it does not contain both a blocked action and a corresponding blocked coaction, and if it does not contain an unblocked action or ambient inside a blocked ambient. The set of deadlock-free terms is denoted by $\mathrm{CAM}^{\surd}$ and is defined by placing constraints on the set of machine terms CAM (see Definition 8.14).

Once the set of deadlock-free terms has been defined in this way, it is possible to prove that the machine is deadlock-free. Lemma 8.1 ensures that reduction cannot cause a term to deadlock. The lemma states that if the machine reduces a deadlock-free term V to V' then V' is deadlock-free.

$$
\begin{aligned}
V ::= \quad & vn\ V \quad && \text{Restriction} && (8.65)\\
\mid \quad & A \quad && \text{List} && (8.66)\\
A,B,C ::= \quad & [\,] \quad && \text{Empty} && (8.67)\\
\mid \quad & \alpha.P :: C \quad && \text{Action, } \alpha.P = x(m).P \Rightarrow C \not\succ a\,\boxed{x^\uparrow\langle n\rangle.P :: A} :: C' && (8.68)\\
\mid \quad & a\,\boxed{A} :: C \quad && \text{Ambient, } a = \underline{a} \Rightarrow (A \not\succ \alpha.P :: A' \wedge A \not\succ b\,\boxed{B} :: A') && (8.69)
\end{aligned}
$$

$$A \succ \overline{\mathrm{out}\,x}.Q :: A' \Rightarrow A' \not\succ b\,\boxed{\mathrm{out}\,x.P :: B} :: A''$$

$$A \succ b \cdot x\langle n\rangle.P :: A' \Rightarrow C \not\succ b\,\boxed{x^\uparrow(m).Q :: B} :: C'$$

$$A \succ x^\uparrow(m).P :: A' \Rightarrow C \not\succ b\,\boxed{a \cdot x\langle n\rangle.Q :: B} :: C'$$

$$A \succ x^\uparrow\langle n\rangle.P :: A' \Rightarrow C \not\succ x(m).Q :: C'$$

$$A \succ \underline{\mathrm{in}\,b \cdot x}.P :: A' \Rightarrow C \not\succ b\,\boxed{\overline{\mathrm{in}\,x}.Q :: B} :: C'$$

$$A \succ \overline{\mathrm{in}\,x}.P :: A' \Rightarrow C \not\succ b\,\boxed{\mathrm{in}\,a \cdot x.Q :: B} :: C'$$

DEFINITION 8.14   *Syntax of Deadlock-Free Terms* $\mathrm{CAM}^{\surd}$

LEMMA 8.1 *(Deadlock-Free Reduction)*
$$\forall V.V \in \text{CAM}^\surd \wedge V \longrightarrow V' \Rightarrow V' \in \text{CAM}^\surd$$

PROOF   By induction on Definition 8.12 of reduction in CAM.   ☐

The main property of deadlock-free terms is that they should be able match all reductions of the corresponding calculus process. In order to prove this, it is first necessary to prove certain properties about the relationship between reduction, decoding, and encoding.

Lemma 8.2 ensures that machine terms with the same decoding can perform corresponding reductions. The lemma states that if terms U, V are deadlock-free and have the same decoding, and if V can reduce to V' then U can reduce to a term that has the same decoding as V'.

LEMMA 8.2   *(Decoding Reduction)*
$$\forall U, V.U, V \in \text{CAM}^\surd \wedge [\![U]\!]=[\![V]\!] \wedge V \longrightarrow V' \Rightarrow \exists U'.U \longrightarrow^* U' \wedge [\![U']\!]=[\![V']\!]$$

PROOF   By induction on Definition 8.12 of reduction in CAM.   ☐

Lemma 8.3 ensures that a process is structurally congruent to the decoding of its encoding.

LEMMA 8.3   *(Decoding Encoding)*
$$\forall P.P \in \text{CA} \Rightarrow [\![(P)]\!] \equiv P$$

PROOF   Follows by induction on the definition of construction in CAM.   ☐

Lemma 8.4 ensures that a term and the encoding of its decoding both have the same decoding. Note that by Definition 8.13, if two terms have the same decoding then they are equal up to blocking of actions and ambients.

LEMMA 8.4   *(Encoding Decoding)*
$$\forall V.V \in \text{CAM} \Rightarrow [\![([\![V]\!])]\!] = [\![V]\!]$$

PROOF   By induction on Definition 8.13 of decoding in CAM.   ☐

Once these properties have been proved for reduction, decoding, and encoding, it is possible to prove the liveness of the machine. Theorem 8.4 ensures that the machine can match all reductions of the calculus. The theorem states that if a given term V is deadlock-free and the decoding of V can reduce to P' then V can reduce to a term V' that decodes to P', up to structural congruence.

THEOREM 8.4   *(Liveness)*
$$\forall V.V \in \mathrm{CAM}^{\surd} \wedge [\![V]\!] \longrightarrow P' \Rightarrow \exists V'.V \longrightarrow^* V' \wedge [\![V']\!] \equiv P'$$

PROOF   By Lemmas 8.2 through 8.4 and by Theorem 8.3. $\square$

8.3.9   Proving Termination to Prevent Livelocks

Termination ensures that a given machine term will always terminate, provided the corresponding calculus process also terminates. This ensures that the machine does not livelock when executing a given term. According to Theorem 8.5, if the decoding of a given term V cannot reduce, then V will be unable to reduce after a finite number of steps.

THEOREM 8.5   *(Termination)*
$$\forall V \in \mathrm{CAM}.[\![V]\!] \not\longrightarrow \Rightarrow \exists V'.V \longrightarrow^* V' \wedge V' \not\longrightarrow$$

PROOF   The reduction rules of the machine can be divided into two types of rules: blocking rules $V \longrightarrow_b V'$ and interactive rules $V \longrightarrow_i V'$. From the proof of Theorem 8.2 it can be shown that

$$V \longrightarrow_i V' \Rightarrow [\![V]\!] \longrightarrow [\![V']\!]$$
$$V \longrightarrow_b V' \Rightarrow [\![V]\!] \equiv [\![V']\!]$$

Therefore, if $[\![V]\!] \not\longrightarrow$ then either $V \not\longrightarrow$ or $V \longrightarrow_b V'$. Furthermore, if $V \longrightarrow_b V'$ then $[\![V]\!] \equiv [\![V']\!]$ and $[\![V']\!] \not\longrightarrow$. By definition, a given term can only perform a finite number of consecutive blocking reductions, since each term can only contain a finite number of actions or ambients to block. Therefore, by induction V will be unable to reduce after a finite number of steps. $\square$

Note that the termination property does not hold in the case where actions to a remote site are prevented from blocking. This is because a given action will continue polling until a synchronization occurs, which may be indefinitely.

8.4   Channel Ambient Runtime

This section presents a runtime for executing processes of the CA calculus, known as the Channel Ambient Runtime. The runtime is implemented by defining a mapping from the CA Machine to functional program code, using the OCaml language [21].

8.4.1 Local Runtime Implementation

The CA Machine can be used to implement a local runtime, which executes a given calculus process on a single physical device.

The architecture of the local runtime is described in Figure 8.7. The main components of the runtime are an *encoder* and an *interpreter*, where a given process of the CA calculus is executed by the local runtime as follows. First, the process is encoded to a corresponding runtime term by the encoder. The resulting term is then executed by the interpreter in steps, according to a reduction relation. At each step, the interpreter transforms the initial term into an updated term. Execution continues until no more reductions are possible.

The implementation of the local runtime is almost a direct mapping from the CA Machine to functional program code. Full details of the mapping are given in Ref. [31].

8.4.2 Distributed Runtime Implementation

The CA Machine can also be used to implement a distributed runtime, which executes a given calculus process over multiple devices in a hierarchical TCP/IP network. A runtime term containing multiple sites is executed by mapping each site in the term to a separate distributed runtime. For example, the following term is executed using three separate runtimes, one for each of the sites s_0, s_1, s_2:

$$v\tilde{z}\, s_0 \boxed{S_0 :: s_1 \boxed{S_1} :: s_2 \boxed{S_2} :: [\,]} :: [\,]$$

The top-level restricted values are assumed to span all of the distributed runtimes and do not need to be implemented explicitly, since they form part of an implicit set of global values in the network.

The terms inside separate distributed runtimes can interact with each other using standard network protocols. This is achieved by configuring each distributed runtime to act as a server on the address of the site it is executing. The tree structure of the network is preserved by adding a link from each site to its parent. These links are implemented by placing each site inside a proxy of its parent, using the local address of the parent. If no parent is specified, then a default root parent is used. For example,

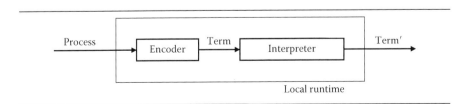

FIGURE 8.7:   Architecture of the local runtime.

the above term is implemented by executing each of the following three terms on a separate distributed runtime:

$$root\,{}^{m_0}_{s_0}\boxed{S_0}::[]::[],\quad m_0\,s_1\boxed{S_1}::[]::[],\quad m_0\,s_2\boxed{S_2}::[]::[]$$

The site s_0 is placed inside a proxy of the default root parent, and each of the sites s_1 and s_2 is placed inside a separate proxy m_0 of the site s_0. The name m_0 of the proxy corresponds to the local address of s_0, which is used for receiving messages and agents from sites that are logically contained in s_0. Although each proxy site is initially empty, during the course of execution it can be used to temporarily store and execute agents that are in transit between sites. Thus, in addition to providing a link to the parent site, a proxy can also be used to decentralize the execution of the contents of the parent. For example, the following runtime term represents a site s with contents S, inside a proxy site m with child agents $g_1,\dots,g_N$:

$$m\,s\boxed{S}::g_1\boxed{G_1}::\dots::g_N\boxed{G_N}::[]::[]$$

Although the child agents $g_1,\dots,g_N$ have already left s, they can still temporarily remain on the runtime inside the proxy m while in transit to their next destination. If one of the agents needs to perform a local interaction inside the parent then it will be moved to the parent on demand. Otherwise, it will remain inside the proxy until it moves to its next destination.

    The architecture of the distributed runtime is described in Figure 8.8. The main components of the runtime are an encoder, an interpreter, a daemon and shared data, and channel buffers. A given process of the CA calculus is executed by the distributed runtime as follows:

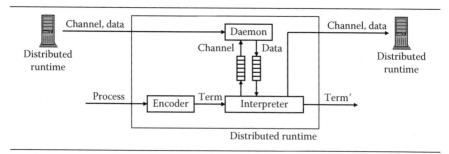

FIGURE 8.8:   Architecture of the distributed runtime.

1. First, the process is encoded to a corresponding runtime term by the encoder. The process is assumed to be of the form $m \boxed{s \boxed{P}}$, where s is the main site to be executed by the runtime, P is contents of the site, and m is a proxy of the parent site. The process is encoded to a runtime term of the form $m \boxed{s \boxed{P : []} :: []} :: []$.

2. The resulting term is then executed by the interpreter in steps, according to a reduction relation. At each step, the interpreter transforms the initial term into an updated term.

3. Since the interpreter is single-threaded, a separate daemon is needed in order to receive data from distributed runtimes. The data can be either a message received by the daemon or a mobile agent accepted by the daemon over a given channel. Any data received by the daemon is added to the data buffer. After each execution step, the interpreter checks this buffer for incoming data, which is added to the site s.

4. During execution, the interpreter can instruct the daemon to receive data on a given channel. The type of data that can be received is determined by the type of the channel, where different channel types are used to distinguish between incoming messages and agents. The interpreter can also send data to distributed runtimes directly over a given channel.

5. Execution continues until no more reductions are possible, after which the interpreter goes into a blocked state, waiting for an interrupt from the data buffer to signal the arrival of new data. As soon as any data arrives it is added to the site s, allowing the interpreter to resume executing.

The reduction rules of the CA Machine are implemented in a distributed setting by making a distinction between sites and agents, as described in Section 8.2, and by taking into account the constraints described in Section 8.3. Although the changes required to go from the local reduction rules to the distributed reduction rules are relatively minor, the corresponding changes in the implementation are more substantial. In particular, the CA Machine defines a single reduction rule for performing a sibling output, parent output, or enter to an ambient, but the distributed implementation needs to distinguish between performing these actions to a remote site or to a local agent. Similarly, a single reduction rule is defined for blocking an external input or accept inside an ambient, but the distributed implementation needs to distinguish between blocking these actions inside a remote site or inside a local agent. As a result, additional functions are required to enable distributed runtimes to interact over a network. Full details are provided in Ref. [31].

8.4.3 Enhanced Runtime Implementation

The distributed runtime has been enhanced in a number of ways in order to improve the efficiency of process execution, while conserving network bandwidth:

- The runtime terms are implemented using a map data structure instead of a list, in order to improve the efficiency of lookups when searching for a blocked coaction.

- A deterministic selection algorithm is used in order to improve the efficiency of runtime execution, while ensuring a suitable notion of fairness.

- Network masks are used to conserve bandwidth, by preventing spurious attempts to interact with inaccessible network addresses.

Full details of the optimizations are described in Ref. [31].

An indication of runtime performance can be obtained by characterizing the efficiency of the enhanced runtime execution algorithm. Essentially, each site in a distributed application is executed by a separate runtime, which can be stopped and started independently. This allows a large number of parallel runtimes to be used in a given application with low initialization overhead. Each individual runtime executes in a loop, as described in Section 8.4.2, where the state A of the runtime consists of a queue of active actions, a queue of blocked actions and a queue of ambients, recursively:

$$\alpha_1.P_1 ::: \ldots ::: \alpha_i.P_i \; , \; \underline{\alpha_j.P_j} ::: \ldots ::: \underline{\alpha_k.P_k} \; , \; a_1 \boxed{A_1} ::: \ldots ::: a_N \boxed{A_N}$$

Each blocked or active action corresponds to a separate, self-contained thread, allowing a large number of threads to be executed in parallel with low overhead. The runtime scheduling algorithm simply picks the first action in the active queue, which takes constant time, and then checks whether there is a corresponding blocked coaction. The check will depend on the type of coaction that is required. If the coaction is in a separate runtime, the action is either blocked immediately or sent over the network to the runtime. If the coaction is in the same runtime, it will either be inside the same ambient, the parent, a child, or a sibling. For a coaction inside the same ambient or inside the parent, the lookup time is $O(N)$, where N is the number of blocked coactions inside the ambient. In practice, the machine state is further refined by grouping blocked coactions of the same type (e.g., all the $in\,a \cdot x$ actions to a given ambient a on a given channel x) into a map data structure, so that the lookup time is $O(\log N)$, where N is the number of different types of blocked coactions inside the ambient. For coactions inside a sibling or child ambient, the lookup time is $O(N_A \times \log N)$, where N_A is the number of sibling or child ambients. This gives us an upper bound on the cost of computing a given interaction. If no coaction is found then the chosen action is blocked. If no actions are left in the active queue then an action inside the first active ambient in the ambient queue is chosen, recursively. In the worst case, the time to schedule an action will be $O(N_T)$, where N_T is the total number of active ambients in the runtime. If no active actions are found anywhere inside an ambient or its children then the ambient is blocked, allowing the ambient to be skipped in future scheduling. If all the actions or ambients in a runtime are blocked, then the entire runtime goes into a blocked state, waiting for messages from the network. This combination of lookup and scheduling will determine the cost of executing a given program, and will vary depending on the number of ambients and actions in the program. As an example, the full agent tracker application in

Section 8.5 took 0.136 s when all the sites were executed as local agents on the same runtime. This includes the disk I/O time for writing the initial and final states of the runtime to separate HTML files. When each site is executed on a separate runtime, the execution time is dominated by the properties of the underlying network.

8.4.4 Channel Ambient Language

The CA calculus has been used as the basis for a programming language for mobile applications, known as the Channel Ambient Language (CA Language). The language extends the calculus with various programming constructs such as data structures, process definitions and system calls, and the execution rules of the language are based largely on the reduction rules of the CA Machine. Although the language itself is merely a prototype, it gives a useful indication of how next-generation programming languages for mobile applications can be based on a formal model. The language and runtime system are available from Ref. [30].

8.4.4.1 Execution

The CA Runtime can be used to execute a file *source.ca* by typing the command *cam.exe source.ca* in a console. The file *source.ca* contains the code for a single site with a given network address. Before the runtime executes the program, it checks to see whether the program code is well typed. The type system for the CA Language is defined in Ref. [31] and ensures that only values of the correct type can be sent and received over channels. When the runtime executes a given program, its internal state is modified with each execution step, according to the reduction rules of the CA Machine.

The execution state of the runtime is regularly streamed to the file *state.html*, which can be viewed in a web browser and periodically refreshed to display the latest state information. For reference, the initial state of the runtime is stored in the file *start.html*.

The state of the runtime looks very much like a source file, and contains any program code that is currently being executed by the runtime. Each currently executing thread is displayed next to a *thread* icon ⊞, where icons of blocked threads are underlined ⊞. Each currently executing ambient is enclosed in a box with an *open* folder icon ⊟ next to the ambient's name, where icons of blocked ambients are underlined ⊟. The state of the ambient can be hidden by clicking on this icon, which collapses the contents of the ambient and displays a *closed* folder icon ⊟ next to the ambient's name. If the entire contents of the runtime are blocked, then execution is suspended until the runtime receives an interrupt from the network, announcing the arrival of new messages or ambients to be executed.

8.4.4.2 System Calls

A given runtime executes a single site, where the top-level process inside the site is specified by the user, but new agents can enter and leave the site dynamically.

The security model only allows the top-level process inside the site to perform a system call, by doing a parent output on one of the predefined *System* channels. These include channels for printing on a console, reading and writing to files, executing external programs, and forwarding messages to remote sites. If a parent output on a *System* channel is performed inside a nested agent, then it is treated as an ordinary parent output. These parent outputs can be forwarded to the parent site through the use of private forwarding channels. Where necessary, a separate forwarding channel can be defined for each agent or group of agents, allowing fine-grained access permissions to be implemented for each system call. If no forwarding channels are implemented, then none of the agents inside a site will be able to perform a system call.

The runtime also allows files to be stored in an ambient in binary form and sent over channels like ordinary values. This can be used to program a wide range of applications. For example, it can be used to program an ambient that moves to a remote site, retrieves a Java class file, and then moves to a new site to execute the retrieved file. A similar approach can be used to coordinate the execution of multiple prolog queries on different sites, or coordinate the distribution and retrieval of multiple HTML forms.

8.4.5   Resource Monitoring Application

The CA Language can be used to program the resource monitoring example of Section 8.2. The code for the client and server is given in Figure 8.9. The syntax of the code is similar to the syntax of the calculus, with minor variations such as using a semicolon instead of a dot for action prefixes. The code also contains type

---

```
192.168.0.3:3000
[ let server = 192.168.0.2:3001 in
  let client = 192.168.0.3:3000 in
  let register =
      register:<site,<<migrate>>> in
  let logout = logout:<migrate> in
  let Monitor() = print<monitor> in
  let Continue() = print<continue> in
( new ack:<<migrate>>
  server.register<client,ack>;
  ack^(x);
  monitor
  [ out logout;
    in server.x;
    Monitor<>
  ]
| -out logout;
  Continue<>
) ]
```

```
192.168.0.2:3001
[ let Report() = print<report> in
  let register =
      register:<site,<<migrate>>> in
  !register^(c,k);
    new login:<migrate>
    c.k<login>;
    -in login;
    Report<> ]
```

FIGURE 8.9:   Program code for the resource monitoring application.

annotations of the form $n : T$, where n is a variable name and T is a type expression, and value declarations of the form *let* $n = V$ *in* P, where V is a value expression. Site values are of the form $IP : i$, where IP is an IP address and i is a port number. Channel values are of the form $n : \langle T \rangle$, where n is a name and T is the type of values carried by the channel. The additional type information helps to preserve type safety when remote machines interact over global channels, since two channels are only equal if both their names and types coincide.

The server runs on port 3001 of IP address 192.168.0.2, while the client runs on port 3000 of IP address 192.168.0.3. In this example the *Report*, *Monitor*, and *Continue* processes are defined as simple outputs. In general they can be defined as arbitrarily complex processes.

The two programs are executed on separate client and server sites by separate runtimes. Initially, each runtime parses the program code and substitutes any value, type, or process definitions. Top-level parallel compositions are expanded accordingly and the client creates a new acknowledgment channel, where ack_1 is an abbreviation for a private channel name. All occurrences of the name ack are substituted with the ack_1 inside the client. The initial internal state of the client and server runtimes are shown in Figure 8.10. These states correspond to HTML output that is automatically generated by the runtimes after each execution step. The final state of the client and server runtimes are shown in Figure 8.10. The monitor agent has entered the server on the login channel and executes a process to monitor the resource on the server. In parallel, the server executes a process to forward information about the resource to the monitor. Further examples can be downloaded from Ref. [30], along with debugging information about the initial and final states of the runtime.

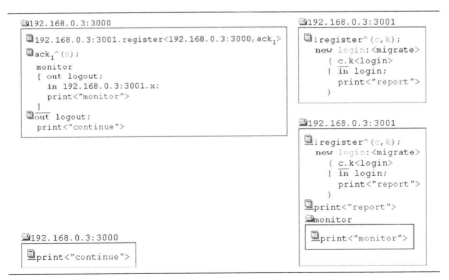

FIGURE 8.10: Initial (top) and final (bottom) states for the server and client defined in Figure 8.9.

8.5   Agent Tracker Application

The CA Language can be used to develop an agent tracker application, which keeps track of the location of registered client agents as they move between trusted sites in a network. The application is inspired by previous work on location-independence, studied in the context of the Nomadic π-calculus [41] and the Nomadic Pict programming language [42]. Algorithms for reliably tracking the location of agents are fundamental in many distributed applications. One example is in the area of information retrieval, where multiple agents visit specialized data repositories to perform computation-intensive searches, periodically communicating with each other to update their search criteria based on high-level goals. A simple example of an information retrieval application is the organization of a conference using mobile agents. Each agent is given a dedicated task, such as flight and train reservations, hotel reservations, sightseeing tours, restaurant bookings, etc. The various agents then move between dedicated sites in order to achieve their specific goals, periodically communicating with each other to resolve conflicts in scheduling or availability. In general, algorithms for tracking the location of migrating agents are a key feature of mobile agent platforms, and are part of the Mobile Agent System Interoperability Facility standard for mobile agent systems [35]. Many agent platforms rely on tracking algorithms to forward messages between migrating agents, including, for example, the JoCaml system. In this respect, the CA Language can be used to develop secure, extensible platforms for mobile agents, using tracker algorithms similar to the one presented.

This section describes a decentralized version of the agent tracker algorithm given in Ref. [40]. The algorithm uses multiple *home servers* to track the location of mobile clients, and relies on a locking mechanism to prevent race conditions. The locking mechanism ensures that messages are not forwarded to a client while it is migrating between sites, and that the client does not migrate while messages are in transit.

The agent tracker application is specified using the *Server*, *Tracker*, *Site*, and *Client* processes defined in Figure 8.11. The corresponding program code for these definitions is also presented. The *Site* process describes the services provided by each trusted site in the network. An agent at a trusted site can receive a message m on channel x from a remote agent via the *child* channel. Likewise, it can forward a message m on channel x to a remote agent a at a site s via the *fwd* channel. Visiting agents can enter and leave a trusted site via the *login* and *logout* channels, respectively. An agent at a trusted site can check whether a given site s is known to be trusted by sending an output on channel s. The *Server* process describes the behavior of a home server that keeps track of the location of multiple clients in the network. A *client* agent can register with a server site via the *register* channel, which creates a new *tracker* agent to keep track of the location of the client. The *Tracker* and *Client* processes describe the services provided by the tracker and client agents, respectively.

Figure 8.12 describes a scenario in which a client registers with its home site. The scenario uses a simplified version of the graphical representation for the CA calculus

$Server(s_i) \triangleq$

$!register(client, ack).vtracker\ vsend\ vmove\ vdeliver\ vlock$

   $(tracker\ \boxed{Tracker\langle client, send, move, deliver, lock \rangle}$

   $|\ client/ack\langle tracker, send, move, deliver, lock \rangle$

   $|\ tracker/lock\langle s_i \rangle)$

$Tracker(client, send, move, deliver, lock) \triangleq$
$(\ !send^{\uparrow}(x, m).lock^{\uparrow}(s).fwd^{\uparrow}\langle s, client, deliver, (s, x, m) \rangle$
$|\ !move^{\uparrow}(s').s'^{\uparrow}\langle \rangle.lock^{\uparrow}(s)fwd^{\uparrow}\langle s, client, lock, s' \rangle)$

$Site() \triangleq$
$(\ !child^{\uparrow}(a, x, m).a/x\langle m \rangle$
$|\ !fwd(s, a, x, m).s \cdot child\langle a, x, m \rangle$
$|\ !\overline{in}\,login\ |\ !\overline{out}\,logout$
$|\ !s_0()\ |\ ...\ |\ !s_N())$

$Client(home, tracker, deliver, lock) \triangleq$
$(!lock^{\uparrow}(s).out\,logout.in\,s \cdot login.fwd^{\uparrow}\langle home, tracker, lock, s \rangle.moved\langle s \rangle$
$|\ !deliver^{\uparrow}(s, x, m).fwd^{\uparrow}\langle home, tracker, lock, s \rangle.x\langle m \rangle)$

```
let Server(home:site)=
!register(client,ack); new tracker:agent  new move:<site>
  new deliver:<site,<'a>,'a>  new lock:<site>
  ( tracker[ Tracker<client,move,deliver,lock> ]
  | client/ack<tracker,move,deliver,lock>
  | tracker/lock<home> )

let Tracker(client:agent,move:<site>,deliver:<site,<'a>,'a>,lock:<site>)=
( !send^(x,m); lock^(s); forward^<s,client,deliver,(s,x,m)>
| !move^(s1); s1^<>; lock^(s); forward^<s,client,lock,s1>)

let Site()=
( !child^(a,x,m); a/x<m>
| !forward(s,a,x,m); s.child<a,x,m>
| !-in login | !-out logout)

let Client(home:site,tracker:agent,deliver:<site,<'a>,'a>,lock:<site>)=
( !lock^(s);out logout;in s.login;forward^<home,tracker,lock,s>;moved<s>
| !deliver^(s,x,m); forward^<home,tracker,lock,s>; x<m>)
```

FIGURE 8.11: Agent tracker specification (top) and implementation (bottom)

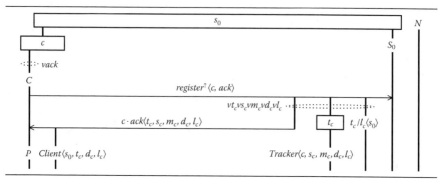

FIGURE 8.12: Tracker registration.

presented in Section 8.2, in which the vertical lines represent parallel processes, the boxes represent ambients, the horizontal arrows represent interaction and the flow of time proceeds from top to bottom. The client c sends a message to its home site s_0 on the *register* channel, consisting of its name and an acknowledgment channel ack. The site creates a new agent name t_c and new send, move, deliver, and lock channels s_c, m_c, d_c, l_c, respectively. It then sends these names to the client on channel ack, and in parallel creates a new tracker agent t_c for keeping track of the location of the client. The tracker is initialized with the *Tracker* process and the current location of the client is stored as an output to the tracker on the lock channel l_c. When the client receives the acknowledgment it spawns a new *Client* process in parallel with process P.

Figure 8.13 describes a scenario in which a tracker agent sends a message to its client. A message can be sent to the client by sending a request to its corresponding tracker agent on the home site s_0. The request is sent to the tracker agent t_c on the send channel s_c, asking the tracker to send a message n to its client on channel x. The tracker then inputs the current location s_i of its client via the lock channel l_c,

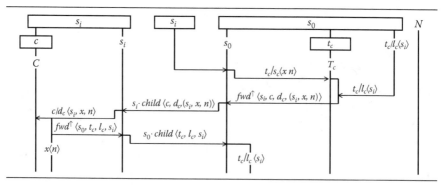

FIGURE 8.13: Tracker delivery.

thereby acquiring the lock and preventing the client from moving. The tracker then forwards the request to the deliver channel d_c of the client. When the client receives the request, it forwards its current location to the tracker on the lock channel, thereby releasing the lock, and locally sends the message n on channel x.

When the client wishes to move to a site s_j, it forwards the name s_j to the tracker agent on the move channel m_c. The tracker agent first checks whether this site is trusted by trying to send an output on channel s_j. If the output succeeds, the tracker inputs the current location of its client on the lock channel, thereby acquiring the lock and preventing subsequent messages from being forwarded to the client. It then forwards the name s_j to the client on the lock channel, giving it permission to move to site s_j. When the client receives permission to move, it leaves on the *logout* channel and enters site s_j on the *login* channel. It then forwards its new location to the tracker agent on the lock channel, thereby releasing the lock.

The above definitions can be used to specify a distributed instance of the tracker application, in which multiple client agents communicate with each other as they move between trusted sites in a network. The following specification describes three client agents a, b, c which start out on a home site s_0.

$$s_0 \boxed{Server \langle s_0 \rangle \mid Site \langle \rangle \mid a \boxed{A} \mid b \boxed{B} \mid c \boxed{C}} \mid s_1 \boxed{Site \langle \rangle} \mid s_2 \boxed{Site \langle \rangle} \mid s_3 \boxed{Site \langle \rangle}$$

The processes A, B, C are used to describe how the clients register with the home site s_0, locally exchange tracker names and then embark on their respective journeys through the network. The clients are able to communicate with each other as they move between the trusted sites $s_0, ..., s_4$ by sending messages to their respective tracker agents on the home site s_0. The trackers then forward the messages to the appropriate client. This example was programmed in the CA Language by executing the following four sites on four different runtimes:

```
192.168.0.2:3010
[ Site<> | Server<192.168.0.2:3010>
| alice[Alice<>] | bob[Bob<>] | chris[Chris<>]
]
192.168.0.3:3011[ Site<> ]
192.168.0.4:3012[ Site<> ]
192.168.0.5:3013[ Site<> ]
```

The full code for the application is available online at Ref. [30]. During execution, each runtime periodically produced an HTML output of its execution state. The state of site s_1 at the beginning and end of execution is shown in Figure 8.14. Initially, the site executes code to provide a number of basic services, but does not contain any agents. By the end of the program execution, the site has been visited by both the bob and chris agents, and the bob agent is still present on the site. Multiple runs of the application show that all of the messages are reliably delivered to the respective agents as they move around between sites.

```
192.168.0.3:3011

( !child^(a,x,m);
      a/x<m>
 | !forward(s,a,x,m);
      route^<s,child,(a,x,m)>
 | !in login
 | !out logout
 | !println(s);
      system.println<s>;
      print<s>
)
```

```
192.168.0.3:3011

!out logout
!in login
!child^(a,x,m);
    a/x<m>
!println(s);
    system.println<s>;
    print<s>
!forward(s,a,x,m);
    route^<s,child,(a,x,m)>
print<"bob received message from alice">
print<"chris was here">
bob

    !lock~10^(s:site);
        out logout;
        in s.login;
        forward^<192.168.0.2:3010,tracker~7,lock~10,s>;
        moved<s>
    !deliver~9^(s:site,x:<'a>,m:'a sub);
        forward^<192.168.0.2:3010,tracker~7,lock~10,s>;
        println^<((show bob + " received message from ") + show m)>;
        x<m>
    from_chris<chris>
    from_alice<alice>
```

FIGURE 8.14: Initial and final Runtime states for site s_1 at address 192.168.0.3: 3011.

8.6 Related Work

The CA calculus is inspired by previous work on calculi for mobility, including the Ambient calculus, the Nomadic π-calculus, and variants of the Boxed Ambient calculus. The main differences with Boxed Ambients are that ambients in CA can interact using named channels and that sibling ambients can communicate directly. Sibling communication over channels is inspired by the Nomadic π-calculus and channel communication is also used in the Seal calculus [11], although sibling seals cannot communicate directly. The use of channels for mobility is inspired by the mechanism of passwords, first introduced in Ref. [24] and subsequently adopted in Ref. [7]. The main advantage of CA over the Nomadic π-calculus is its ability to regulate access to an ambient by means of named channels, and its ability to model computation within nested locations, both of which are lacking in Nomadic π. In the Safe Ambient calculus [22], coactions are used to allow an ambient to enter, leave, or open another ambient. However, there is no way of controlling which ambients are allowed to perform the corresponding actions. In Ref. [41], the authors of Nomadic π argue that some form of local synchronous communication is fundamental for programming mobile applications. In contrast, Boxed Ambient calculi typically require all communication between sibling agents to take place via the parent. However, there is nothing to prevent one or both of these agents from migrating while the message is still in transit, leaving the undelivered message stuck inside the parent.

The CA Machine is inspired by the Pict abstract machine [38], which is used as a basis for implementing the asynchronous π-calculus. Like the Pict machine, CAM uses a list syntax to represent the parallel composition of processes. In addition, CAM extends the semantics of the Pict machine to provide support for nested ambients and ambient migration. Pict uses channel queues in order to store blocked inputs and outputs that are waiting to synchronize. In CAM these channel queues are generalized to a notion of blocked processes, in order to allow both communication and migration primitives to synchronize. A notion of unblocking is also defined, which allows mobile ambients in CAM to re-bind to new environments. In addition, CAM requires an explicit notion of restriction in order to manage the scope of names across ambient boundaries. The Pict machine does not require such a notion of restriction, since all names in Pict are local to a single machine. By definition, the Pict abstract machine is deterministic, and, although it is sound with respect to the π-calculus, it is not complete. In contrast, CAM is nondeterministic, and is both sound and complete with respect to the CA calculus.

A number of abstract machines have also been defined for variants of the Ambient calculus. In Ref. [8], an informal abstract machine for Ambients is presented, which has not been proved sound or complete. In Ref. [18], a distributed abstract machine for ambients is described, based on a formal mapping from the Ambient calculus to the Distributed Join calculus. However, it is not clear how such a translation can be applied to the CA calculus, which uses more high-level communication primitives. In addition, the translation is tied to a particular implementation language (JoCaml), whereas the CA Machine uses more a low-level approach that can be implemented in any language with support for function definitions. Furthermore, the abstract machine in Ref. [18] separates the logical structure of ambients from the physical structure of the network, whereas the CA Machine uses ambients to directly model the hierarchical topology of the network. This approach assumes that the network topology is part of the application specification, which leads to a simplified implementation. In Ref. [36], a distributed abstract machine for Safe Ambients is presented, which uses logical forwarders to represent mobility. Physical mobility can only occur when an ambient is opened. However, such an approach is not applicable to CA, where the *open* primitive is nonexistent.

The Nomadic Pict runtime [42] is an extension of the Pict runtime with support for the migration of mobile agents between runtimes over a network. Unlike the CA Runtime, the Nomadic Pict runtime is not based on a formal abstract machine, although such a machine could in principle be defined using the approach described in Section 8.3. Another feature of the Nomadic Pict runtime is that all the agents in a given application are typically launched from a single runtime, and then migrated to other runtimes in the network. This is because the constructs for agent creation require each agent to be created with a unique identity that is only known within a limited scope.

The JoCaml runtime [15] is implemented based on a high-level abstract machine for the Join calculus [17] and a compilation of the calculus to a suitable target language [20]. In JoCaml, migrating agents communicate by sending messages to each other irrespective of their location, which requires complete network transparency. Therefore, in order to fully implement the semantics of the Join calculus, sophisticated distributed algorithms are required, such as the information retrieval infrastructure described in this chapter. In general, the JoCaml runtime relies on a small set of powerful primitives that provide complete network transparency, whereas the CA Runtime relies on a small set of low-level primitives that can be directly implemented above standard network protocols, together with an easy way of encoding more expressive primitives on top.

In Ref. [41], Nomadic Pict is used to program an infrastructure for reliably forwarding messages to mobile agents, and a centralized version of this algorithm is proved correct in Ref. [40]. This chapter shows how similar applications can be programmed using the Ambient paradigm. One of the advantages of Ambients is that they can directly model computation within nested locations, which is not possible in Nomadic Pict. In addition, the CA Language provides constructs for regulating access to a location by means of named channels, whereas Nomadic Pict assumes that all locations are freely accessible.

8.7   Conclusion

The agent tracker application for information retrieval relies on distributed algorithms previously presented in Ref. [41], which also describes a variety of algorithms for fault tolerance, load-balancing, large-scale parallel computation, and event-driven mobility. In future, some or all of these algorithms could be specified in the CA calculus and used to implement a range of mobile applications in the CA Language.

From a security perspective, perhaps the most interesting area for future work lies in the use of Ambient Logics for reasoning about the security properties of mobile applications. Such logics are a powerful tool that can express a much broader range of properties than process equivalences [9]. In general, there has been a significant amount of research on security mechanisms for Ambient calculi. Although only a fraction of this research has been applied to the CA calculus, much more could be applied in future. For the time being, this chapter focuses on ensuring that, if a given security property holds for the calculus specification of an application, then it will also hold for its implementation.

From an implementation perspective, a number of improvements can be made to both the CA Runtime and the CA Language. Thanks to the completeness of the CA Machine, a number of optimizations can be introduced for the scheduling of processes, while still preserving the correctness of the runtime. The CA Language is merely a prototype, and a number of extensions to the language can be envisaged. The Pict language demonstrated how high-level encodings of functions, procedures, and data structures could be readily embedded in a π-calculus language. Since the CA calculus is an extension of the π-calculus, a full embedding of a functional language can be incorporated into the CA Language in a similar way.

From a specification perspective, a number of extensions to the calculus can also be envisaged, including a notion of time and nondeterministic choice. A promising approach to modeling time in the π-calculus is presented in Ref. [2], which can also be readily applied to Ambient calculi. A possible extension for modeling nondeterministic choice is described in the Bio Ambient calculus [34]. The combination of these two extensions could be readily used to program timeouts and other exception mechanisms. For example, $ack^{\uparrow}(k).P + \tau_n.Q$ could be used to model a choice ($+$) between receiving an acknowledgment k on channel ack and then executing P, or waiting for time n and then executing Q. In this example, if time n elapses before the acknowledgment is received, the process Q is executed. Such an extension would be relatively straightforward—for comparison, a notion of stochastic choice with timeout has already been implemented in Ref. [32]. So far, the CA calculus has been used to specify mobile applications for networks that support the TCP/IP. In principle, the calculus could also be used to specify applications on a range of networks types. For example, wireless networks with mobile devices can be readily modeled using the synchronization primitives of the calculus. Indeed, mobile wireless networks were some of the first network types to be targeted by process calculi. For example, [25] uses the π-calculus to describe a simple protocol for switching mobile phones between base stations.

This chapter investigates to what extent a variant of the Ambient calculus can be used for specifying and implementing secure mobile applications. In particular, the chapter investigates whether ambients can be used as the basis for a distributed, mobile programming language. Over the years there has been substantial theoretical research on the Ambient calculus and its many variants, but there has been comparatively little research on the implementation of ambients. In this regard, a number of lessons can be learned from the Nomadic π-calculus, for which the underlying theory [39] and implementation [41] were developed in concert, in order to produce the beginnings of a programming language and runtime system for developing real mobile applications [42]. Ambient calculi have matured substantially over the years, and it is perhaps time to seriously address how the theory of ambients can be turned into practice, in order to reap the benefits of the last decade of theoretical research.

This follows on from one of the early papers on ambients [10], which states "On this foundation, we can envision new programming methodologies, programming libraries and programming languages for global computation."

References

[1] M. Abadi and A. D. Gordon. A calculus for cryptographic protocols: The Spi calculus. *Journal of Information and Computation*, 148(1):1–70, 1999.

[2] M. Berger. Basic theory of reduction congruence for two timed asynchronous π-calculi. In *Concurrency Theory (CONCUR'04), Lecture Notes in Computer Science*, 3170:115–130, Springer, January 2004.

[3] L. Bettini, R. De Nicola, and R. Pugliese. Xklaim and klava: Programming mobile code. *Electronic Notes in Computer Science*, 62, Elsevier, 2002.

[4] K. Bhargavan, C. Fournet, and A. D. Gordon. A semantics for web services authentication. In *POPL '04: Proceedings of the 31st ACM SIGPLAN-SIGACT Symposium on Principles of Programming Languages*, ACM Press, New York, 2004, pp. 198–209.

[5] M. Bugliesi, G. Castagna, and S. Crafa. Boxed ambients. In *Theoretical Aspects of Computer Science (TACS'01), Lecture Notes in Computer Science*, 2215:38–63, Springer, 2001.

[6] M. Bugliesi, G. Castagna, and S. Crafa. Reasoning about security in mobile ambients. In *Concurrency Theory (CONCUR'01). Lecture Notes in Computer Science*, 2154:102–120, Springer, 2001.

[7] M. Bugliesi, S. Crafa, M. Merro, and V. Sassone. Communication interference in mobile boxed ambients. In *Foundations of Software Technology and Theoretical Computer Science (FSTTCS'02). Lecture Notes in Computer Science*, 2556:71–84, Springer, 2002.

[8] L. Cardelli. Mobile ambient synchronization. Technical Report SRC-TN-1997-013, Digital Equipment Corporation, July 25, 1997.

[9] L. Cardelli and A. D. Gordon. Anytime, anywhere: Modal logics for mobile ambients. In *Principles of Programming Languages (POPL '00)*, pp. 365–377, ACM Press, January 2000.

[10] L. Cardelli and A. D. Gordon. Mobile ambients. *Theoretical Computer Science*, 240(1):177–213, 2000. An extended abstract appeared in *FoSSaCS '98*: 140–155.

[11] G. Castagna, J. Vitek, and F. Zappa. The seal calculus. 2003. Available from ftp://ftp.di.ens.fr/pub/users/castagna/seal.ps.gz.

[12] D. Chess, C. Harrison, and A. Kershenbaum. Mobile agents: Are they a good idea? Technical Report RC 19887 (December 21, 1994—Declassified March 16, 1995), Yorktown Heights, NY, 1994.

[13] A. Church. *The Calculi of Lambda Conversion.* Princeton University Press, Princeton, NJ 1941.

[14] K. G. Coffman and A. M. Odlyzko. Internet growth: Is there a "Moore's Law" for data traffic? In Panos M. Pardalos and Mauricio Resende, (Eds.), *Handbook of Massive Data Sets,* Kluwer, Norwell, MA 2002, pp. 47–93.

[15] S. Conchon and F. Le Fessant. Jocaml: Mobile agents for Objective-Caml. In *First International Symposium on Agent Systems and Applications (ASA'99)/Third International Symposium on Mobile Agents (MA'99),* Palm Springs, CA, 1999.

[16] S. Crafa, M. Bugliesi, and G. Castagna. Information flow security for boxed ambients. In *Foundations of Wide Area Network Computing (F-WAN'02). Electronic Notes in Computer Science,* 66(3), 2002.

[17] C. Fournet and G. Gonthier. The reflexive chemical abstract machine and the join-calculus. In *Principles of Programming Languages (POPL '96),* pp. 372–385, ACM Press, January 1996.

[18] C. Fournet, J.-J. Lévy, and A. Schmitt. An asynchronous distributed implementation of mobile ambients. In *Exploring New Frontiers of Theoretical Informatics. Lecture Notes in Computer Science,* 1872:348–364, Springer, August 2000.

[19] ITU-TS. *Recommendation Z.120: Message Sequence Chart (MSC).* Geneva, 1996.

[20] F. Le Fessant and L. Maranget. Compiling join-patterns. In Uwe Nestmann and Benjamin C. Pierce, (Eds.), *Proceedings of HLCL '98,* ENTCS, 16.3, Elsevier Science Publishers, New York, 1998.

[21] X. Leroy, D. Doligez, J. Garrigue, D. Remy, and J. Vouillon. The Objective Caml system, release 3.08, Documentation and user's manual, INRIA. Available at http://caml.inria.fr/ocaml/, 2004.

[22] F. Levi and D. Sangiorgi. Controlling interference in ambients. In *Principles of Programming Languages (POPL'00),* pp. 352–364, ACM Press, 2000.

[23] M. Merro and M. Hennessy. Bisimulation congruences in safe ambients. In *Principles of Programming Languages,* pp. 71–80, ACM Press, 2002.

[24] M. Merro and V. Sassone. Typing and subtyping mobility in boxed ambients. In *Concurrency Theory (CONCUR'02). Lecture Notes in Computer Science,* 2421:304–320, 2002.

[25] R. Milner. *Communicating and Mobile Systems: The π-Calculus.* Cambridge University Press, New York, May 1999.

[26] R. Milner, J. Parrow, and D. Walker. A calculus of mobile processes, part I/II. *Journal of Information and Computation*, 100:1–77, September 1992.

[27] R. Milner, M. Tofte, R. Harper, and D. MacQueen. *The Definition of Standard ML (Revised)*. MIT Press, Cambridge, MA, 1997.

[28] A. M. Odlyzko and K. G. Coffman. Growth of the Internet. In T. Li and I. P. Kaminow, (Eds.), *Optical Fiber Telecommunications IV B: Systems and Impairments*, Academic Press, New York, 2002. pp. 17–56.

[29] L. L. Peterson and B. S. Davie. *Computer Networks: A Systems Approach*, Elsevier, 4th ed., 2007.

[30] A. Phillips. *The Channel Ambient System*, 2005. Runtime and documentation available from http://research.microsoft.com/˜aphillip/cam/.

[31] A. Phillips. Specifying and implementing secure mobile applications in the channel ambient system. PhD thesis, Imperial College London, April 2006.

[32] A. Phillips and L. Cardelli. Efficient, correct simulation of biological processes in the stochastic pi-calculus. In *Computational Methods in Systems Biology (CMSB'07). Lecture Notes in Computer Science*, 4695:184–199, Springer, September 2007.

[33] A. Phillips, N. Yoshida, and S. Eisenbach. A distributed abstract machine for boxed ambient calculi. In *European Symposium on Programming (ESOP'04). Lecture Notes in Computer Science*, 2986:155–170, Springer, April 2004.

[34] A. Regev, E. M. Panina, W. Silverman, L. Cardelli, and E. Shapiro. Bioambients: An abstraction for biological compartments. *Theoretical Computer Science*, 325(1):141–167, September 2004.

[35] D. S. Milojicic, M. Breugst, I. Busse, J. Campbell, S. Covaci, B. Friedman, K. Kosaka, D. Lange, K. Ono, M. Oshima, C. Tham, S. Virdhagriswaran, and J. White. MASIF: The OMG Mobile Agent System Interoperability Facility. In Kurt Rothermel and Fritz Hohl, (Eds.), *Proceedings of the Second International Workshop on Mobile Agents (MA'98), Stuttgart (Germany), September 1998*, volume 1477 of *LNCS*, Springer-Verlag, New York, 1999, pp. 50–67.

[36] D. Sangiorgi and A. Valente. A distributed abstract machine for safe ambients. In *International Colloquium on Automata, Languages and Programming (ICALP'01). Lecture Notes in Computer Science*, 2076:408–420, 2001.

[37] P. Sewell, P. Wojciechowski, and B. Pierce. Location independence for mobile agents. In *International Conference on Computer Languages (ICCL'98)*, 1998.

[38] D. N. Turner. *The Polymorphic Pi-Calculus: Theory and Implementation*. PhD thesis, University of Edinburgh, June 1996. CST_126_96 (also published as ECS_LFCS_96_435).

[39] A. Unyapoth. Nomadic π-calculi: Expressing and verifying communication infrastructure for mobile computation. Technical Report UCAM-CL-TR-514, University of Cambridge, Computer Laboratory, Cambridge, June 2001.

[40] A. Unyapoth and P. Sewell. Nomadic Pict: Correct communication infrastructures for mobile computation. In *Principles of Programming Languages (POPL'01)*, 2001, pp. 116–127.

[41] P. T. Wojciechowski. *Nomadic Pict: Language and Infrastructure Design for Mobile Computation*. PhD thesis, University of Cambridge, June 2000.

[42] P. T. Wojciechowski. *The Nomadic Pict System*, 2000. Available electronically as part of the Nomadic Pict distribution http://www.cs.put.poznan.pl/pawelw/npict/.

Part III

Embedded Systems

Chapter 9

Calculating Concurrency Using *Circus*

Alistair A. McEwan

Contents

9.1   A Brief History

> We must try to get a better understanding of the processes of computing
> and their description in programming languages. In computing we have
> what I believe to be a new field of mathematics. (Christopher Strachey,
> 1965)

In the early days of Computer Science, Christopher Strachey held the belief that programming should be treated as a mathematical activity, in order to be given a sound theoretical basis. The above quote is taken from Ref. [34], containing some of his previously unpublished lecture notes in an edition dedicated to his memory. Perhaps an even greater dedication is that much work since his pioneering days has been dedicated to realizing that goal; and today it is still a fundamental aim of computer science research to understand, describe, and solve the problems of programming using sound mathematical techniques.

In the decades following Strachey's work, methodologies have become much more sophisticated, whilst maintaining the original goal of systematic and orderly development of computer systems. However, systems are also becoming vastly more complex, and the distinction between hardware and software increasingly blurred. The tasks which computer systems are asked to undertake are becoming increasingly fundamental to our everyday lives. Indeed, it may be argued that the demands placed on these systems are rapidly outstripping our ability to reason about them. With this in mind, it is clear that it is becoming ever more important to strive to achieve Strachey's early insight.

One of the practical challenges in designing and understanding modern systems centers around the increasingly concurrent nature of systems. Concurrency is commonplace, ranging from coprocessors in desktop personal computers, to massively parallel systems, which, even only a few years ago seemed infeasible. Customizable, reconfigurable hardware is becoming more commonplace, and with it comes the potential for even more concurrency.

Reasoning about concurrent systems is inherently more difficult than reasoning about sequential ones—in part due to the increased number of states a concurrent system can be in, but also due to the increased behaviors enjoyed by concurrent systems. These can be classified in the following ways:

- Nondeterminism: A system is *nondeterministic* if behaviors may be different given exactly the same sets of inputs in different experiments. Nondeterminism can often arise through, for instance, *race conditions*. Depending on the circumstances, nondeterminism may be a desirable property or it may not: either way it should be detectable.

- Deadlock: A system is *deadlocked* if no components can make any further progress. This is often caused by a chain of dependency between processes waiting for communications, and one, or more of the components refusing to engage in a communication.

- Livelock: A system is *livelocked*, or *divergent*, if it is partaking in a (potentially) infinite sequence of internal events over which the environment has no control. Externally, the view of a livelocked system is similar to that of a deadlocked one; although operationally, and semantically, the two are very different.

9.1.1 Recent Developments: Linking Theories

The challenge to computer science is to provide a methodology and notation that can be used to describe all of the properties of a system together, in a clear, and concise manner, while allowing as much of the life cycle of a system to be contained within the same notation.

The development of a Unifying Theory of Programming [14] has fueled interest in integrating different languages and notations to achieve this. The aim is generally to offer notations that allow the designer to reason about more aspects of systems and specifications, in the ways described above. The desire to combine reactive behavior of systems with functional properties has led to several combinations of state-based formalisms and process algebras being proposed. In this chapter, we are concerned with one such integration, *Circus*.

9.1.2 Contribution

Previous works exist documenting the case study presented in this chapter, an Internet packet filter, and how it can be implemented on an FPGA [19,20], and attempts to verify the implementation in Ref. [39] The major contributions of this chapter can be summarized by the following points:

1. An introduction to *Circus* with examples

2. A demonstration of the calculation of concurrency from a sequential specification using laws of *Circus*

3. The presentation of a top-down design strategy from requirements and a verification strategy for a development using Z/Eves and FDR

4. The presentation of design patterns for refining *Circus* processes into *Handel-C* models, incorporating a model of synchronous clock timing

5. Evidence that the laws of *Circus* allow the exploration of valid refinements, guided by engineering intuition where requirements may not have been explicit in the specification

The chapter is structured as follows: Section 9.2 presents an introduction to *Circus*. This introduction assumes the reader has some previous knowledge of CSP and Z, although readers without this experience will benefit from studying the examples. Section 9.3 shows how laws of *Circus* can be used to calculate a concurrent implementation from a sequential specification. Sections 9.4 through 9.10 present the case study that forms the majority of this chapter. The purpose of this case study is to motivate this calculational approach, by starting with an abstract specification that may be reasoned about, and end up with a highly concurrent implementation tailored to the hardware upon which it is to be deployed. Background to the case study is given in Section 9.4 and the abstract specification in Section 9.5. The specification is refined into different components in Section 9.6, and into the implementation architecture in Sections 9.7 through 9.9; and the final implementation given in Section 9.10. We summarize the conclusions and achievements of this work in Section 9.11.

9.2   About *Circus*

Circus is the most modern approach to combining notations, integrating CSP, Z, the Guarded Command Language and the Z Refinement Calculus, with a central notion of refinement. The semantics of the language have been defined [38]; while refinement, consisting of simulation rules for processes, and a development calculus, is presented in Ref. [37]. The approach is the most modern because the semantic model of the combination is built on Hoare and He's Unifying Theory [14].

The Z notation [4,30,31] is a well-established formal specification language used to model the behavior of systems in terms of state. It offers rich facilities to describe state-based properties of a system, its data structures, and operations upon it. Z has been applied in many case studies, for instance [9,33]; a wide range of tools are available such as the type-checker Fuzz [32], and the theorem provers Balzac [8], Z/Eves [27], and ProofPower [15]; and several courses and textbooks are available [7,25,40].

Central to the Z notation is the notion of *abstract data types*. An *abstract data type* comprises a space of states, and an indexed collection of operations. Any use of the type begins with an initialization, proceeds with some sequence of operations, and ends with a finalization. Furthermore, data refinement and simulation rules allow development of abstract data types in a system, from abstract, nondeterministic specifications into deterministic implementations; however, it has no concept of operation evaluation order, no potential for the system to refuse to engage in an operation—or indeed to nondeterministically select one operation over another, nor of concurrency.

These concepts are all offered by the theory of Communicating Sequential Processes (CSP) [12,13,26]. CSP is presented by Hoare [13], although it was originally presented in earlier works [12]. It is further built on by Roscoe in [26]. It is a mathematical formalism for reasoning about interacting, concurrent, communicating entities in a system. Each entity in the system is represented by a *process*, which engages in atomic actions, called *events*. An event is an observable action in the life of a process, and acts as a communication, or synchronization between cooperating processes. An *alphabet* is a set of events. The role of alphabets in the semantic domain differs between Refs. [13] and [26]. In the former, processes have alphabets, while in the latter, it is the process combinators that are alphabetized. *Circus* is concerned with the alphabetized operator version of Ref. [26].

Complex descriptions of behavior can be built up as networks of processes, using the operators of CSP. It is a process algebra: descriptions of processes can be manipulated in accordance with sets of algebraic laws; the correctness of these laws being justified within the semantic models associated with CSP.

Failures-Divergences Refinement (FDR) [17] and *Probe* [18] are the most popular tools for investigating networks of CSP processes, and offer model-checking and simulation support, respectively.

A *Circus* program is a sequence of paragraphs: Z paragraphs, channel definitions, channel set definitions, and process definitions. The basic form of process definition describes the process state and operations on that state—as in a Z specification. Operations are called *actions* and can be specified using schemas, CSP events, and guarded commands. Every process has a single unnamed action, starting with •, in its definition—this is the main action of the process.

A *channel definition* declares the channels to which processes can refer: it gives the name of each channel and the type of values it can communicate. The channels over which a process communicates constitute the external interface of that process. A *process definition* gives a specification of the process. The simplest form of process specification consists of a set of process paragraphs, and the main process action, delimited by a **begin** and **end**.

Example 9.1 A *Circus buffer of length 1*

> **channel** *in*, *out* : $\mathbb{N}$
> **channel** *init*, *reset*
>
> **process** *Buffer* $\hat{=}$
> **begin**
> $\quad USt \hat{=} [\, p : \mathbb{N} \,]$
> $\quad Init \hat{=} [\, USt' \mid p' = 0 \,]$
> $\quad Input \hat{=} [\, \Delta USt, i? : \mathbb{N} \mid p' = i? \,]$
> $\quad Reset \hat{=} [\, \Delta USt \mid p' = 0 \,]$
>
> $\quad Run \hat{=} \mu X \bullet (in?i \to Input;\ out!p \to X) \ \Box\ (reset \to Reset;\ X)$
>
> $\quad \bullet\ init \to Init;\ Run$
> **end**

Example 9.1, similar to that in Ref. [5], declares four channels. Two of these, *in* and *out*, communicate natural numbers; the other two, *init* and *reset*, are for synchronization. This *Buffer* process synchronizes with the environment on startup via event *init*, and performs process *Init* to initialize *p* of the user state (*USt*) to 0. After this, The *Run* process either inputs a value, which is stored in local state by process *Input*, after which it insists on outputting a value, or it resets the buffer. The event *in?i* introduces the variable *i*, which has scope over the remainder of the action, and corresponds to the *i?* in *Input*. In the output action, *out!p*, *p* is in scope as it is contained in the user state; this value is communicated and the system recurses. To reset the system, the world synchronizes with the *reset* event.

9.2.1   Program Development: Refinement

One of the major aims of *Circus* is to establish a calculational approach to implementing specifications. Fundamental to this idea is the notion of refinement. Refinement is common to many formal methods, although different methods adopt different approaches, for instance [1,36] differ in the treatment of guards and preconditions. Much work has also been done in showing how specifications can be refined to executable code [2,3,6,21,22].

In the Unifying Theory, refinement is expressed as implication: an implementation *P* satisfies a specification *S* if $[P \Rightarrow S]$, where the brackets denote universal quantification over the alphabets of *P* and *S*, which must be equivalent. Reference [28] presents the meaning of refinement in *Circus*—process refinement is given in terms of action refinement, as in Definitions 9.1 and 9.2.

DEFINITION 9.1   Action refinement

Suppose that A and B are actions on the same state space. Action A is refined by action B if, and only if, every observation of B is permitted by A as well: $A \sqsubseteq_A B \mathrel{\widehat{=}} [B \Rightarrow A]$.

DEFINITION 9.2   Process refinement

Assume P.st, P.init, P.act, Q.st, Q.init, and Q.act are the states, initializations, and actions, of processes P and Q, respectively. $P \sqsubseteq_P Q$ is defined to mean that process P is refined by process Q if, and only if,

$$(\exists P.st, P.st' \bullet P.init \wedge P.act) \sqsubseteq_A (\exists Q.st, Q.st' \bullet Q.init \wedge Q.act)$$

Data refinement techniques require a formalization of the link between abstract and concrete states—relations describing this link are usually referred to as *forwards* and *backwards simulation* [10,11,40].* A forwards simulation (Figure 9.1) is usually used to prove the correctness of a development step when the implementation side

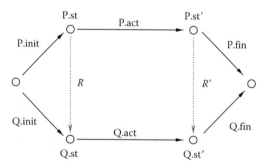

FIGURE 9.1: Forwards (or downwards) simulation.

---

* References [10,11] actually use the term downwards simulation for forwards simulation, and upwards simulation for backwards simulation.

resolves some nondeterminism in the specification; while a backwards simulation is used to prove the correctness of the step when the implementation *postpones* some nondeterminism present in the specification. Rules for forwards simulation in *Circus* have been proposed in Ref. [28], and these are recounted in Definition 9.3. Process refinement in *Circus* corresponds to the *failures-divergences* refinement ordering in CSP.

DEFINITION 9.3   *Forwards simulation of Circus processes*
   A forwards simulation between actions A and B of processes P and Q is a relation R of type Q.st $\longleftrightarrow$ P.st, satisfying

$$[\forall Q.st \bullet Q.init \Rightarrow (\exists P.st \bullet P.init \wedge R)] \qquad\qquad [initialization]$$
$$[\forall P.st; \; Q.st; \; Q.st' \bullet R \wedge B \Rightarrow (\exists P.st' \bullet R' \wedge A)] \qquad\qquad [correctness]$$

A forwards simulation between P and Q is a forwards simulation between their main actions.

9.3   Calculating Concurrency Using *Circus*

A major goal of *Circus* is to make development of concurrent programs as systematic as that of sequential ones. In this section, the focus is on laws that may be applied to a specification in order to calculate a concurrent refinement. The section opens by presenting the main law for introducing concurrency into a specification: *process splitting*. This is followed by a law known as *process indexing*, used for replicating, indexing, and interleaving collections of like components. Finally, a new law and associated operator *generalised chaining* that relaxes some of the restrictions in process splitting is presented.

9.3.1   Process Splitting

The first law in this section, *process splitting*, applies to processes whose state components can be partitioned in such a way that each partition has its own set of process paragraphs. The result of applying this law to a process of the correct form is two new processes, each containing their own paragraphs; and together they exhibit behaviors that are a refinement of the original.

DEFINITION 9.4   *Process forms applicable to process splitting*

    process $P \mathrel{\widehat{=}}$
    begin
        $State \mathrel{\widehat{=}} Q.st \wedge R.st$
        $Q.pps \uparrow R.st$
        $R.pps \uparrow Q.st$
        $\bullet F(Q.act, R.act)$
    end

Let *pd* stand for the process declaration in Definition 9.4, where *Q.pps* and *R.pps* stand for the paragraphs of processes *Q* and *R*; and *F* for an arbitrary context (function on processes). The operator $G \uparrow S$ takes a set of paragraphs *G* and schema expressions *S*, and joins the paragraphs of *G* with ΞS. For the expression to be well-formed, the paragraphs in *G* must not change state in *S*. This is the general form of processes to which process splitting laws apply.

The state of *P* is defined as the conjunction of schemas *Q.st* and *R.st*. The actions of *P* are $Q.pps \uparrow R.st$ and $R.pps \uparrow Q.st$. They must handle the partitions of the state separately. In $Q.pps \uparrow R.st$ each schema expression in *Q.pps* is conjoined with $\Xi R.st$. Similar comments apply to $R.pps \uparrow Q.st$.

LAW 9.1 *Process splitting*

$$pd = (\mathbf{process}P \stackrel{\frown}{=} F(Q.act, R.act))$$

provided Q.pps and R.pps are disjoint with respect to R.st and Q.st.

Two sets of process paragraphs *pps* and *pps'* are said to be disjoint with respect to states *s* and *s'* if and only if $pps = pps' \uparrow s'$ and $pps' \uparrow s$, and no action in *pps* refers to components of *s'* or to paragraph names in *pps'*; further, no action in *pps'* refers to components of *s* or to paragraph names in *pps*. Informally, this may be summarized by saying that *pps* should not depend upon anything in *s'*, and *pps'* should not depend upon anything in *s*.

Example 9.2 is a process to which this law is applicable. The process represents a very simple operating system that allows users to log in, and for files to be written to. Initialization of user state has been omitted, as the technique for dealing with initialization is the same as for the other actions.

Example 9.2 *A very simple operating system*

> **process** *OperatingSystem*
> **begin**
> $USt \stackrel{\frown}{=} [\, file : \mathbb{N} \rightarrowtail File; \; users : \mathbb{P} \, Person \,]$
> $WriteFile \stackrel{\frown}{=} [\, \Delta USt; \; i? : 1..maxfiles; \; k? : Key; \; d? : Data \, |$
> $\qquad\qquad file \, i?' = file \, i \oplus \{k? \mapsto d?\} \wedge users' = users \,]$
> $AddUser \stackrel{\frown}{=} [\Delta USt; \; p? : Person \, |$
> $\qquad\qquad p? \notin users \wedge users' = users \cup \{p?\} \wedge file' = file \,]$
>
> $Run \stackrel{\frown}{=} \mu X \bullet (write?i?k?d \rightarrow WriteFile \; \Box \; add?p \rightarrow AddUser); \; X$
>
> $\bullet \, Run$
> **end**

First, the state of the process may be considered. Inspection shows that *WriteFile* accesses one part of the state, while *AddUser* accesses the other; both operations guarantee to leave the complementary state unchanged. The user state can therefore be rewritten as in Example 9.3. Using the schema calculus, the correctness of splitting the user state can be shown by conjoining each operation with an unaltered version of the partner state and proving that it is equal to the original: for instance, *WriteFile* = *WriteFile$_1$* $\wedge$ Ξ*USt$_2$*. A proof obligation such as this may typically be automated using Z/Eves.

Example 9.3   *Separating operating system state*

$USt_1 \,\widehat{=}\, [\, file : \mathbb{N} \,\nrightarrow\, File \,]$
$USt_2 \,\widehat{=}\, [users : \mathbb{P} \, Person \,]$

$WriteFile_1 \,\widehat{=}\, [\, \Delta USt_1; \; i? : 1..maxfiles; \; k? : Key; \; d? : Data \,|$
$\qquad\qquad\qquad file\, i?' = file\, i \oplus \{k? \mapsto d?\} \,]$

$AddUser_1 \,\widehat{=}\, [\Delta USt_2; \; p? : Person \,|\, p? \notin users \wedge users' = users \cup \{p?\} \,]$

The second task is to separate the actions of the process, such that there are two new actions, each of which accesses only one part of the process state. This is shown in Example 9.4.

Example 9.4   *Separating operating system actions*

$Run_1 \,\widehat{=}\, \mu X \bullet write?i?k?d \rightarrow WriteFile_1; \; X$
$Run_2 \,\widehat{=}\, \mu X \bullet add?p \rightarrow AddUser_1; \; X$

The splitting of the actions may be shown to be correct by checking the correctness of the implementation using FDR: for instance, $Run \sqsubseteq (Run_1 \,|||\, Run_2)$. The obligation that the paragraphs are truly disjoint can be met by showing that each action leaves the opposing state unchanged ($Q.pps \uparrow R.st$): for instance $Run_1 \,\widehat{=}\, \mu X \bullet write?i?k?d \rightarrow WriteFile_1 \wedge \Xi USt_2; \; X$. There is no direct tool support for this obligation; however, it typically follows trivially if the previous two are met. Applying the law now gives rise to two new processes, given in Example 9.5.

Example 9.5   *The split operating system processes*

process$Users \,\widehat{=}$	**process**$FileSystem \,\widehat{=}$
begin	**begin**
$Ust_1 \,\widehat{=}\, [\, ... \,]$	$Ust_2 \,\widehat{=}\, [\, ... \,]$
$Run_1 \,\widehat{=}\, ...$	$Run_2 \,\widehat{=}\, ...$
$\bullet \, Run_1$	$\bullet \, Run_2$
end	**end**

$OperatingSystem \sqsubseteq_P (FileSystem \,|||\, Users)$

9.3.2   Process Indexing

Specifications in Z are often structured using *promotion* [40]—where global state is defined in terms of local state. In this section, a refinement law of *Circus* called *process indexing* is presented. This law relies on promotion to produce an indexed array of concurrent processes from a sequential specification. An important requirement for applying this law is that the promotion must be free, and that the states of the indexed processes must be disjoint.

Process indexing laws refine a specification structured using free promotion to an indexed family of processes, each one representing an element of the local type. In Definition 9.5, L stands for a local process, G for the global process, and *Promote* for the promotion schema. A local element is identified in the global state as $f(i)$. Promotion of a schema operation is as in Z, while promoting *Skip*, *Stop*, and *Chaos* leaves them unchanged.

The promotion of a communication $c.e$ requires the addition of the identifier of the value in the collection. Therefore, for every channel c there is a corresponding promoted channel pc that communicates a pair formed by an identifier and its value.

DEFINITION 9.5   *Promoting state, actions, and communications*

promote$(SExp) \cong \exists \Delta L.st \bullet SExp \wedge Promote$

promote$(A) \cong A \text{ for } A \in \{Skip, Stop, Chaos\}$

promote$(c.e \to A) \cong pc.\textbf{promote}(e) \to \textbf{promote}(A)$

Promotion distributes through the action operators. For a guarded action, the guard must be promoted; for parallelism and hiding, channels are replaced with corresponding promoted channels. Let pd stand for the process declaration in Definition 9.6. Indexing laws apply to processes of this form.

DEFINITION 9.6   *Processes applicable to indexing*

process $P \cong$
begin
$State \cong [f : Range \rightarrowtail L.st \mid pred]$
$L.action_k \upharpoonright State$
$L.act \cong \mu X \bullet F(L.action_k);\ X$
$Promote \cong [\ \Delta L.st;\ \Delta State, i? : Range \mid$
$\qquad\qquad i? \in \operatorname{dom} f \wedge \theta L.st = f(i?) \wedge$
$\qquad\qquad \theta State' = \theta State \oplus \{i? \mapsto \theta L.st'\}\ ]$

$action_k = \textbf{promote}(L.action_k)$
$\bullet\ \mu X \bullet F(action_k);\ X$
end

The global state component is a function from *Range* to a local state *L.st*. Actions *L.action* over the local state do not affect the global state. The main local action is defined recursively, as is the main global action. Both have the same structure, but the former uses the actions on the local states, and the latter the corresponding promoted actions. Considering the indexed process:

process $IL \mathrel{\widehat{=}} i : Range \odot L[L.c\_i := pc]$

the process $i : Range \odot L$ acts on indexed channels c_i, where L acts on a channel c. Like the promoted channels pc used in P, they communicate pairs of values—the index and the original value. In the process above, each c_i was renamed pc. In this way, *IL* may be used in the refinement of P.

LAW 9.2 *Process indexing*

$$pd = (\textbf{process}\, P \mathrel{\widehat{=}} ||| \; i : Range \odot IL[i])$$

provided L.pps and pps are disjoint with respect to L.state and state.

In the new definition of P, local state is available through indexed processes *IL*. Interleaving ensures there is no interference among the individual elements.

The file system example in the previous section may be used to demonstrate an application of process indexing. A file is a relation between a set of keys and some data, and actions controlling these—given by the process *File* of Example 9.6. For simplicity the only operation allowed is writing to the file.

Example 9.6 *A file*

process*File* $\mathrel{\widehat{=}}$
begin
$File \mathrel{\widehat{=}} [\; contents : Key \twoheadrightarrow Data \;]$
$Init \mathrel{\widehat{=}} [\; contents' = \varnothing \;]$
$Write \mathrel{\widehat{=}} [\; \Delta File;\; k? : Key;\; d? : Data \;|$
$\qquad\qquad k? \in \mathrm{dom}\, contents \wedge contents' = contents \oplus \{k? \mapsto d?\} \;]$
$Run \mathrel{\widehat{=}} \mu X \bullet wrt?k?d \rightarrow Write;\; X$
$\bullet\; Init;\; Run$
end

The global view of a collection of files is given by the schema *GlobalFS*; and the relationship between a *File* and a *GlobalFS* by *Promote*.

Example 9.7   *The global view and promotion schema*

$$GlobalFS \mathrel{\widehat{=}} [\ filesystem : \mathbb{N} \ \rightarrowtail\ File \mid \text{dom } filesystem = 1..maxfiles\ ]$$

$Promote$
$\Delta File$
$\Delta GlobalFS$
$i? : 1..maxfiles$

$i? \in \text{dom } filesystem$
$\theta File = filesystem\ i?$
$filesystem' = filesystem \oplus \{i? \mapsto \theta File'\}$

Now, the *WriteFile* operation of Example 9.2 can be defined using promotion. If the promotion of the channel *wrt* is considered to be the channel *write*, then the result of promoting the action *Run* of Example 9.6 is the action *Run'*.

Example 9.8   *The promoted operation and action*

$$FileWrite \mathrel{\widehat{=}} \exists\, DeltaFile \bullet Write \wedge Promote$$
$$Run' = \mu X \bullet write?i\,?k?d \to FileWrite;\ X$$

The final step is to implement each file as an independent *Circus* process as the result of applying process indexing to *FileSystem*.

Example 9.9   *Indexing individual files*

process $IndexedFile \mathrel{\widehat{=}} (i : 1..maxfiles \odot File)[wrt \leftarrow write]$
process $FileSystem \mathrel{\widehat{=}} \,|||\ i : 1..maxfiles \odot IndexedFile$

9.3.3 Generalised Chaining

The previous laws relied on processes not containing references to nonlocal data. However, sometimes this restriction is a hindrance: for instance, pipelines pass values between neighboring processes. In this section, this restriction is relaxed and a new operator is defined, applicable to processes meeting requirements for *process indexing* with the exception that neighboring state may be shared in the unchained processes.

DEFINITION 9.7   *Generalised chaining*

$$[c_1 \longleftrightarrow c_2, S]n \bullet P \;\widehat{=}$$
$$\mathbf{var}\, v : \mathbb{N}_1 \bullet v := n; \; \mu X \bullet v = 1 \;\&\; P$$
$$\square$$
$$v > 1 \;\&\; v := v - 1; \; (P[m_v := c_1]$$
$$\llbracket \{m_v\} \cup S \rrbracket$$
$$X[m_v := c_2]) \setminus \{m_v\}$$

Definition 9.7 presents the *generalised chaining* operator. This operator relates an indexed set of processes, where each process may engage in communications with adjacent processes. Additionally, synchronizations with the environment may also occur. Given that the major tool for removing shared state is to introduce new events communicating state, it can be seen that this is the role played by synchronizations with adjacent processes. A process declared in terms of this operator can be seen as analogous to one declared using indexed processes such as *FileSystem* of Example 9.9, with the operator also handling the task of communicating and synchronizing neighboring state.

P is the process to be replicated and chained; n copies are requested. Each copy of P will synchronize with the environment on all events in the (possibly empty) set S. The channel c_1 is indexed and renamed m_n in process n, as is c_2 in process $n+1$; processes n and $n+1$ are concurrently composed, synchronizing on m_n, which is internal to the system. External observations are c_1 in the leftmost (lowest indexed) process in the chain, and c_2 in the rightmost (highest indexed) process in the chain. This operator is called *generalised chaining*, as it permits a set of external synchronizations S, in addition to the chained communications, to be specified. In the case where no synchronizations with the environment are required, $S = \emptyset$.

In Figure 9.2, the process $Q \;\widehat{=}\; [\,\{\!|c_1|\!\} \longleftrightarrow \{\!|c_2|\!\}, \{\!|S|\!\}\,]\, 3 \bullet P$ consists of three copies of P, with internal communications along the channels c_1 and c_2 renamed and indexed to m_i. The channels in $\{\!|S|\!\}$ are not renamed, but are placed in the synchronization set for all processes. The result is the process Q, which has the channels c_1, c_2, and $\{\!|S|\!\}$, and behaves as a 3-place buffer.

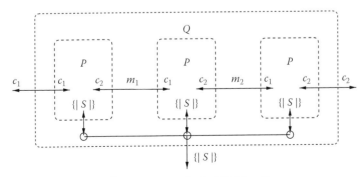

FIGURE 9.2:   Process $Q \;\widehat{=}\; [\,\{\!|c_1|\!\} \longleftrightarrow \{\!|c_2|\!\}, \{\!|S|\!\}\,]\, 3 \bullet P$.

9.3.4 Calculated Introduction of Generalised Chaining

Generalised chaining is most applicable when the abstract process has a structure corresponding to a Z promotion, but where a promoted operation may rely on some nonlocal value. Processes suitable for indexing are also suitable for generalised chaining. However, by insisting on free promotion, factoring the processes out works as before—because the dependency on the global view is removed by introducing the new chained events.

Global state is a function from *Range* to local state. Local actions do not affect global state, and the global action is defined in terms of the promoted local action. For each channel c used by a local action, the global action uses a corresponding promoted channel pc.

LAW 9.3   *Process chaining*

$$pd = (\textbf{process } P \mathrel{\widehat{=}} [c_1 \longleftrightarrow c_2, S]n \bullet L)$$

provided L.pps and pps are disjoint with respect to L.state and state.

9.4 Using *Circus*: A Case Study

In the remainder of this chapter, a case study where a security device is specified, designed, and implemented using *Circus* is investigated. The interesting aspect of this case study is that the device is to be implemented in hardware, and some of the design requirements are dependent on the hardware on which it is deployed. The development of the device is guided by the laws of *Circus*, and proof obligations for development steps center around proving the correctness of the resulting refinement on each major development phase. The case study in question is an Internet packet filter [19,20], a device which sits on a network, monitoring traffic passing through it, and watching for illegal traffic on the network. Typically, these devices can be employed to monitor, route, or prevent traffic on different parts of networks: in all of these cases, but particularly in the case of prevention, confidence in the correctness of the implementation is necessary if network security is to be assured.

The remainder of this section presents the problem domain, and relevant background material. This is followed by a formalization of requirements and an abstract specification in Section 9.5. In Sections 9.6 through 9.9, different components of the implementation are developed and verified. Section 9.10 presents the implementation as a final *Handel-C* model.

9.4.1 Packet Filters

An Internet packet filter is an application that monitors network traffic, and performs actions based on criteria concerning the traffic it observes. In this case study, the packet filter monitors traffic on a local section of Ethernet, flagging observations of predetermined source/destination address pairs. An important property of a monitoring device such as this one is it must not interfere with traffic of no concern to it: essentially, its presence should be effectively unobservable unless it is required to take action. This manifests itself in a number of ways:

- It should not block legitimate traffic.

- It should either flag, or prevent, illegal traffic before it leaves the filter.

- It should not constrain network bandwidth or throughput.

- It should not introduce unnecessary, inconsistent, or unexpected delays.

The packet filter assumes traffic is transmitted using the Internet protocol (IP) version 4 [16,24]. In IP v4, a packet consists of a *header* and a *payload*. The header contains accounting information about the packet, whilst the payload contains the information itself.

Traffic is assumed to be transmitted as a byte-stream. The application should passively observe this byte-stream, identify when sections of the byte-stream corresponds to a packet header, and investigate the addresses contained within that header. This is a nontrivial task: the stream is passing at a rapid rate, and the vast majority of the stream will be payload data. The device must be able to identity a packet header, including performing necessary checksum calculations, extract source, and destination address from the header, compare it to a dictionary of known addresses, and return the result of this comparison before the full header has passed through the stream; and this must be done with the minimum amount of interference to the stream.

9.4.2 Identifying Packets

Packet headers are identified in the stream by performing a checksum calculation, which should be equal to 0 in ones complement, checking the IP version number, which should be equal to 4, checking the least significant bit, which should always be 0, and checking the protocol number, which in this case should be 6, representing TCP/IP. When IP packets are identified, they are examined to identify whether or not the packet requires action. In this case study, the filter simply observes and acknowledges source/destination address pairs predetermined to be of interest.

9.4.3 Content Addressable Memories

A Content Addressable Memory (CAM), also known as an *associative memory*, is a device in which search operations are performed based on content rather than on address. Retrieval of data in a CAM is done by comparing a search term with the contents of the memory locations. If the contents of a location match the supplied data, a match is signaled. Typically, searching in a CAM can be performed in

Calculating concurrency using circus

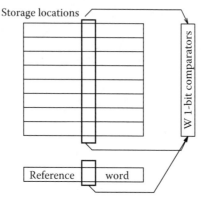

FIGURE 9.3:   Word parallel, bit serial CAM.

a time independent of the number of locations in the memory, which is where it differs from, for instance, a hash table. Various CAM architectures have been proposed [23,29,35]. A CAM needs storage for data, circuitry to compare the search term with the terms in memory, and a mechanism to deliver the results of comparisons. It also needs a mechanism to add and delete the data in memory if the dictionary is not fixed.

Conventional CAMs require circuitry to perform a *word parallel bit parallel* search on words in memory. While this offers the fastest lookups because of the fully parallel nature of the search, it has very high hardware costs.

Reference [20] shows how the packet filter application benefits from content-based lookups, as fast, constant time lookups are required over an arbitrary data set in the form of a pipeline data stream. The design of CAM adopted is called a *Rotated ROM CAM* [19,20], shown in Figure 9.3. Each dictionary word stored has an associated comparator, and this comparator iterates along the word, comparing the relative positions in the dictionary with the search term word. Its simplicity makes it an ideal CAM architecture to implement on an FPGA because the reconfigurable nature of the FPGA means that the ROM can be designed to be as wide as the number of words in the CAM dictionary and as deep as the width of the words in the dictionary.

This design is chosen as it exploits the architecture of the FPGA, allowing a trade-off between hardware costs and the speed of lookups. The trade-off is in the way that the words in memory are not searched in a fully parallel manner, thus reducing the amount of hardware required to search the dictionary, bringing with it an increase in the time it takes to search the dictionary.

9.4.4   Implementation Architecture

The application consists of a number of discrete systems: the packet detector, the search engine, and a process to output results. The packet detector is responsible for monitoring input data, and signaling when a subset of this data constitutes an IP packet. When a packet is detected, the search engine decides if it is a packet of interest by comparing source and destination addresses with those stored in the CAM.

9.4.5 Approach to Verification

In this chapter, many laws of *Circus* are used. Each law typically has a proof obligation associated with it. Discharging all these proof obligations by hand in a large development is infeasible. The technique adopted in this case study is much more pragmatic, and is intended to work more as a realistic industrial development. Laws of *Circus* are repeatedly applied in order to manipulate the actions into the required form. Instead of proving the application of each law, the proof is to show that the final result is a valid refinement of the starting point. This proof obligation is met by model-checking. The CSP_M used in verification for each stage may be easily reconstructed.

Some laws require manipulation of user state, encapsulated by Z schemas. Where there are proof obligations associated with schema and abstract data type manipulation, these are discharged using Z/Eves. The source input for Z/Eves is not included, as these proofs are relatively routine.

The complete development, including all steps and the tool applied to proof obligations at each step, is given graphically in Figures 9.4, 9.5, and 9.9. Figure 9.4 represents the decomposition of the abstract specification into the three major components. Figure 9.5 represents the decomposition into a chained pipeline and a CAM

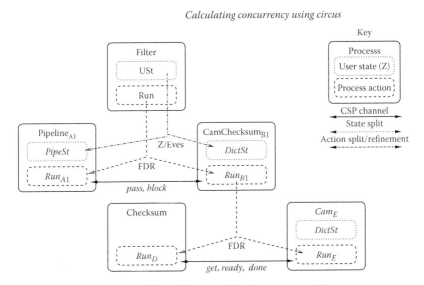

FIGURE 9.4: Development/proof strategy for first system refinements.

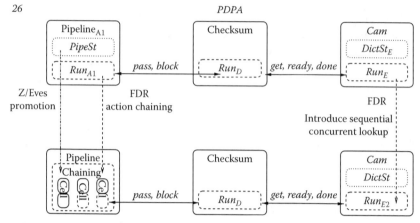

FIGURE 9.5: Development/proof strategy for second system refinements.

with sequentialized lookups. Figure 9.9 represents the refinement of this model into a clocked *Handel-C* implementation.

9.5   An Abstract Specification

In this section, an abstract sequential specification of the packet filter is given. The packet filter reads in a set of bytes from the network, looking to spot addresses. At this stage of development it is not necessary to know the details about representation of data types, so they are specified as given sets.

DEFINITION 9.8   *Given sets and axiomatic definitions*

[*Byte, Addr, BitPair*]
RESULT ::= *yes* | *no*

$$
\begin{array}{|l}
dictsize : \mathbb{N} \\
pipesize : \mathbb{N} \\
addr : (\mathbb{N} \nrightarrow Byte) \nrightarrow (Addr \times Addr) \\
chk : (\mathbb{N} \nrightarrow Byte) \longrightarrow \mathbb{B} \\
\hline
dictsize \geq 1 \\
pipesize = 20 \\
\end{array}
$$

Addresses of interest are to be stored in a dictionary, while the bytes currently being examined will be stored in a pipeline. At this stage of development it is not yet known how many addresses will be stored in the dictionary when it is finally deployed, so this constant is left loose. An IP header is 20 bytes long: this constant is the length of the pipeline.*

The partial function *addr* takes a sequence of bytes and returns a pair of addresses, while the function *chk* takes a sequence of bytes and returns a boolean result. The former will later be used to extract addresses from a packet, and the latter to identify a packet.

The local state of the abstract system, given in Definition 9.9, has two components: a pipeline and a dictionary. As the pipeline will be built in hardware, the state invariant does not allow its size to ever change. The size of the dictionary is also constant. Unusually, an initialization operation has not been specified for these. In the case of the pipeline, this is because initially the values are meaningless: when hardware is powered up, registers have an arbitrary value. In the case of the dictionary, this is because the specification is purposefully being left loose with regards to the addresses of interest. When the final implementation is deployed, these would be given.[†]

DEFINITION 9.9   *The abstract packet filter*

process *Filter* $\widehat{=}$
begin
$USt \widehat{=} [\ pipe : \mathbb{N} \nrightarrow Byte;\ dict : \mathbb{N} \nrightarrow (Addr \times Addr)\ |$
$\qquad\qquad \text{dom}\ pipe = 1..pipesize \wedge \text{dom}\ dict = 1..dictsize\ ]$

$$
\begin{array}{l}
\underline{UpdateAll}\underline{\qquad\qquad\qquad\qquad\qquad\qquad\qquad\qquad\qquad\qquad} \\
\Delta USt \\
b? : Byte \\
\hline
pipe' = \\
\quad \{x : 1..pipesize - 1;\ y : Byte\ | \\
\qquad x \mapsto y \in pipe \wedge x + 1 \in \text{dom}\ pipe \bullet x + 1 \mapsto y\} \cup \{1 \mapsto b?\} \\
dict' = dict
\end{array}
$$

---

* The development could have started with a more abstract description of the problem, with a data refinement between it and the concrete. However, in this chapter the concern is developing the concrete data model into a concurrent implementation.

[†] Where and when these were given would depend upon the target implementation environment. For instance, for a purpose built hardware CAM, they could be included in the state invariant; whilst for a software implementation such as a hash table, it may be done by adding a new operation to initialize the dictionary variable.

$$Run \mathrel{\widehat{=}} \mathbf{var}\, c : \mathbb{N};\ h : (Addr \times Addr);\ \mu X \bullet$$
$$(in?b \rightarrow SKIP \mathrel{|||} out!last(pipe) \rightarrow SKIP);$$
$$\neg\, chk(pipe)\ \&\ UpdateAll;\ X$$
$$\square$$
$$chk(pipe)\ \&$$
$$h := addr(pipe);\ c := 1;\ \mu Y \bullet UpdateAll;$$
$$c < \#pipe - 1\ \&\ (in?b \rightarrow SKIP$$
$$\mathrel{|||} out!last(x) \rightarrow SKIP);$$
$$c := c + 1;\ Y$$
$$\square$$
$$c = \#pipe - 1\ \&\ (in?b \rightarrow SKIP$$
$$\mathrel{|||} out!last(x) \rightarrow SKIP$$
$$\mathrel{|||} match!(h \in dict)) \rightarrow SKIP);\ X$$

$\bullet$ *Run*

end

An operation to update the local state exists. This operation takes the state of the pipeline, and a new byte as input. It adds the new byte to the head of the pipeline, and drops the last byte in the pipeline. The dictionary remains unchanged. There are no outputs.

The main action of the process reads an input into the local variable b and outputs the last element in the pipeline. Then the predicate *chk* returns *true* or *false* indicating whether or not it believes the pipeline to correspond to a packet header. If not, the new data is stored and shifted. If it does then the address pair in the pipeline is recorded in the local variable h before the shift. Another *pipesize–2* shift cycles are permitted before a result is output on the channel *match* indicating whether or not the addresses were known to the dictionary. The condition on *pipesize* ensures that the result is known to the environment before the data has fully left the pipeline.

9.6   Refining into a Pipeline and a Checksum

Ultimately, the abstract specification must be split into three components: a pipeline, a CAM, and a checksum calculation. This decomposition is shown in Figure 9.4, including the channels that will be inserted and the strategy for proof of refinement.

Inspecting Definition 9.9 suggests a development strategy. The main action may be split into two: one that acts on the dictionary, and the other that acts on the pipeline; the requirement is that the dictionary state and the pipeline state must also be split. To split the action *Run*, it must be manipulated into a suitable form.

The implementation is to contain a pipeline of cells, storing data from the network. As these are to be implemented as concurrent registers, user state in each cell must be disjoint from the next—the requirement for *Process splitting* (Law 9.1).

Counterintuitively, therefore, the first development step is to separate the *checksum calculation* from the rest of the application.

This is nonobvious as a first step, but is vitally important. The checksum, when implemented, will require to inspect the value of a number of pipeline cells, and given that the states must be disjoint it *cannot* be implemented as a global predicate over these cells. The specification must be manipulated such that the checksum inspects a local copy of the pipeline.

The goal of this development phase is to split the component of the action that maintains the pipeline from the component that calculates the checksum and performs the lookup. This is done by replicating the behavior in two actions, and then removing from each the behaviors that are not required in each. In doing so, there are several synchronous properties of the specification that care must be taken to preserve. First, *in* and *out* occur pairwise. Second, when a match occurs, it must interleave the correct pair of *in* and *out* events. Third, the address to be looked up must be stored and made available on the correct *in–out* cycle. To achieve this, two new events are introduced. The event *block* is used to delimit *in–out* cycles after each *UpdateAll* operation. On each iteration a second new event *pass* is introduced that communicates the values held in the pipeline to those who may desire read access. In this, and following definitions the actions that pass local state across are factored out in Definition 9.10 for presentation.

DEFINITION 9.10 *Passing pipeline state*

$$Pass \mathrel{\widehat{=}} in?b \rightarrow SKIP$$
$$\mathbin{|||} out!last(pipe) \rightarrow SKIP$$
$$\mathbin{|||} (\mathbin{|||} i : \operatorname{dom} pipe \bullet pass.i!pipe(i) \rightarrow SKIP)$$

$$Pass' \mathrel{\widehat{=}} in?b \rightarrow SKIP$$
$$\mathbin{|||} out!last(pipe) \rightarrow SKIP$$
$$\mathbin{|||} (\mathbin{|||} i : \operatorname{dom} pipe \bullet pass.i?copy(i) \rightarrow SKIP)$$

$$Match \mathrel{\widehat{=}} Pass \mathbin{|||} match!(h \in dict) \rightarrow SKIP$$

$$Match' \mathrel{\widehat{=}} Pass' \mathbin{|||} match!(h \in dict) \rightarrow SKIP$$

The checksum behavior must also be factored out. This step relies on the property that when P is deterministic, $P = P \parallel P$. *Run* is deterministic, so may be placed in parallel with a second copy of itself. However, although the two actions synchronize on their events—preserving the deterministic property—the operation schemas and state variable assignments do not. If this were disregarded the action would, for instance, execute *UpdateAll* twice every time it was intended to execute once. The second copy of this action therefore has its own copies of local variables c and h; and does not write to global state. In fact, this development step goes one stage further: it introduces a new variable *copy* to the new action that has the same type as the pipeline, and each *pass* cycle updates the local copy. This step further relies on laws for local variable introduction, and for introducing a direction in the *pass* communication. The replicated action is given in Definition 9.11.

DEFINITION 9.11   *Introducing internal events and a concurrent action*

$Run_A \cong \textbf{var} \, c : \mathbb{N}; \ h : (Addr \times Addr); \ \mu X \bullet$
        $Pass;$
                $\neg \, chk(pipe) \ \& \ UpdateAll; \ block \rightarrow X$
                $\square$
                $chk(pipe) \ \& \ h := addr(pipe); \ c := 1; \ \mu Y \bullet$
                        $UpdateAll; \ block \rightarrow SKIP;$
                                $c < \#pipe - 1 \ \&$
                                        $Pass;$
                                                $c := c + 1; \ block \rightarrow Y$
                                $\square$
                                $c = \#pipe - 1 \ \&$
                                        $Match;$
                                                $UpdateAll;$
                                                        $block \rightarrow X$

$Run_B \cong \textbf{var} \, copy : \mathbb{N} \rightarrowtail Byte \mid \text{dom} \, copy = 1..pipesize;$
        $c : \mathbb{N}; \ h : (Addr \times Addr); \ \mu X \bullet$
        $Pass';$
                $\neg \, chk(pipe) \ \& \ block \rightarrow X$
                $\square$
                $chk(pipe) \ \& \ h := addr(copy); \ c := 1; \ \mu Y \bullet$
                        $block \rightarrow SKIP;$
                                $c < \#pipe - 1 \ \&$
                                        $Pass';$
                                                $c := c + 1;$
                                                        $block \rightarrow Y$
                                $\square$
                                $c = \#pipe - 1 \ \&$
                                        $Match; \ block \rightarrow X$

$Run \cong (Run_A \ [\![\, \{\!\mid in, out, match, pass, block \mid\!\} \,]\!] \ Run_B)$

Next we separate concerns between the actions, relying on properties of synchronization and distributed cotermination. Run_A is to form the pipeline; therefore, Run_B

should not engage in *in* or *out*. Given that *block* was introduced to delimit the *in–out* sequences, *in* and *out* can safely be dropped from Run_B and the synchronization. The same argument holds for *match*: the pipeline should not be aware of matches; therefore, it can be dropped from Run_A. This allows a further simplification: the variable h no longer plays a role in Run_A, so its scope may be restricted to Run_B. Moreover, the role played by c was to implement a loop causing a number of shifts before a *match*—this is no longer necessary in Run_A. The two actions are given in Definitions 9.12 and 9.13, where Run_{A1} inputs, outputs, and passes data to Run_{B1}, which records the data and indicates lookup results as appropriate.

DEFINITION 9.12 *Removing replicated behaviors in the pipeline*

$$Run_{A1} \;\widehat{=}\; \mu X \bullet Pass; \; UpdateAll; \; block \to X$$

DEFINITION 9.13 *Removing replicated behaviors*

$$
\begin{aligned}
&Run_{B1} \;\widehat{=}\; \textbf{var}\, copy : \mathbb{N} \;\nrightarrow\; Byte \mid \mathrm{dom}\, copy = 1..pipesize; \\
&\qquad\qquad c : \mathbb{N};\; h : (Addr \times Addr);\; \mu X \bullet \\
&\quad (\,\|\|\; i : \mathrm{dom}\, copy \bullet pass.i?copy(i) \to SKIP); \\
&\qquad \neg\, chk(copy) \;\&\; block \to X \\
&\qquad\quad \square
\end{aligned}
$$

$$
\begin{aligned}
&\quad chk(copy) \;\&\; \\
&\qquad c := 1;\; h := addr(copy);\; \mu Y \bullet \\
&\qquad\quad block \to SKIP; \\
&\qquad\qquad c < \#\,\mathrm{dom}\, copy - 1 \;\&\; \\
&\qquad\qquad\quad (\,\|\|\; i : \mathrm{dom}\, copy \bullet pass.i?copy(i) \to SKIP); \\
&\qquad\qquad\qquad c := c + 1;\; block \to Y \\
&\qquad\quad \square \\
&\qquad c = \#\,\mathrm{dom}\, copy - 1 \;\&\; \\
&\qquad\quad (\,\|\|\; i : \mathrm{dom}\, copy \bullet pass.i?x \to SKIP \\
&\qquad\quad \|\|\; match!(h \in dict) \to SKIP\,); \\
&\qquad\qquad block \to X
\end{aligned}
$$

Run_{A1} and Run_{B1} are disjoint with respect to user state. There is no associated proof obligation—but if this were false, further development would fail. However, there is an obligation to show the new actions have been correctly derived. Theorem 9.1 states that their parallel combination is a refinement of the specification.

THEOREM 9.1   *The action refinement is correct*

$$Run \sqsubseteq_A (Run_{A1} [\![\{pass, block\}]\!] Run_{B1}) \setminus \{pass, block\}$$

This is proved using the equivalent refinement assertion in FDR: *assert Run $\sqsubseteq_{FD}$ (Run_{A1} [[$\{pass, block\}$]] Run_{B1}) \ $\{pass, block\}$.*

9.6.1   Partitioning the Global User State

Although the actions have now been split, they both still reference the single global user state: Run_{A1} accesses the pipeline state, while Run_{B1} accesses the dictionary. In order to show that this process meets the form applicable to process splitting, these states and operations must be disjoint. In this section, the disjoint states are calculated and verified using Z/Eves. By relying on the properties of schema conjunction, the original user state can be split in two, and *Update* acts on the pipeline, leaving the dictionary unchanged.

DEFINITION 9.14   *Partitioned user state and Update*

$$PipeSt \cong [\ pipe : \mathbb{N} \ \nrightarrow Byte \mid \text{dom} \ pipe = 1..pipesize \ ]$$
$$DictSt \cong [\ dict : \mathbb{N} \ \nrightarrow (Addr \times Addr) \mid \text{dom} \ dict = 1..dictsize \ ]$$

UpdateAll' _____
$\Delta PipeSt$
$b? : Byte$
_____
$pipe' = \{x : 1..\#\text{dom} \ pipe; \ y : Byte \mid$
$\qquad\qquad x \mapsto y \in pipe \wedge x + 1 \in \text{dom} \ pipe \bullet x + 1 \mapsto y\}$
$\qquad \cup \{1 \mapsto b?\} \ ]$

THEOREM 9.2   *Partitioned states and update are correct*

$$Ust \Leftrightarrow (PipeSt \wedge DictSt)$$
$$UpdateAll \Leftrightarrow (UpdateAll' \wedge \Xi DictSt)$$

9.6.2   First Application of Process Splitting

The actions and state are now of the correct form for *Process Splitting* to be applied. The abstract pipeline is a very simple process, containing an abstract data type holding all pipeline data. On each iteration, it reads in a new value, outputs the oldest value, informs the environment of the current state of the pipeline, and then shifts all the data. The abstract checksum and lookup process contains the dictionary. The main action of this process performs the checksum calculation on the local copy of the state that is passed to it on each pipeline shift, and outputs the result of the lookup accordingly.

DEFINITION 9.15   *The abstract pipeline and CAM/checksum processes*

process $Pipeline_{A1} \cong$ **process** $CamChecksum_{B1} \cong$
begin   **begin**
$PipeSt \cong$ $DictSt \cong$
   $[pipe : \mathbb{N} \nrightarrow Byte \mid$ $[dict : \mathbb{N} \nrightarrow (Addr \times Addr)]$
      $\text{dom } pipe = 1..pipesize]$

$UpdateAll' \cong$ Definition 9.14
$Run_{A1} \cong$ Definition 9.12 $Run_{B1} \cong$ Definition 9.13
• Run_{A1} • Run_{B1}
end   **end**

The complete packet filter is the pipeline in parallel with the checksum synchronising on the channel used to pass pipeline state across. Both also synchronize on the channel *block*, thus maintaining the synchronous nature of the behavior between the two components.

DEFINITION 9.16   *The split packet filter*

$Filter' \cong (CamChecksum_{B1} \mathbin{[\![} \{pass, block\} \mathbin{]\!]} Pipeline_{A1}) \setminus \{pass, block\}$

9.7   Implementing the Pipeline as Concurrent Cells

The next development stage, shown in Figure 9.5, is to implement the pipeline as an array of concurrent cells, using the generalised chaining operator of Definition 9.7. Reference [28] shows how a Z-style promotion distributes through a *Circus* process. By factoring out a promotion from the abstract pipeline, a single cell is exposed. However, $UpdateAll'$ describes a shift of the entire pipeline: therefore we must rewrite this in terms of a single element. The schema $Update$ defines an update on a single pipeline element, identified by the input variable i.

DEFINITION 9.17   *A single update*

$Update \cong [\ \Delta PipeSt;\ b? : Byte;\ i? : \mathbb{N} \mid pipe' = pipe \oplus \{i? \mapsto b?\}\ ]$

For this operation to implement a complete pipeline shift, it must act upon all elements of the pipeline. This may be achieved by replicating and interleaving *pipeline* copies of the operation. It is necessary to store the initial state of the pipeline in a local variable to ensure that each interleaved *Update* acts on the correct initial value of its predecessor element.

DEFINITION 9.18   *Interleaved updates*

$IUpdate \,\widehat{=}$
   var $copy : \mathbb{N} \, \longrightarrow\!\!\!\!\!+ \, Byte \mid copy = pipe \bullet$
      $|||\, i : 1..pipesize \bullet (i \neq 1 \wedge b = copy\,(i - 1); \; Update)$

Now local state may be described as a promotion. A local element is simply a *Byte*, while the global state of the system is a function from natural numbers (the index of each element in the pipeline) to the elements. The update operation changes an individual element to the value of the input variable. The global view in terms of local elements is given by the schema *Promote*.

DEFINITION 9.19   *Local and global views, and the promotion schema*

$PipeCell \,\widehat{=}\, [\, elem : Byte\,]$
$GlobalPipe \,\widehat{=}\, [\, pipe : \mathbb{N} \, \longrightarrow\!\!\!\!\!+ \, PipeCell\,]$
$UpdateCell \,\widehat{=}\, [\, \Delta PipeCell;\; b? : Byte \mid elem' = b?\,]$

$$
\begin{array}{|l}
\hline
\,Promote \,\underline{\hspace{8cm}}\\
\,\Delta GlobalPipe\\
\,\Delta PipeCell\\
\,i? : \mathbb{N}\\
\hline
\,i? \in \mathrm{dom}\, pipe\\
\,\theta PipeCell = pipe(i?)\\
\,pipe' = pipe \oplus \{i? \mapsto \theta PipeCell'\}\\
\hline
\end{array}
$$

For this promotion to be factored out, it must be *free*: that there is no global constraint that cannot be expressed locally. If this were not the case, then each local element of state would need to be aware of other local elements of state: something that is not permitted if processes are to be concurrent. Theorem 9.3 captures this with a typical free promotion proof obligation.

THEOREM 9.3   *The promotion is free:*

$$\exists\,PipeCell' \bullet \exists\,GlobalPipe' \bullet Promote \Rightarrow$$
$$\forall\,PipeCell' \bullet \exists\,GlobalPipe' \bullet Promote$$

9.7.1   A Local Process Implementing a Cell

A single pipeline cell is a process that encapsulates the local data, with an unpromoted input and output action.

process *Cell* $\hat{=}$
begin
PipeCell $\hat{=}$ [*elem* : *Byte*]
UpdateCell $\hat{=}$ [$\Delta PipeCell$; *b*? : *Byte* | *elem'* = *b*?]

$Run_C \hat{=} \mu\,X \bullet$
    $(in?b \rightarrow SKIP \;|||\; out!elem \rightarrow SKIP \;|||\; pass.i!elem \rightarrow SKIP)$;
        $UpdateCell$; $block \rightarrow X$
$\bullet\,Run_C$
end

It seems possible to compose a number of cells to form a pipeline using process indexing, Law 9.2. However, this is insufficient as it requires no interference between local processes. This is precisely not the case here, where each process needs the local value of its numeric predecessor in the pipeline—the role played by the local variable *copy* in *IUpdate* (Definition 9.18). Furthermore, only the first input *in* and the last output *out* are externally visible.

9.7.2   Composing Local Processes

This is exactly the scenario achieved by generalised chaining. The promotion of each cell is the function from local to global state, and states and local operations are disjoint as the promotion is free. Promoting *in*, last *out*, *tick*, and *tock* gives the global action with all the internal communications hidden. The index of each process *i* is taken from the promotion schema.

The operator composes *pipesize* copies of *Cell* concurrently. A parameter given to the operator is a pair of events that are to chain—to synchronize—numerically adjacent processes, uniquely renamed for adjacent processes accordingly. This synchronization communicates the initial value of a cell to its neighbor. Internal events are hidden. The *ripple effect* of nearest neighbor communication ensures inputs and outputs are ordered correctly. This construction is demonstrated for a pipeline of three cells in Figure 9.6. Theorem 9.4 states that this implementation is correct

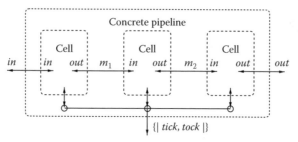

FIGURE 9.6: Process $[in \longleftrightarrow out, \{| tick, tock |\}] \ 3 \bullet Cell$.

with respect to the specification, and this is proved using FDR with the assertion *assert Pipeline$_{A1}$ $\sqsubseteq_{FD}$ Pipeline*.

DEFINITION 9.20   *The pipeline implementation*

$$Pipeline \,\widehat{=}\, [in \longleftrightarrow out, \{|block|\}]pipesize \bullet Cell$$

THEOREM 9.4   *The pipeline cell implementation is correct*

$$Pipeline_{A1} \sqsubseteq_P Pipeline$$

9.7.3   Modeling the Local Processes as *Handel-C* Variables

The description of a single cell is very close to *Handel-C*. The remaining task is to include the model of the clock. Assignments take place on a rising clock edge, modeled using an event *tick*. All assignments in all processes must happen before a clock cycle completes, modeled using an event *tock*. In adapting the description of a cell to behave as a *Handel-C* process this clock model must be included. It is trivial in this case: each process performs a single assignment after its communications; therefore, each iteration in *Run$_{C1}$* is a single clock cycle. The event *block* ensured the synchronous behavior of pipeline cells on each iteration: this function is now achieved by the clock, so *block* may be dropped.[*] This description is now only a syntactic step away from the implementation of each cell as a simple variable.

---

[*] Alternatively, this step could be regarded as renaming *block* to *tock* and including the *tick* event to globally synchronize and separate communications from state updates.

DEFINITION 9.21 *A Handel-C model of cell implementation*

> **process** *ClockedCell* $\widehat{=}$
> **begin**
> *PipeCell* $\widehat{=}$ [*elem* : *Byte*]
> *UpdateCell* $\widehat{=}$ [$\Delta PipeCell$; *b*? : *Byte* | *elem′* = *b*?]
>
> *Run*$_{C1}$ $\widehat{=}$
> $\mu X \bullet$ (*in*?*b* $\rightarrow$ *SKIP* ||| *out*!*elem* $\rightarrow$ *SKIP* ||| *pass.i*!*elem* $\rightarrow$ *SKIP*);
>   *tick* $\rightarrow$ *UpdateCell*;
>   *tock* $\rightarrow$ *X*
> $\bullet$ *Run*$_{C1}$
> **end**

Adapting the chained processes is trivial: all instances share the same clock, achieved by placing the events in the global (shared) synchronization set.

DEFINITION 9.22 *The clocked* Handel-C *pipeline implementation*

> *ClockedPipeline* $\widehat{=}$ [*in* $\longleftrightarrow$ *out*, \{*tick*, *tock*\}] *pipesize* $\bullet$ *ClockedCell*

In the above definition, the clock may be hidden if no further processes are to be introduced, or if they use a separate clock; however, it is left visible in this implementation as the CAM will share the same global clock. Theorem 9.5 states that the clocked implementation is correct and can be proved with the FDR assertion **assert** *Pipeline* \ \{*block*\} $\sqsubseteq_{FD}$ *ClockedPipeline* \ \{*tick*, *tock*\}.

THEOREM 9.5 *The implementation of the clocked pipeline is correct*

> *Pipeline* \ \{*block*\} $\sqsubseteq_P$ *ClockedPipeline* \ \{*tick*, *tock*\}

9.8 Separating the Checksum and the CAM

The checksum is a standard calculation used in IP packet identification. If the checksum calculation returns the same result as that contained within the header, and several other sanity checks also hold, then the state of the pipeline represents an IP packet header. The next development stage is to separate this out from the process that performs the inspection on the addresses.

The first steps in separating the checksum and the CAM follow a similar pattern as before. A *match* is prefixed with *ready* and followed with *done*. A new second action agrees to this *ready-match-done* cycle. As before, *match* may now be dropped from the original. This is shown in Definition 9.23; some parts of the definition not relevant to this design step have been elided (...).

DEFINITION 9.23   *Adding the partner action*

$$Run_{B2} \;\widehat{=}\; \mathbf{var}\, h : (Addr \times Addr)... \;\bullet$$
$$\mu X \bullet ...; \; c < \#\operatorname{dom} pipe - 2 \;\&\; ...; \; X$$
$$\square$$
$$c = \#\operatorname{dom} pipe - 2 \;\&$$
$$(...$$
$$\||| \; ready \rightarrow$$
$$match!(h \in dict) \rightarrow$$
$$done \rightarrow SKIP); \; X$$
$$\|[\{ready, match, done\}]\|$$
$$\mu Y \bullet ready \rightarrow match!(h \in dict) \rightarrow done \rightarrow Y$$

If Run_{B2} is to be split, it must not share the global variable h. The main tool for removing such dependencies is the introduction of a new event to communicate state: *get* is introduced for this purpose. Now restricting the scope of h (and of *copy* and c to the first action) is trivial, and Run_{B2} may be rewritten as two separate actions Run_D and Run_E.

DEFINITION 9.24   *The checksum and CAM actions*

$$Run_D \;\widehat{=}\; \mathbf{var}\, copy : \mathbb{N} \rightarrowtail Byte \mid \operatorname{dom} copy = 1..pipesize; \; c : \mathbb{N}; \; \mu X \bullet$$
$$(\||| \; i : \operatorname{dom} copy \bullet pass.i?copy(i) \rightarrow SKIP);$$
$$\neg \, chk(copy) \;\&\; block \rightarrow Check$$
$$\square$$
$$chk(copy) \;\&$$
$$c := 1; \; get!addr(copy) \rightarrow SKIP \bullet \mu Y \bullet block \rightarrow SKIP;$$

$$c < \operatorname{dom} copy - 1 \;\&$$
$$(\||| \; i : \operatorname{dom} copy \bullet pass.i?copy(i) \rightarrow SKIP);$$
$$block \rightarrow c = c + 1; \; Y$$
$$\square$$
$$c = \#\operatorname{dom} copy - 1 \;\&$$
$$ready \rightarrow$$
$$(\||| \; i : \operatorname{dom} copy \bullet pass.i?copy(i) \rightarrow SKIP);$$
$$done \rightarrow block \rightarrow X$$

$$Run_E \;\widehat{=}\; \mu X \bullet get?term \rightarrow ready \rightarrow match!(term \in dict) \rightarrow done \rightarrow X$$

9.8.1 Splitting the Checksum and CAM

The two actions are now of a form that allows the process to be split. The correctness of this development step can be verified by proving the refinement relation that holds between the main action of the abstract CAM and checksum process of Definition 9.15 and the newly split actions (Definitions 9.25 and 9.26) with the FDR assertion *assert* $Run_{B1} \sqsubseteq_{FD} (Run_D \, [\![\, get, ready, done \,]\!] \, Run_E) \setminus \{\!| get, ready, done |\!\}$.

THEOREM 9.6 *The split actions are a refinement*

$$Run_{B1} \sqsubseteq_A (Run_D \, [\![\, get, ready, done \,]\!] \, Run_E) \setminus \{\!| get, ready, done |\!\}$$

DEFINITION 9.25 *The abstract CAM process*

> **process** $Cam_E \;\widehat{=}$
> **begin**
> $DictSt \;\widehat{=}\; [dict : \mathbb{N} \;\nrightarrow\; (Addr \times Addr) \mid \mathrm{dom}\, dict = 1..dictsize]$
> $Run_E \;\widehat{=}\;$ Definition 9.24
> • *Run*
> **end**

In the checksum, variables local to the action may be encapsulated as the user state of the process.

DEFINITION 9.26 *The checksum implementation*

> **process** $Checker \;\widehat{=}$
> **begin**
> $USt \;\widehat{=}\; [copy : \mathbb{N} \;\nrightarrow\; Byte; \; c : \mathbb{N} \mid \mathrm{dom}\, copy = 1..pipesize]$
> $Run_D \;\widehat{=}\;$ Definition 9.24
> • Run_D
> **end**

The process monitoring the checksum calculation is now ready for implementation, and may have the program clock introduced. The interleaved *pass* events are actually emulating read access to the pipeline process, so they must appear before a *tick*—there is no value to be latched in. Other assignments, such as to the local variable c, must be latched between the *tick* and the *tock*. The clock now additionally performs the role of making sure that each *pass* cycle is synchronous with respect to pipeline shifts (as the clock is shared with the pipeline), so *block* may be dropped.

Another less obvious role of the clock comes from the fact that the events *ready* and *done* occur exactly $\# \mathrm{dom}\, copy - 2$ clock cycles after a *get* is issued (counting starts at 1). These events were introduced to ensure that the CAM would output results at the correct time. As the CAM is to be implemented on the same FPGA, with the same global clock, the assumption that the *match* output will happen # dom *copy*–2 clock cycles after it receives a *get* can be made. As long as the development of the CAM respects this assumption, *ready* and *done* no longer play a significant role in the behavior of the checksum and the CAM, and may be dropped. This step has not been made as a result of a direct application of any of laws and is discussed further in Section 9.9.

DEFINITION 9.27   *The clocked checksum action*

$$Run_{D1} \cong \mu X \bullet$$
$$(||| \, i : \text{dom } copy \bullet pass.i?copy(i) \rightarrow SKIP);$$
$$\neg \, chk(copy) \, \&$$
$$tick \rightarrow tock \rightarrow Check$$
$$\square$$
$$chk(copy) \, \&$$
$$get!addr(copy) \rightarrow tick \rightarrow c := 1; \, tock \rightarrow SKIP \bullet \mu Y \bullet$$
$$c < \# \text{dom } copy - 1 \, \&$$
$$(||| \, i : \text{dom } copy \bullet pass.i?copy(i) \rightarrow SKIP);$$
$$tick \rightarrow c = c + 1; \, tock \rightarrow Y$$
$$\square$$
$$c = \# \text{dom } copy - 1 \, \&$$
$$(||| \, i : \text{dom } copy \bullet pass.i?copy(i) \rightarrow SKIP);$$
$$tick \rightarrow tock \rightarrow X$$

DEFINITION 9.28   *Handel-C model of the checksum*

process *ClockedChecker* $\cong$
begin
$USt \cong [copy : \mathbb{N} \nrightarrow Byte; \, c : \mathbb{N} \mid \text{dom } copy = 1..pipesize]$
$Run_{D1} \cong$ Definition 9.27
$\bullet \, Run_{D1}$
end

Theorem 9.7 states that the clocked implementation is correct and may be proved with the FDR assertion *assert Checker* \ $\{ready, done, block\} \sqsubseteq_{FD} ClockedChecker$ \ $\{tick, tock\}$.

THEOREM 9.7   *The clocked checksum is correct*

$$Checker \setminus \{ready, done, block\} \sqsubseteq_P ClockedChecker \setminus \{tick, tock\}$$

9.9   Implementing the Content Addressable Memory

The size of an IP packet header, combined with the clock cycle requirements of the components, is a useful piece of information in designing the CAM. It allows splitting combinatorial logic over multiple clock cycles and a more serial implementation of a CAM, which may reuse comparand registers and reduce area requirements.

The *Rotated ROM* design of Refs. [19,20] consists of ROMs of depth 16 bits, and width 2 bits, giving 32-bit words, each one of which corresponds to an IP address. The search circuitry compares 2 bits at a time, meaning that 16 comparisons are required to compare the search term with a word in the dictionary. The circuitry assigns a value to a flag indicating whether a word matches the search term or not. Figure 9.7 shows the area costs of this design on a sample FPGA in terms of configurable logic blocks. In Figure 9.8, the clock speeds attainable for increasing CAM sizes are shown under two conditions: when the FPGA synthesis tools place and route the circuit using no special constraints (lower line), versus when the tools are told to optimize the speed (upper line). These experimental results confirm that a Rotated ROM CAM will permit the FPGA to be clocked at a sufficiently fast speed to allow the rest of the application to monitor a network in real time.

9.9.1   Making the Dictionary Tight

Initially, the definition of the dictionary state was left loose. This section begins by making it tight—given in the replacement definition of *DictSt*. The state in the

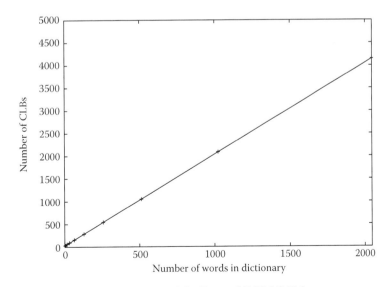

FIGURE 9.7:   Example area costs of the Rotated ROM CAM.

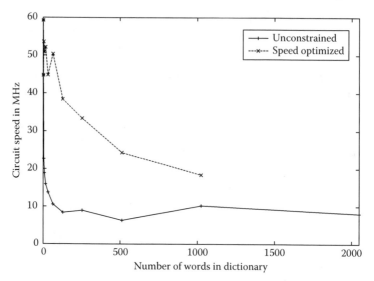

FIGURE 9.8:   Example clock speeds of the Rotated ROM CAM.

schema *Dict* is the source/destination addresses given as 2×16 *BitPair* representations. As the dictionary is being implemented in hardware, the values are invariant (statically determined)—there is no need for an initialization operation, as they never change.

DEFINITION 9.29   *The state of the dictionary*

DictSt

$dict : \mathbb{N} \nrightarrow \text{seq } BitPair$

---
dom *dict* $= 1..4$
$dict\ 1 = \langle (1,0), (1,0), (0,0), (1,1), (0,0), (0,0), (0,0), (0,1),$
$\qquad\qquad (0,0), (0,1), (1,0), (1,1), (1,1), (0,0), (0,0), (0,0) \rangle$
...*etc*...

User state is now a dictionary consisting of four concrete addresses, implemented as four *BitPair* sequences of length 16. In implementing the lookup, it is tempting to specify this by defining simple equality tests over dictionary elements; however, the design goal is directed by area and speed concerns—optimally, a single, iterative comparator for each word in the dictionary.

In the Rotated ROM CAM, local variables maintain lookup results and index iterations. The definition of the main CAM action below reflects this. The local array *result* stores the boolean result of comparing each dictionary entry with the search term. The *match* line outputs a result indicating if comparisons returned true—implemented as the disjunction over the elements in the result array. Rather than a simple comparison over each word, it is implemented as an iteration over comparing each *BitPair* with the relevant corresponding place in the search term, meaning there are 16 comparisons performed for each word, with the *n*th comparison for each word being performed in parallel. When a comparison fails, this is recorded. Implementation of a 2-bit comparator is cheap: this way area and combinatorial costs have been drastically reduced. Events *ready* and *done* ensure that the implementation still meets the constraints that the lookup should be complete before the word has left the pipeline.

DEFINITION 9.30   *Sequencing the comparators*

$$Run_{E1} \mathrel{\widehat{=}} \textbf{var}\ result : \mathrm{dom}\ dict \;\longrightarrow\; \mathbb{B};\ \mu X \bullet$$
$$get?term \rightarrow c := 0;$$
$$\mathrm{ran}\ result := true;\ \mu Y \bullet$$
$$c = 16\ \&$$
$$ready \rightarrow match!(true \in \mathrm{ran}\ result) \rightarrow done \rightarrow X$$
$$\square$$
$$0 \le c < 16\ \&$$
$$(\forall i : \mathrm{dom}\ result \bullet$$
$$result(i) = result(i) \wedge dict.i(c) = head(term));$$
$$term := tail(term);\ c := c + 1;\ Y$$

The universal quantifier in Definition 9.30 may be expanded—resulting in the conjunction of assignments indexed by *i* shown in Definition 9.31. As these assignments are disjoint with respect to state, they may be implemented in an arbitrary order.

DEFINITION 9.31   *Concurrent words*

$$Run_{E2} \mathrel{\widehat{=}} \textbf{var}\ result : \mathrm{dom}\ dict \;\longrightarrow\; \mathbb{B};\ \mu X \bullet$$
$$get?term \rightarrow c := 0;$$
$$\mathrm{ran}\ result := true;\ \mu Y \bullet$$
$$c = 16\ \&$$
$$ready \rightarrow match!(true \in \mathrm{ran}\ result) \rightarrow done \rightarrow X$$
$$\square$$

$$c \ge 0 \wedge c < 16\ \&$$
$$(|||\ i : \mathrm{dom}\ result \bullet$$
$$result(i) = result(i) \wedge dict.i(c) = head(term));$$
$$term := tail(term);\ c := c + 1;\ Y$$

DEFINITION 9.32   *The CAM implementation*

> **process** *Cam* $\widehat{=}$
> **begin**
> *DictSt* $\widehat{=}$ Definition 9.29
> *Run*$_{E2}$ $\widehat{=}$ Definition 9.31
> • *Run*$_{E2}$
> **end**

Unlike the other components in this case study, concurrency is not introduced using process splitting—it has been introduced with concurrent assignments to user state. To attempt process splitting is not useful: although each entry in the dictionary and the comparisons are disjoint, the result array is not. To split these processes—and therefore the result array—and collate results using further communications would mean that either the clock cycle constraint is not met, or that the implementation is less serial.

As the implementation is to be in *Handel-C* on the same FPGA as the other components, the global clock may now be introduced, in Definition 9.33.

DEFINITION 9.33   *The clocked CAM action*

$$Run_{E3} \widehat{=} result : \mathrm{dom}\, dict \longrightarrow \mathbb{B};\ \mu X \bullet$$
$$tick \rightarrow tock \rightarrow X$$
$$\Box$$
$$get?term \rightarrow tick \rightarrow c := 0;$$
$$\mathrm{ran}\, result := true;\ tock \rightarrow SKIP;\ \mu Y \bullet$$
$$c = 16\ \&$$
$$match!(true \in \mathrm{ran}\, result) \rightarrow tick \rightarrow tock \rightarrow X$$
$$\Box$$
$$c \geq 0 \wedge c < 16\ \&$$
$$(\|\ i : \mathrm{dom}\, result \bullet$$
$$\|\{tick, tock\}\,\|\ result(i) = result(i)$$
$$\wedge\ dict.i(c) = head(term));$$
$$tick \rightarrow term := tail(term);$$
$$c := c + 1;$$
$$tock \rightarrow Y$$

The first clock cycle reads in and assigns the value of the search term. The last clock cycle for any given lookup is occupied by outputting the result. In between, there are 16 clock cycles available for the lookup—this is the assumption that was made in the final development step for the clocked checksum process in Definition 9.28 that allowed the events *ready* and *done* to be dropped. The lookup itself actually consists of a sequence of assignments to the result array, where each element in the array is assigned concurrently. As an assignment in *Handel-C* is latched in during a clock cycle, each assignment in the sequence is a single clock cycle. As there are 16 assignments, one for each *BitPair* in an address, there are 16 clock cycles.

It is important that the clock does not block when the process is waiting for input—a communication on *get* will not happen on every clock cycle. If the extra choice containing *tick* and *tock* were not included, this would be a modeling error that is not apparent in the implementation—in reality, a *Handel-C* process cannot block the clock from ticking. This modeling error would prevent accurate verification of the implementation.

DEFINITION 9.34   *The clocked CAM*

> **process** *ClockedCam* $\widehat{=}$
> **begin**
> *DictSt* $\widehat{=}$ *Definition* 9.29
> *Run$_{E3}$* $\widehat{=}$ *Definition* 9.33
> • *Run$_{E3}$*
> **end**

THEOREM 9.8   *The CAM implementation is valid*

> *Cam* \ {|*ready*, *done*|} $\sqsubseteq_P$ *ClockedCam* \ {|*tick*, *tock*|}

Verification of the clocked implementation proves to be interesting. If the CSP_M assertion corresponding to Theorem 9.8 is checked, it is found to fail. In this assertion, as with all the other checks of clocked processes, the clock must be hidden to ensure the alphabets of the processes match. The reason for the failure is that a divergence exists in *ClockedCam*. This divergence corresponds to the state where it is waiting for an input: it may perform an infinite series of clock ticks, and a *get* never occurs.

Verification therefore needs more care: it must ignore this divergence. A simplistic way of achieving this would be to disallow this possibility and recheck; however, this may not always be possible for more general examples. Instead, the technique is to show that the divergence did not result from the *real* activities of the process: it is shown to be divergence free when all events other than *tick* and *tock* are hidden. The divergence, therefore, must have come from the clock. If the process can then be shown to be a failures refinement then the implementation is correct. The FDR assertions given in Definition 9.35 achieve this.

DEFINITION 9.35   *The equivalent CSP$_M$ assertions*

> *assert ClockedCam* \ $\lbrace\! \lbrace get, match \rbrace\! \rbrace$: *divergence free*
> *Cam* \ $\lbrace\! \lbrace ready, done \rbrace\! \rbrace \sqsubseteq_{FD}$ *ClockedCam* \ $\lbrace\! \lbrace tick, tock \rbrace\! \rbrace$

9.10   Final Implementation

The final implementation is the parallel combination of each of the processes developed above, as shown in Figure 9.9 and given in Definition 9.36.

DEFINITION 9.36   *The final clocked implementation*

> **process** *ClockedFilter* $\widehat{=}$
>   ((*ClockedPipeline* $[\![\ \lbrace\! \lbrace tick, tock, pass \rbrace\! \rbrace\ ]\!]$ *ClockedChecker*) \ $\lbrace\! \lbrace pass \rbrace\! \rbrace$
> $[\![\lbrace\! \lbrace get, done, tick, tock \rbrace\! \rbrace]\!]$
>   *ClockedCam*) \ $\lbrace\! \lbrace get, done, tick, tock \rbrace\! \rbrace$

THEOREM 9.9   *The Handel-C implementation is correct*

> *Filter* $\sqsubseteq_P$ *ClockedFilter* \ $\lbrace\! \lbrace tick, tock \rbrace\! \rbrace$

PROOF   From the monotonicity of refinement.   ☐

From the development strategy of Figures 9.4, 9.5, 9.9, and monotonicity of refinement, confidence in the correctness of the implementation is assured. In any case, an industrial development will be too large to model-check, and monotonicity of refinement will be relied upon.

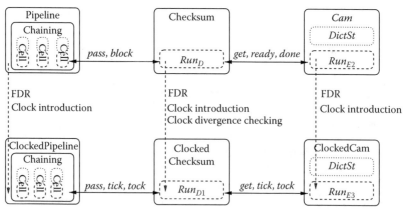

FIGURE 9.9:   Development/proof strategy for final clocked system.

9.10.1 Final Twist

In Definition 9.37, the action $BadRun_D$ is presented because it highlights an interesting error that may be made in the assumptions.

DEFINITION 9.37 *A bad checksum process*

$$BadRun_D \cong \mu X \bullet (\parallel\parallel i : \text{dom } copy \bullet pass.i?x \to SKIP);$$
$$\neg chk(copy) \& tick \to tock \to X$$
$$\Box$$
$$chk(copy) \& get!addr(copy) \to tick \to tock \to X$$

In the correct checksum process, when the calculation returns true the pipeline is not tested again until that packet has left. The above definition evaluates the checksum on each iteration. When it returns true, it passes the address to the CAM. If the CAM is in the middle of a lookup it refuses the communication—therefore, the system may deadlock. This version is not a valid refinement: a counter-example evidencing the deadlock can be found. The error was present in early versions of the application. In hardware, leaving a test condition such as this one *permanently running* often appears to be a natural, and inconsequential, assumption. The code was extensively tested, both in simulation and on real networks and the deadlock was not discovered.

The assumption is the checksum will not produce a false positive: if it does, it may result in *apparent* headers overlapping. In reality, the chance of it doing so is extremely remote, and this is why testing did not highlight the issue.

Not only does this highlight the value of the formal development in *Circus*, but provides an interesting starting point for requirements checking and investigating the real security offered by the device. While real networks may not present the possibility for false positives, formal development has shown that the device does not function if they did actually happen—and this may form the starting point for a malicious attack on the device.

9.11 Summary

The case study began with a high level abstract specification of a network packet filter. Through a series of design steps—each one of which was guided by domain knowledge—an implementation corresponding to a *Handel-C* program was calculated. The correctness of each design step was verified using Z/Eves and FDR. Calculating concurrency in the specification proved to be relatively straightforward; however, some manipulations relied heavily on *post hoc* verification. The structure, and rigor, of the development is the most advanced recorded in the literature for *Handel-C* and *Circus*.

The intention was to capture the level of rigor and applicability of domain expertize that may be adhered to in an industrial development, and show that this level of rigor is both feasible and sufficient for large projects. This was achieved: in fact, an erroneous assumption in the original design was uncovered that testing alone had not exposed.

Due to its simplicity, the implementation has a natural mapping onto *Handel-C*; although, as a formal semantics for *Handel-C* has not yet been approved, this final step is not as formal as that of Ref. [6] or Ref. [21]. Due to the nature of refinement in *Circus*, some of the traditional problems in *Handel-C* were naturally avoided: for instance, it should not normally be possible to derive a program where two processes attempt to assign to a variable concurrently. This leads to an interesting artifact in the model: although the *Handel-C* code may share access to variables—in particular, read access—the *Circus* model may not. An idiom involving regular updates of local state was appealed to in order to emulate this read access. However, in the final code, there is no need to copy the state of the pipeline in the checksum process—it is *Circus* that requires this. This is clearly an important consideration for hardware area constraints, and is a problem in need of further attention.

Decisions about the design of the device, and where and how concurrency was introduced and exploited, was governed by domain knowledge and empirical evidence rather than solely by laws of *Circus*. The necessity of supporting application of domain knowledge is important. Significant gains in the end product were made by targeting design steps at features of *Handel-C* and the FPGA. A different correct implementation could have been developed without this knowledge; but it may not have met the speed and area requirements which only become apparent after hardware has been built and tested. Experiences, gained from empirical experiments, guided these judgments.

The achievements of this chapter are severalfold. First, it introduced *Circus*. Second, it introduced operators, and laws for the introduction of those operators, that are relevant to many a typical hardware development. Third, it evidenced that an algebraic, refinement-based approach to systems development in *Circus* is both feasible for industrially sized case studies, and beneficial in terms of verification, and uncovering errors and assumptions. Perhaps the most significant achievement of the case study is that the requirements of the device have been met by drawing on expert domain knowledge; and that the correctness of applying this domain knowledge has been verified at every stage by drawing on formal techniques. This is a significant demonstration in the applicability of formal techniques to a typical engineering process and shows the relevance of formal techniques.

References

[1] R. J. R. Back and K. Sere. Stepwise refinement of action systems. *Structured Programming*, 12:17–30, 1991.

[2] R. J. R. Back and J. von Wright. Refinement calculus, part I: Sequential nondeterministic programs. In J. W. de Bakker, W. P. de Roever, and G. Rozenberg, (Eds.), *Stepwise Refinement of Distributed Systems*, Mook, the Netherlands. *Lecture Notes in Computer Science*, 430:42–66, May/June 1989. Springer-Verlag.

[3] R. J. R. Back and J. von Wright. Refinement calculus, part II: Parallel and reactive programs. In J. W. de Bakker, W. P. de Roever, and G. Rozenberg, (Eds.), *Stepwise Refinement of Distributed Systems*, Mook, the Netherlands. *Lecture Notes in Computer Science*, 430:67–93, 1989.

[4] S. M. Brien and J. E. Nicholls. Z base standard, version 1.0. Technical monograph TM-PRG-107, Oxford University Computing Laboratory, U.K., November 1992.

[5] M. Broy. Specification and refinement of a buffer of length one. In M. Broy, (Ed.), *Proceedings of the NATO Advanced Study Institute on Deductive Program Design*, Marktoberdorf, Germany, *Computer and System Sciences*, 152:273–304, 1996.

[6] A. L. C. Cavalcanti. *A Refinement Calculus for Z*. DPhil thesis, The University of Oxford, Oxford, U.K., 1997.

[7] A. Diller. *An Introduction to Formal Methods*. John Wiley & Sons, 2nd ed., New York, 1994.

[8] W. T. Harwood. Proof rules for Balzac. Technical Report, WTH/P7/001, Imperial Software Technology, Cambridge, U.K., 1991.

[9] I. Hayes, (Ed.). *Specification Case Studies*. 2nd ed., Prentice-Hall International Series in Computer Science, Prentice-Hall, Hertfordshire, U.K., 1993.

[10] Jifeng He, C. A. R. Hoare, and J. W. Sanders. Data refinement refined. In B. Robinet and R. Wilhelm, (Eds.), *ESOP 86, European Symposium on Programming*, SaarbrÄucken, Federal Republic of Germany. *Lecture Notes in Computer Science*, 213:187–196, March 17–19, 1986. Springer.

[11] C. A. R. Hoare. Proof of correctness of data representations. *Acta Informatica*, 1:271–281, 1972.

[12] C. A. R. Hoare. Communicating Sequential Processes. *Communications of the ACM*, 21(8):666–677, 1978.

[13] C. A. R. Hoare. *Communicating Sequential Processes*. Prentice-Hall International Series in Computer Science. Prentice-Hall, Hertfordshire, U.K., 1985.

[14] C. A. R. Hoare and Jifeng He. *Unifying Theories of Programming*. Prentice-Hall Series in Computer Science. Prentice-Hall, Hertfordshire, U.K., 1998.

[15] R. B. Jones. ICL ProofPower. *BCS FACS FACTS*, Series III,1(1):10–13, Winter 1992.

[16] S. Kent and R. Atkinson. IP authentication header. Technical Report RFC-2401, The Internet Society, November 1998.

[17] Formal Systems (Europe) Ltd. FDR: User manual and tutorial, version 2.28. Technical Report, Formal Systems (Europe) Ltd., 1999.

[18] Formal Systems (Europe) Ltd. Probe user manual. Technical Report, Formal Systems (Europe) Ltd., 1999.

[19] A. McEwan, J. Saul, and A. Bailey. A high speed reconfigurable firewall based on parameterizable FPGA based Content Addressable Memories. In *Proceedings of the International Conference on Parallel and Distributed Processing Techniques and Applications*, Vol. 2, CSREA Press, Las Vegas, NV, June 1999, pp. 1138–1144.

[20] A. A. McEwan and J. Saul. A high speed reconfigurable firewall based on parameterizable FPGA-based content addressable memories. *The Journal of Supercomputing*, 19(1):93–105, May 2001.

[21] Carroll Morgan. *Programming from Specifications*. International Series in Computer Science. Prentice-Hall, Hertfordshire, U.K., 1990.

[22] Carroll Morgan and Trevor Vickers, (Eds.). *On the Refinement Calculus*. Springer-Verlag, London, 1992.

[23] Behrooz Parhami. Architectural tradeoffs in the design of VLSI-based associative memories. *Journal of Microprocessing and Microprogramming*, 38:27–41, 1993.

[24] J. Postel. Internet protocol. Technical Report RFC-791, The Internet Society, September 1981.

[25] B. F. Potter, J. E. Sinclair, and D. Till. *An Introduction to Formal Specification and Z*, 2nd ed. Prentice-Hall, New York, 1996.

[26] A. W. Roscoe. *The Theory and Practice of Concurrency*. Prentice Hall Series in Computer Science. Prentice Hall, Hertfordshire, U.K., 1998.

[27] Mark Saaltink. The Z/EVES system. In J. P. Bowen, M. G. Hinchey, and D. Till, (Eds.), *ZUM'97: Z Formal Specification Notation*, Reading, U.K., *Lecture Notes in Computer Science*, 1212:72–85, 1997.

[28] A. Sampaio, J. Woodcock, and A. Cavalcanti. Refinement in *Circus*. In Lars-Henrik Eriksson and Peter Alexander Lindsay, (Eds.), *FME 2002: Formal Methods—Getting IT Right*, Springer-Verlag, New York, 2002, pp. 451–470.

[29] K. J. Schultz and P. G. Gulak. Architectures for large capacity CAMs. *INTEGRATION, the VLSI Journal*, 18:151–171, 1995.

[30] J. M. Spivey. *Understanding Z: A Specification Language and Its Formal Semantics*. Cambridge University Press, New York, 1988.

[31] J. M. Spivey. *The Z Notation: A Reference Manual*. Prentice-Hall, 2nd ed., New York, 1992.

[32] J. M. Spivey. *The Fuzz Manual*. The Spivey Partnership, Oxford, U.K., 1995.

[33] S. Stepney, D. Cooper, and J. Woodcock. An electronic purse: Specification, refinement, proof. Programming Research Group Monograph, PRG-126 ISBN 0-902928-41-4, Oxford University Computing Laboratory, U.K., July 2000.

[34] C. Strachey. Fundamental concepts in programming languages. *Higher Order and Symbolic Computation*, 13:11–49, 2000.

[35] D. Tavangarian. Flag-oriented parallel associative architectures and applications. In *IEEE Proceedings*, 136:5, pp. 357–365, 1989.

[36] M. Waldéen and K. Sere. Refining action systems with B-tool. In M.C. Gaudel and J. C. P. Woodcock, (Eds.), *FME'96: Industrial Benefit and Advances in Formal Methods*, Oxford. *Lecture Notes in Computer Science*, 1051:85–104, 1996.

[37] J. C. P. Woodcock and A. L. C. Cavalcanti. A concurrent language for refinement. In *5th Irish Workshop on Formal Methods*, Dublin, Ireland, July 16–17, 2001.

[38] J. C. P. Woodcock and A. L. C. Cavalcanti. The semantics of *Circus*. In Didier Bert, Jonathan P. Bowen, Martin C. Henson, and Ken Robinson, (Eds.), *Formal Specification and Development in Z and B*, ZB 2002, Springer-Verlag, New York, 2002, pp. 184–203.

[39] J. C. P Woodcock and Alistair A. McEwan. An overview of the verification of a *Handel-C* program. In *Proceedings of the International Conference on Parallel and Distributed Processing Techniques and Applications*. CSREA Press, Las Vegas, NV, 2000.

[40] J. Woodcock and J. Davies. *Using Z: Specification, Refinement, and Proof.* International Series in Computer Science. Prentice-Hall, Hertfordshire, U.K., 1996.

Chapter 10

PARS: A Process Algebraic Approach to Resources and Schedulers

MohammadReza Mousavi, Michel A. Reniers, Twan Basten, and Michel Chaudron

Contents

10.1 Introduction

Scheduling theory has a rich and long history. In addition, process algebras have been studied as a formal theory of system design and verification since the early 1980s. However, these two separate worlds have not been connected until recent years and the connection is not yet complete. In other words, using the models and algorithms of scheduling theory in a process algebraic design is still involved with many theoretical and practical complications. In this chapter, building upon previous attempts in this direction, we propose a process algebra, called *PARS* for Process Algebra with Resources and Schedulers, for the design of scheduled real-time systems. Previous attempts to incorporate scheduling algorithms in process algebra either did not have an explicit notion of schedulers [5,15,16] (thus, coding the scheduling policy in the process specification) or scheduling is treated for restricted cases that only support single-processor scheduling [6,12].

Our approach to modeling scheduled systems is depicted in Figure 10.1. In this approach, process specification (including aspects such as causal relations of actions, their timing and resource requirements) is separated from the specification of schedulers. Then one can apply schedulers to processes to obtain scheduled systems

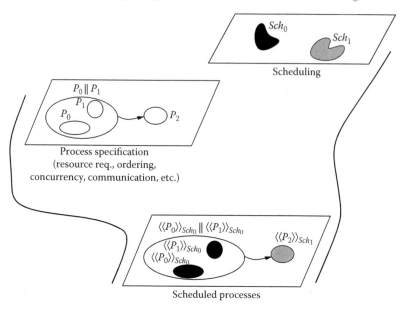

FIGURE 10.1: Schematic view of the *PARS* approach.

and further compose scheduled systems together. A distinguishing feature of our process algebra is the possibility of specifying schedulers as process terms (similar to resource-consuming processes). Another advantage of the proposed approach is the separation between process specification and scheduler specification that provides a separation of concerns, allows for specifying generic scheduling strategies and makes it possible to apply schedulers to systems at different levels of abstraction. Common to most process algebraic frameworks for resources, the proposed framework provides the possibility of extending standard schedulability analysis to the formal verification process.

10.1.1 Related Work

Several theories of process algebra with resources have been proposed recently. Our approach is mainly based on dense-time Algebra of Communicating Shared Resources (ACSR) of Ref. [5]. ACSR [15,16] is a process algebra enriched with possibilities to specify priorities and resources. Several extensions to ACSR have been proposed over time for which Ref. [16] provides a summary. The main shortcoming of this process algebra is the absence of an explicit scheduling concept. In this approach, the scheduling strategy is coded by means of priorities inside the process specification domain. Due to the absence of a resource provision model, some other restrictions are also imposed on resource demands of processes. For example, two parallel processes are not allowed to call for the same resource or they deadlock.

Our work has also been inspired by Ref. [6]. In Ref. [6], a process algebraic approach to resource modeling is presented and the application of scheduling to process terms is investigated. This approach has an advantage over that of ACSR in that scheduling is separated from the process specification domain. However, first, there is no structure or guideline to define schedulers in this language (as Ref. [16] puts it, the approach looks like defining a new language semantics for each scheduling strategy) and second, the scheduling is restricted to a single resource (single CPU) concept.

Scheduling algebra of Ref. [14] defines a process algebra that has processes with interval timing. In order to have an efficient scheduling, actions are supposed to be scheduled without delay or only after another action terminates (so-called anchor points). The semantics of the process algebra takes care of defining and extending anchor points over process structure. Since scheduling algebra abstracts from resources, the notion of scheduling is also very abstract and comes short of specifying examples of scheduling strategies such as those that we specify in the remainder of this chapter.

RTSL of Ref. [12] defines a discrete-time process algebra for scheduling analysis of single-processor systems. The basic process language allows for specifying tasks as sequential processes, and a system language takes care of composition of tasks (using parallel composition) and selecting the higher priority active tasks. Furthermore, the approach studies the issue of exception handling in case of missed deadlines. Similar to Ref. [6], there is no need for an explicit notion of resources in RTSL, since the only shared resource is the single CPU. The restriction of tasks, in this approach, to sequential processes makes the language less expressive than ours (for example, in the process language a periodic task whose execution time is larger than its period cannot be specified). Also, coding the scheduling policy in terms of a priority function may make specification of scheduling more cumbersome (similar to Ref. [6]).

Asynchrony in timed parallel composition (interleaving of relative timed-transitions) has received little interest in timed process algebras. Semantics of parallel composition in ATP [20] and different versions of timed-ACP [2], timed-CCS [8,10] and timed-CSP [11] all enforce synchronization of timed- transitions such that both parallel components evolve in time. The *cIPA* of Ref. [1] is among few timed process algebras that contain a notion of timed asynchrony. In this process algebra nonsynchronizing actions are forced to make asynchronous (interleaving) time transitions and synchronizing actions are specified to perform synchronous (concurrent) time transition. We do not see this distinction as necessary, since nonsynchronizing actions may find enough resources to execute in true concurrency and synchronizing actions may be forced to make interleaving time transitions due to the use of shared resources (e.g., scheduling two synchronizing actions on a single CPU). In other words, making the resource model explicit and separating it from process specification allows us to delay such kinds of design decisions and reflect them in the scheduling strategy and scheduled systems semantics.

A related (but different) issue in this regard is laziness and eagerness of actions (see Ref. [9] for a detailed account of the issue) that can lead to semantics similar to what we call an abstract parallel composition, which implements timed asynchrony. In general, in presence of only lazy actions, the difference between asynchronous and synchronous time (called abstract and strict, respectively in this chapter) parallel compositions vanishes and the two types of composition behave the same. However in our case, actions are not absolutely lazy and do not fit in the general framework proposed in Ref. [9]. They can only idle if another process in their abstract parallel context can perform an action. We abstract from this implicit idling by allowing time transitions to be done in an asynchronous (interleaving) as well as synchronous (concurrent) manner. This abstraction comes in handy when taking resource contention into account (where actions may be prevented from executing concurrently due to resource contention). Strict, i.e., synchronous time, parallel composition is differentiated from abstract, i.e., asynchronous time, parallel composition in this context by separating the resource concerns of its two arguments and forcing parallelism between them.

This chapter can be considered as a continuation of our previous work regarding separation of concerns in the design of embedded systems [17,18]. There, we propose separation of functionality, timing, and coordination aspects, which are represented here in the process specification language. In the *PARS* approach, we add the aspect of scheduling to the above set by assuming a (composed) model of timed functionality and coordination underneath our theory.

The main ideas of the *PARS* approach (a separation of processes and schedulers) are used in the CARAT tool for analyzing performance and resource use of component-based embedded systems [4]. Given a set of component models, the CARAT tool builds a system model that closely resembles a *PARS* model. Subsequently, different analysis techniques can be applied to that system model. In order to reduce the analysis time of system models, mostly scenario-based analyses have been performed to get indications of best-case and worst-case behaviors of a system [3].

This chapter is a revised and extended version of the extended abstract that appeared as Ref. [19]. The semantics of the process algebraic formalism presented in this chapter are improved substantially compared to those in Ref. [19], by introducing worst-case execution time (WCET) and deadline predicates. Consequently, a congruence result for strong bisimilarity is obtained here, which was impossible to obtain in Ref. [19]. We also give a number of sound axioms with respect to strong bisimilarity for our process algebraic formalism and give more elaborate examples.

10.1.2 Structure of the Chapter

The chapter is organized as follows. We define the syntax and semantics of *PARS* in three parts. In Section 10.2, we build a process algebra with asynchronous relative dense time (i.e., with the possibility of interleaving timing transition) for process specifications that have a notion of resource consumption. As explained before, we consider time asynchrony relevant and helpful in this context, since it models possible delays due to resource contentions (such as implementation of multiple parallel compositions on a single or a few CPUs). Then, in Section 10.3, a similar process algebraic theory is developed for schedulers as resource providers. Subsequently, in Section 10.4, application of a scheduler to a process and composition of scheduled systems is defined. To illustrate the usage of our method, we give some examples from the literature in each section. Finally, Section 10.5 concludes the results and presents future research directions.

10.2 Process Specification in *PARS*

A specification in *PARS* consists of three parts: a process specification that represents the usual process algebraic design of the system together with resource requirements of its basic actions, a scheduler specification that specifies availability of resources and policies for providing different resources, and a system specification that applies schedulers to processes and composes scheduled systems.

In our framework, resources are represented by a set R. The amount of resources required by a basic action is modeled by a function $\rho : R \rightarrow R^{\geq 0}$. The resource requirement is assumed to be constant at any point of time during the action execution. However, extending this to a function of time, i.e., action duration, is a straightforward extension of our theory. The resources provided by schedulers are modeled using a function $\overline{\rho} : R \rightarrow R^{\leq 0}$. Active tasks that require or provide resources are represented by multisets of such tasks in the semantics. We assume that scheduling strategies can address process identifiers, the execution time of processes, and their deadlines as typical and most common parameters for scheduling. Nevertheless, the extension of our theory with other parameters is orthogonal to our theory.

As a notational convention, we refer to the set of all multisets as M (we assume that the type of elements in the multiset is clear from the context). To represent a multiset extensionally (using its elements) we use the notation $[a, b, c, \ldots]$. The empty multiset is denoted by $\emptyset$ and $+$ and $-$ are overloaded to represent addition and subtraction of multisets, respectively.

The syntax of process specification in *PARS* is presented in Figure 10.2. It resembles a relative dense-time process algebra (such as relative dense-time ACP of Ref. [2]) with empty process ($\epsilon(0)$) and deadlock (δ). The main difference with such a theory is the attachment of resource requirements to basic actions (most process algebras abstract from resource requirements by assuming abundant availability of shared resources).

Basic action $\epsilon(t)$ represents a nonaction or idling lasting for t time which does not require any resource. Other basic actions $(a, \rho)(t)$ are pairs of actions from the set A together with the respective resource requirement function ρ and the timing t during which the required resources should be available for the action.

$$P ::= \delta \mid p(t) \mid P : P \mid P \parallel P \mid P \parallel\mid P \mid P + P \mid$$

$$\sigma_t(P) \mid \mu X.P(X) \mid \int_{x \in T} P(x) \mid \partial_{Act}(P) \mid id : P$$

$$p \in (A \times \rho) \cup \{\epsilon\}, \ t \in R^{\geq 0}, \ X \text{ is a recursive variable,}$$
$$Act \subseteq A, \ x \in V_t, \ T \subseteq R^{\geq 0}, \ id \in N$$

FIGURE 10.2: Syntax of *PARS*, part 1: process specification.

Example 10.1 Portable tasks

Suppose that the task a can be run on different platforms, either a RISC processor on which it will take 2 units of time and 100 units of memory (during those 2 time units) or on a CISC processor for which it will require 4 units of time and 70 units of memory (over the 4 time units). This gives rise to using the basic actions

$$(a, \{RISC \mapsto 1, Mem \mapsto 100\})(2)$$

and

$$(a, \{CISC \mapsto 1, Mem \mapsto 70\})(4).$$

☐

Terms $P \; ; \; P$, $P \parallel P$, $P \parallel\parallel\parallel P$, $P + P$ represent sequential composition, abstract parallel composition, strict parallel composition, and nondeterministic choice, respectively. Abstract parallel composition refers to cases where the ordering (and possible preemption) of actions has to be decided by a scheduling strategy. In practice, it is used when two processes have no causal dependency and thus *can* run in parallel but they do not necessarily have to. Particularly, when not enough resources are provided, two components that are composed using the abstract parallel composition may be scheduled sequentially. Strict parallel composition is similar to standard parallel composition in timed process algebra in that it forces concurrent execution of the two operands. Resource-consuming processes composed by this operator should be provided with enough resources to run concurrently or they will all deadlock. Both strict and abstract parallel composition allow for synchronization of actions using the synchronization function γ, i.e., when $\gamma(a, b)$ is defined, actions a and b may synchronize resulting in $\gamma(a, b)$.

Example 10.2 Portable tasks

The following is a description of the two possibilities of executing task a from the previous example:

$$P \doteq (a, \{RISC \mapsto 1, Mem \mapsto 100\})(2) + (a, \{CISC \mapsto 1, Mem \mapsto 70\})(4)$$

☐

Example 10.3 (Abstract and strict parallel composition)

Consider the following two tasks a and b which can be run independently from each other:

$$P \doteq (a, \{CPU \mapsto 1\})(2) \parallel (b, \{CPU \mapsto 1\})(3)$$

If there is only one CPU available, then the two tasks must be scheduled sequentially and thus take 5 time units to run. If two CPU's are available, then process P may complete in 3 time units.

Consider now the following task, which comprises two concurrent subtasks a and b that respectively need a CPU and a coprocessor to run. The task cannot commence (or continue) if both resources are not provided to the task.

$$P \doteq (a, \{CPU \mapsto 1\})(2) \;|||\; (b, \{COPROC \mapsto 1\})(1)$$

☐

Deadline operator $\sigma_t(P)$ specifies that process P should terminate within t units of time or it will deadlock. Recursion is specified explicitly using the expression $\mu X.P(X)$ where free variable X may occur in process P and is bound by μX. The term $\int_{x \in T} P(x)$ specifies continuous choice of timing variable x over set T. Similar to recursion, variable x is bound in term P by operator $\int_{x \in T}$. To prevent process P from performing particular actions in set Act (in particular, to force communication among two parallel parties), encapsulation term $\partial_{Act}(P)$ is used. Process terms are decorated with identifiers (natural numbers, following the idea of Ref. [6]) using the $id : P$ construct. Identifiers serve to group processes for scheduling purposes. Note that an atomic action is neither required to have an identifier, nor need its identifier be unique. In other words, there is a (possibly empty) set of identifiers attached to each process term and thus related to its comprising actions.

Precedence of binding among binary composition operators is ordered as $;, |||, ||, +$ where $;$ binds the strongest and $+$ the weakest. Unary operators are followed by a pair of parentheses or they bind to the largest possible term. In this chapter, we are only concerned with closed terms (processes that do not have free recursion or timing variables).

To show how the process specification language is to be used, we specify a few patterns common in scheduling literature [7].

Example 10.4 Periodic tasks

First, we specify a periodic task, consisting of an action a with resource requirements ρ, execution time t, and period t':

$$P_1 \doteq \mu X.(a, \rho)(t) \;|||\; \epsilon(t') \;;\; X$$

Note that strict parallel composition is used here to denote that the arrival of a new task happens concurrently with the execution of an already arrived task. The $\epsilon(t')$ operator in combination with strict parallel composition enforces the duration of the period to exactly t'. The use of parallel composition allows an execution time t greater than the period t'.

Suppose that the exact execution time of a is not known, but that the execution time is within an interval I, then this is specified as follows:

$$P_2 \doteq \mu X. \left(\int_{x \in I}(a, \rho)(x) \right) \;|||\; \epsilon(t') \;;\; X$$

The same technique can be applied to give the scheduling of a new task a variable arrival phase within an interval. Throughout the rest of the chapter, for intervals I, we use the syntactic shorthand $p(I)$ instead of $\int_{x \in I} p(x)$.

Note that in the above examples, the newly arrived task is set in strict parallel composition with the old tasks, which means that simultaneously enabled tasks should be executed concurrently. If this is not desirable, one can use a synchronization scheme, such as the one used in the following process P, to signal the arrival of the task and further put the arrived tasks in abstract parallel composition. This yields a system in which the task arrivals are strictly periodic while tasks execution is subject to resource availability:c

$$P \doteq \partial_{\{s\,Signal, r\,Signal\}} X \;|||\; Y$$
$$X \doteq (\epsilon(t')\;;\;(s\,Signal, \emptyset)(0))\;;\;X$$
$$Y \doteq (r\,Signal, \emptyset)(0)\;;\;((a, \rho)(t)\;||\;Y)$$
$$\gamma(r\,Signal, s\,Signal) \doteq \gamma(s\,Signal, r\,Signal) \doteq signal$$

▯

Example 10.5   Aperiodic tasks

Specification of aperiodic tasks follows a pattern similar to the specification of periodic tasks, with the difference that their period of arrival is not known:

$$S \doteq \mu X.(b, \rho')(t)\;|||\;\epsilon([0, \infty))\;;\;X$$

If the process specification of the system consists of periodic user level tasks and aperiodic system level tasks (e.g., system interrupts) that are to be scheduled with different policies, the specification goes as follows:

$$Sys\,Proc \doteq System : (S)\;||\;User : (P_1)$$

where *System* and *User* are distinct identifiers for these two types of tasks, and where P_1 is as specified in Example 10.4. To prevent the system from deadlocking when the resources are not available, we can compose the system with an idling process as follows:

$$Idle \doteq \mu X.\epsilon([0, \infty))\;;\;X$$
$$Sys\,Proc_{Id} \doteq Idle\;||\;Sys\,Proc$$

▯

The semantics of process specification is given in Figures 10.3 through 10.6 in the style of Structural Operational Semantics of [22]. In this semantics, states are process terms from the syntax (together with an auxiliary operator defined next). Corresponding to the possible events of spending time on actions and committing them, there are two types of transitions in the semantics. First, time passage (by spending time on resources or idling) $\overset{M,t}{\to}$ ($t \in \mathbf{R}^{>0}$), where M is the multiset that represents actions participating in the transition and the amount of resources required by each. Elements of M are of the form (ids, ρ), where ids is a set of identifiers related to the action having resource requirements ρ. Each $\tilde{id} \in ids$ is of the form (id, t_1, t_1') where id is the syntactic identifiers of the process and t_1 and t_1' are,

$$(\text{I0})\frac{}{\epsilon(0)\checkmark} \qquad (\text{I1})\frac{t' \leq t}{\epsilon(t) \stackrel{[(\emptyset,\overline{0})],t'}{\rightarrow} \epsilon(t - t')}$$

$$(\text{A0})\frac{t' \leq t}{(a, \rho)(t) \stackrel{[(\emptyset,\rho)],t'}{\rightarrow} (a, \rho)(t - t')} \qquad (\text{A1})\frac{}{(a, \rho)(0) \stackrel{a}{\rightarrow} \epsilon(0)}$$

$$(\text{S0})\frac{P \stackrel{\chi}{\rightarrow} P'}{P \; ; \; Q \stackrel{\chi}{\rightarrow} P' \; ; \; Q} \qquad (\text{S1})\frac{P\checkmark \quad Q \stackrel{\chi}{\rightarrow} Q'}{P \; ; \; Q \stackrel{\chi}{\rightarrow} Q'} \qquad (\text{S2})\frac{P\checkmark \quad Q\checkmark}{P \; ; \; Q\checkmark}$$

$$(\text{C0})\frac{P \stackrel{\chi}{\rightarrow} P'}{\substack{P + Q \stackrel{\chi}{\rightarrow} P' \\ Q + P \stackrel{\chi}{\rightarrow} P'}} \qquad (\text{C1})\frac{P\checkmark}{\substack{P + Q\checkmark \\ Q + P\checkmark}}$$

$$(\text{D0})\frac{P \stackrel{M,t}{\rightarrow} P' \quad t \leq t_0}{\sigma_{t_0}(P) \stackrel{M,t}{\rightarrow} \sigma_{t_0-t}(P')} \qquad (\text{D1})\frac{P \stackrel{a}{\rightarrow} P'}{\sigma_{t_0}(P) \stackrel{a}{\rightarrow} \sigma_{t_0}(P')} \qquad (\text{D2})\frac{P\checkmark}{\sigma_{t_0}(P)\checkmark}$$

$$(\text{E0})\frac{P \stackrel{a}{\rightarrow} P' \quad a \notin Act}{\partial_{Act}(P) \stackrel{a}{\rightarrow} \partial_{Act}(P')} \qquad (\text{E1})\frac{P \stackrel{M,t}{\rightarrow} P'}{\partial_{Act}(P) \stackrel{M,t}{\rightarrow} \partial_{Act}(P')} \qquad (\text{E2})\frac{P\checkmark}{\partial_{Act}(P)\checkmark}$$

$$(\text{R0})\frac{P(\mu X.P(X)) \stackrel{\chi}{\rightarrow} P'}{\mu X.P(X) \stackrel{\chi}{\rightarrow} P'} \qquad (\text{R1})\frac{P(\mu X.P(X))\checkmark}{\mu X.P(X)\checkmark}$$

$$(\text{CC0})\frac{y_t \in T \quad P(y_t) \stackrel{\chi}{\rightarrow} P'}{\int_{x \in T} P(x) \stackrel{\chi}{\rightarrow} P'} \qquad (\text{CC1})\frac{y_t \in T \quad P(y_t)\checkmark}{\int_{x \in T} P(x)\checkmark}$$

$$(\text{Id0})\frac{P \dagger_{t_1} \quad \eth_{t_1'}(P) \quad P \stackrel{M,t}{\rightarrow} P'}{id : P \stackrel{M \oplus (id,t_1,t_1'),t}{\rightarrow} id : P'} \qquad (\text{Id1})\frac{P \stackrel{a}{\rightarrow} P'}{id : P \stackrel{a}{\rightarrow} id : P'} \qquad (\text{Id2})\frac{P\checkmark}{id : P\checkmark}$$

$$a \in A, \; t, t' \in \mathbb{R}^{>0}, \; t_0 \in \mathbb{R}^{\geq 0}, \; t_1, t_1' \in \mathbb{R}^{\geq 0} \cup \{\infty\}, \; \chi \in (M \times \mathbb{R}^{>0}) \cup A$$

FIGURE 10.3: Semantics of *PARS*, part 1(a): process specification, sequential subset.

respectively, the last deadline and the worst-case execution of the corresponding process as specified later in Figures 10.5 and 10.6. The second type of transitions are action transitions $\stackrel{a}{\rightarrow}$ ($a \in A$) that happen when an action has spent enough time on its required resources such that the remaining time for the action is zero (e.g., rule

$$(\text{P0}) \frac{P \xrightarrow{M,t} P'}{\begin{array}{c} P \parallel Q \xrightarrow{M,t} P' \parallel t \gg Q \\ Q \parallel P \xrightarrow{M,t} t \gg Q \parallel P' \end{array}} \qquad (\text{P1}) \frac{P \xrightarrow{a} P'}{\begin{array}{c} P \parallel Q \xrightarrow{a} P' \parallel Q \\ Q \parallel P \xrightarrow{a} Q \parallel P' \end{array}}$$

$$(\text{P2}) \frac{P \xrightarrow{M,t} P' \quad Q \xrightarrow{M',t} Q'}{P \parallel Q \xrightarrow{M+M',t} P' \parallel Q'} \qquad (\text{P3}) \frac{P \xrightarrow{a} P' \quad Q \xrightarrow{b} Q' \quad \gamma(a,b) = c}{P \parallel Q \xrightarrow{c} P' \parallel Q'}$$

$$(\text{SP0}) \frac{P \xrightarrow{M,t} P' \quad Q \xrightarrow{M',t} Q'}{P \parallel\parallel Q \xrightarrow{M+M',t} P' \parallel\parallel Q'} \qquad (\text{SP1}) \frac{P \xrightarrow{a} P'}{\begin{array}{c} P \parallel\parallel Q \xrightarrow{a} P' \parallel\parallel Q \\ Q \parallel\parallel P \xrightarrow{a} Q \parallel\parallel P' \end{array}}$$

$$(\text{SP2}) \frac{P \xrightarrow{a} P' \quad Q \xrightarrow{b} Q' \quad \gamma(a,b) = c}{P \parallel\parallel Q \xrightarrow{c} P' \parallel\parallel Q'}$$

$$(\text{DS0}) \frac{P \xrightarrow{a} P' \quad P \dagger_{t_0} \quad t \le t_0}{t \gg P \xrightarrow{a} P'} \qquad (\text{DS1}) \frac{P \xrightarrow{M,t'} P' \quad P \dagger_{t_0} \quad t' \le t_0 - t}{t \gg P \xrightarrow{M,t'} t \gg P'}$$

$$(\text{DS2}) \frac{P \surd}{t \gg P \surd}$$

$$(\text{P4}) \frac{P \surd \quad Q \surd}{P \parallel Q \surd} \qquad (\text{SP3}) \frac{P \surd \quad Q \surd}{P \parallel\parallel Q \surd}$$

$$a \in A, \ t, t' \in \mathbf{R}^{>0}, \ t_0 \in \mathbf{R}^{\ge 0}, \ \chi \in (M \times \mathbf{R}^{>0}) \cup A$$

FIGURE 10.4: Semantics of *PARS*, part 1(b): process specification, parallel operators.

(A1) in Figure 10.3). We decided not to combine resource requirements of different actions and keep them separate in a multiset since they may be provided (according to their respective process identifiers) by different scheduling policies. We use $\xrightarrow{\chi}$ as an acronym for either of the two transitions. Without making it explicit in the semantics, we assume maximal progress of actions in that the system can only progress in time whenever no action transition is possible. This assumption can be made explicit by an extra condition, namely absence of action transition, in the premises of all timed-transitions. However, we do not reflect this idea in the formal semantics for sake of readability. Predicate $P \surd$ refers to possibility of successful termination of P. The semantics of process specification is the smallest transition relation (union of the time and action transition relations) satisfying the rules of Figures 10.3 through 10.6.

$$\overline{\delta \dagger_\infty} \quad \overline{\epsilon(t) \dagger_\infty} \quad \overline{(a, \rho)(t) \dagger_\infty}$$

$$\frac{\neg P \surd \quad P \dagger_t}{P : Q \dagger_t} \quad \frac{P \surd \quad P \dagger_t \quad Q \dagger_{t'}}{P : Q \dagger_{\max(t,t')}} \quad \frac{P \dagger_t \quad Q \dagger_{t'}}{P + Q \dagger_{\max(t,t')}}$$

$$\frac{P \dagger_t}{\sigma_{t'}(P) \dagger_{\min(t,t')}} \quad \frac{P \dagger_t}{t' \gg P \dagger_{t-t'}} \quad \frac{P \dagger_t \quad Q \dagger_{t'}}{P \parallel Q \dagger_{\min(t,t')}} \quad \frac{P \dagger_t \quad Q \dagger_{t'}}{P \parallel\parallel Q \dagger_{\min(t,t')}}$$

$$\frac{P(\mu X.P(X)) \dagger_t}{\mu X.P(X) \dagger_t} \quad \frac{}{\int_{x \in T} P(x) \dagger_{\text{Sup}\{t_y \mid P(y) \dagger_{t_y} \wedge y \in T\}}} \quad \frac{P \dagger_t}{\partial_{Act}(P) \dagger_t} \quad \frac{P \dagger_t}{id : P \dagger_t}$$

FIGURE 10.5: Semantics of *PARS*, part 1(c): process specification, deadlines $(t, t' \in \mathbb{R}^{\geq 0} \cup \{\infty\})$.

$$\overline{\mho_0(\delta)} \quad \overline{\mho_t(\epsilon(t))} \quad \overline{\mho_t((a, \rho)(t))} \quad \frac{\mho_t(P) \quad \mho_{t'}(Q) \quad (P : Q) \dagger_{t''}}{\mho_{\min(t'', t+t')}(P : Q)}$$

$$\frac{\mho_t(P) \quad \mho_{t'}(Q) \quad (P + Q) \dagger_{t''}}{\mho_{\min(t'', \max(t,t'))}(P + Q)} \quad \frac{\mho_t(P) \quad \sigma_{t'}(P) \dagger_{t''}}{\mho_{\min(t'', t)}(\sigma_{t'}(P))} \quad \frac{\mho_t(P) \quad t' \gg P \dagger_{t''}}{\mho_{\min(t'', t)}(t' \gg P)}$$

$$\frac{\mho_t(P) \quad \mho_{t'}(Q) \quad (P \parallel Q) \dagger_{t''}}{\mho_{\min(t'', t+t')}(P \parallel Q)} \quad \frac{\mho_t(P) \quad \mho_{t'}(Q) \quad (P \parallel\parallel Q) \dagger_{t''}}{\mho_{\min(t'', t+t')}(P \parallel\parallel Q)}$$

$$\frac{\mho_t(P(\mu X.P(X))) \quad (\mu X.P(X)) \dagger_{t'}}{\mho_{\min(t', t)}(\mu X.P(X))} \quad \frac{\{\mho_{t_y}(P(y)) \mid y \in T\} \quad (\int_{x \in T} P(x)) \dagger_{t'}}{\mho_{\min(t'.\text{Sup}_{y \in T}(t_y))}(\int_{x \in T} P(x))}$$

$$\frac{\mho_t(P)}{\mho_t(\partial_{Act}(P))} \quad \frac{\mho_t(P)}{\mho_t(id : P)}$$

FIGURE 10.6: Semantics of *PARS*, part 1(d): process specification, WCETs $(t, t', t_y \in \mathbb{R}^{\geq 0} \cup \{\infty\})$.

In Figure 10.3, rules (**I0**) and (**I1**) specify termination and a time transition, respectively. In rule (**I1**), $\overline{0}$ is an acronym for the function mapping all resources to zero. Rules (**A0**) and (**A1**) specify how an atomic action can spend its time on resources and after that commit its action, respectively. Rules (**S0**)–(**S2**) present the semantics of sequential composition. Rules (**C0**)-(**C1**) provide a semantics for nondeterministic choice. Our choice operator does not have the property of time-determinism (i.e., passage of time cannot determine choices). The reason is that in *PARS*, spending time on resources can reveal the decision taken for the nondeterministic choice. Semantics of the deadline operator is defined by **(D0)–(D2)**. Note that there is no rule for the case $\sigma_0(P)$ where process P can only do a time step. Absence of a semantic rule for such a case means that this process deadlocks (i.e., missing a deadline will result in a deadlock). Semantics of the encapsulation operator is defined in rules **(E0)–(E2)**. These rules state that the encapsulation operator prevents process P from performing actions in *Act*. This can be quite useful in forcing parallel processes to synchronize on certain actions [2]. Rules **(R0)–(R1)** and **(CC0)–(CC1)** specify the semantics of recursion and continuous choice, respectively. A recursive process can perform a transition, if the unfolded processes can do so. Note that in the continuous choice, the choice is made as soon as the bound term makes a transition. Rules **(Id0)–(Id2)** specify the semantics of *id* by adding $\widetilde{id}$ to the multiset in the transition, where $\widetilde{id}$ is the tuple (id, t_1, t_1') consisting of the syntactic *id*, deadline, and WCET of the process (see Figures 10.5 and 10.6 for the definitions of deadline and WCET, respectively). The operator $\oplus : M \times Type(\widetilde{id}) \to M$, used in the semantic rule **(Id0)**, intuitively merges the new identifier $\widetilde{id}$ with the existing identifiers (associated to a particular resource requirement information) and is formally defined as follows:

$$\emptyset \oplus \widetilde{id} \doteq \emptyset$$
$$([(ids, \rho)] + M) \oplus \widetilde{id} \doteq [(ids \cup \{\widetilde{id}\}, \rho)] + (M \oplus \widetilde{id})$$

In Figure 10.4, abstract parallel composition is specified by rules **(P0)–(P4)** and strict parallel composition is defined by rules **(SP0)–(SP3)**. In rule **(P0)**, $t \gg Q$ is an auxiliary operator (called deadline shift) that is used to specify that Q is getting t units of time closer to its deadline. The semantics of this operator is formally defined by means of the rules **(DS0)–(DS2)**. The semantics for deadline shift takes into account only the deadline of active actions. Active actions are actions that can introduce a task at the moment or already have introduced one. This is in line with the intuition that in scheduling theory only *ready* actions can take part in scheduling and other actions have to wait for their causal predecessors to commit. Function $\gamma(a, b)$ in rules **(P3)** and **(SP2)** specifies the result of a synchronized communication between a and b.

The semantics of abstract parallel composition deviates from standard semantics of parallelism in timed process algebras in that it allows for asynchronous passage of time by the two parties (rule **(P0)**). This reflects the fact that depending on availability of resources and due to scheduling, concurrent execution of tasks can be preempted and serialized at any moment of time.

The latest deadline predicate $P\dagger_t$ is defined by the deduction rules in Figure 10.5. The latest deadline is supposed to indicate the longest period, during which the process requirements should be met or the process will certainly miss a deadline (specified by some σ_t operator) and thus, will turn into a deadlocking situation. Deduction rules defining the concept of the latest deadline for δ, $\epsilon(t)$, and $(a, \rho)(t)$ are self-explanatory; they are all defined to be infinity because they do not contain any deadline operator. In case of sequential composition, we make a case distinction between the case where the first argument does not terminate and where it does. In the former case, the latest deadline of the process is due to its first argument. In the latter case, both arguments may have deadlines and both may continue their execution; thus, the argument which has a later deadline determines the latest possible deadline. The deadline of a process of the form $P + Q$ is determined by the later deadline between that of P and of Q. The deadline of $\sigma_t(P)$ is determined by the minimum of t and the latest deadline of P (since P can wait no more than both t and its latest deadline). Process $t' \gg P$ has already spent t' of its time, so its latest deadline is shifted by t'. Missing a deadline in either of the parallel components results in a missed deadline in the composite process; thus, deadline of both $P \parallel Q$ and $P \parallel\parallel Q$ is defined as the minimum of the deadlines of their arguments. The latest deadline of a recursive process is determined by unfolding its definition. For a continuous choice, the deadline of the process is defined as the supremum, i.e., the least upper bound of the set of deadlines of the alternative choices. Note that the maximum of such deadlines may not exist, e.g., if the deadlines form an open interval. Neither encapsulation nor adding an identifer influence our estimation of the latest deadline.

The WCET predicate $\mho_t(P)$ is defined by the deduction rules in Figure 10.6. As the names suggest, the intuition behind WCET is to be an upper bound estimation of the longest computation time of a process. Most of the deduction rules are self-explanatory; WCET is generally defined by taking the minimum of the deadline of the process and its maximal execution time. The following lemma shows that WCET of each process is less than or equal to its *latest* deadline.

LEMMA 10.1
For each process P and $t \in \mathbb{R}^{\geq 0} \cup \{\infty\}$, if $\mho_t(P)$ holds, then there exists $t' \in \mathbb{R}^{\geq 0} \cup \{\infty\}$ such that $P \dagger_{t'}$ and $t \leq t'$.

PROOF By a case distinction on the last deduction rule in the structure of the proof for $\mho_t(P)$. The lemma holds vacuously for δ, $\epsilon(t)$, and $(a, \rho)(t)$ since for all of them, the latest deadline is ∞. For other deduction rules, there exists a premise of the form $\dagger_{t'} P$ and t is of the form $\min(t', e)$, for some expression e. Thus, it holds that $t \leq t'$.   □

The deadline and WCET are just meant to be estimations of process measures, and any other well-defined performance measure on processes can replace or extend the semantics.

In order to compare processes, e.g., to compare a specification with its implementation, it is customary to define a notion of equivalence or preorder among processes. In this chapter, we adapt the notion of strong bisimilarity to our setting which provides an appropriate theoretical starting point. Other weaker notions of equality and preorder can be analogously adopted to our settings.

DEFINITION 10.1

(Strong bisimulation) A symmetric relation R on process terms is a strong bisimulation for resource requiring processes if and only if for all pairs $(P, Q) \in R$:

1. $P \xrightarrow{\chi} P' \Rightarrow \exists_{Q'} Q \xrightarrow{\chi} Q' \wedge (P', Q') \in R$

2. $P\surd \Rightarrow Q\surd$

3. $P\dagger_t \Rightarrow Q\dagger_t$

4. $\mho_t(P) \Rightarrow \mho_t(Q)$

Two processes P and Q are called strongly bisimilar, denoted by $P \leftrightarrow Q$ if and only if there exists a strong bisimulation relation R such that $(P, Q) \in R$.

THEOREM 10.1   Congruence of strong bisimilarity for the process language

Strong bisimilarity, as defined in Definition 10.1, is a congruence with respect to all operators in our process specification language.

PROOF   The deduction rules are in the PANTH format of Ref. [23]. Furthermore, by counting the number of symbols in each predicate/left-hand side of each transition, we can define a stratification measure which does not increase from the conclusion to the positive premises and decreases from conclusion to negative premises. Thus, the set of SOS rules for our process language specification are complete, i.e., they univocally define a transition relation. Hence, it follows from the metatheorem of Ref. [23] that our notion of strong bisimilarity is a congruence.   ☐

Below we present some properties of the operators introduced in this section. The notation $FV(P)$ denotes the free variables of process P. The proofs are tedious but straightforward and therefore omitted.

THEOREM 10.2   Properties of processes

For arbitrary processes P, Q, and R

$$
\begin{aligned}
P + P &\leftrightarrow P \\
P + Q &\leftrightarrow Q + P \\
(P + Q) + R &\leftrightarrow P + (Q + R) \\
P + \sigma_0(\delta) &\leftrightarrow P \\[6pt]
(P + Q) ; R &\leftrightarrow (P ; R) + (Q ; R) \\[6pt]
\sigma_t(\sigma_{t'}(P)) &\leftrightarrow \sigma_{\min(t,t')}(P) \\
\sigma_t(P + Q) &\leftrightarrow \sigma_t(P) + \sigma_t(Q) \\
\sigma_t(P ; Q) &\leftrightarrow \sigma_t(\sigma_t(P) ; \sigma_t(Q)) \\
\sigma_0(\epsilon(0)) ; P &\leftrightarrow P \\
\mu X.P(X) &\leftrightarrow P(\mu X.P(X))
\end{aligned}
$$

$$\int_{x\in\varnothing} P(x) \quad\quad\quad \underline{\leftrightarrow}\ \sigma_0(\delta)$$
$$\int_{x\in T} P(x) \quad\quad\quad \underline{\leftrightarrow}\ P(t) + \int_{x\in T} P(x) \quad\quad (t \in T)$$
$$\int_{x\in T} P(x) \quad\quad\quad \underline{\leftrightarrow}\ P(x) \quad\quad\quad\quad\quad\quad (x \notin FV(P(x)))$$
$$\int_{x\in T}(P(x) + Q(x)) \ \underline{\leftrightarrow}\ \int_{x\in T} P(x) + \int_{x\in T} Q(x)$$

$$P \parallel Q \quad\quad\quad\quad\quad \underline{\leftrightarrow}\ Q \parallel P$$
$$(P \parallel Q) \parallel R \quad\quad\ \underline{\leftrightarrow}\ P \parallel (Q \parallel R)$$
$$P \parallel \epsilon(0) \quad\quad\quad\ \underline{\leftrightarrow}\ P$$

$$P \parallel\parallel Q \quad\quad\quad\quad\ \underline{\leftrightarrow}\ Q \parallel\parallel P$$
$$(P \parallel\parallel Q) \parallel\parallel R \quad\ \underline{\leftrightarrow}\ P \parallel\parallel (Q \parallel\parallel R)$$

$$t \gg t' \gg P \quad\quad\quad \underline{\leftrightarrow}\ (t + t') \gg P$$

Due to the fact that the definitions of the deadline predicate and the WCET are just approximations, many of the properties that are quite standard for the operators in standard process algebra are not valid in this setting. An example is the associativity of sequential composition.

10.3 Scheduler Specification

The syntax of scheduler specification (Sc) is similar to process specification and is specified in Figure 10.7. Basic actions of schedulers are predicates (*Pred*) mentioning appropriate processes to be provided with resources and the amount of resources ($\overline{\rho}: R \to \mathbb{R}^{\leq 0}$) provided during the specified time. Note that resource provisions are denoted by functions from resources to non-positive integers so that they can cancel the resource requirements when confronted with them in the next section. The

---

$$Sc \quad ::= \delta \mid s(t) \mid Sc\ ;\ Sc \mid Sc \parallel Sc \mid Sc \parallel\parallel Sc \mid Sc + Sc \mid$$

$$\int_{x\in T} P(x) \mid Sc \rhd Sc \mid Sc \rhd^n Sc \mid \|_{t\in T} Sc(t) \mid \|_{t\in T}^n Sc(t) \mid \mu X.Sc(X)$$

$$Pred ::= Id.id\ Op\ Num \mid Id.Dl\ Op\ time \mid Id.WCET\ Op\ time \mid$$

$$Pred \wedge Pred \mid Pred \vee Pred \mid true$$

$$Op ::=< \mid = \mid >$$

$$s \in Pred \times \overline{\rho}, t \in \mathbb{R}_\infty^{\geq 0},\ x \in V_t,\ time \in V_t \cup \mathbb{R}^{\geq 0},\ Id \in V_i,\ Num \in \mathbb{N}$$

FIGURE 10.7: Syntax of *PARS*, part 2: scheduler specification.

predicate can refer to the syntactic identifiers, deadline or WCET for processes. In the syntax of *Pred*, Id is a variable from set V_i (with a distinguished member $\underline{Id}$ and typical members Id_0, Id_1, etc.). $\underline{Id}$ and Id_i refer to the semantic identifier of the particular process receiving the specified resource and its environment processes, respectively. Following the structure of a semantic identifier, Id is a tuple containing syntactic identifier ($Id.id$), deadline ($Id.Dl$), and execution time ($Id.WCET$). As in the process language, the language for predicates can be extended to other metrics of processes. Since we aimed at separating the process specification aspects from scheduling aspects, we did not include constructs such as resource-consuming actions and identifiers in our schedulers language.

A couple of new operators are added to the ones in the process specification language. The preemptive precedence operator $\triangleright$ gives precedence to the right-hand side term (with the possibility of the right-hand side taking over the execution of left-hand side at any point) and continuous preemptive precedence $\rangle_{t\in T}$ which gives precedence to the choice of least possible t. Note that this need not be the least element of T (which may not even exist) but rather it is the predicate with the least possible t that can match a resource requirement. (If no such least possible t exists the result of application of the scheduler to the process should be a deadlock.) The following examples illustrate the use of these operators.

Example 10.6 Precedence operator
Consider the process specification of Example 10.5, where the system consists of two types of processes: user processes and system processes. Suppose that system processes always have priority over user processes in using a single CPU. The following scheduler specifies a general scheduling policy that observes the above priorities:

$$PrSch \doteq (\underline{Id}.id = User, CPU \mapsto -1)([0, \infty)) \triangleright$$
$$(\underline{Id}.id = System, CPU \mapsto -1)([0, \infty))$$

□

Example 10.7 Continuous precedence operator
Assume that our scheduling strategy assigns the only available CPU to the process with shortest deadline for the WCET of the process. The following specification provides us with such a scheduler:

$$CntPrSch \doteq \rangle_{t\in R^{\geq 0}}(\underline{Id}.Dl = t \wedge \underline{Id}.WCET = t', \{CPU \mapsto -1\})(t')$$

In the above scheduler the continuous preemptive precedence operator $\rangle_{t\in R^{\geq 0}}$ in combination with $\underline{Id}.Dl = t$ enforces the process with earliest deadline to be chosen.

□

The nonpreemptive counterparts of the above operators $\triangleright^n$ and $\rangle_{t\in T}^n$ have the same intuition but they do not allow taking over of one side if the other side has already decided to start. The timing variables bound by continuous choice or generalized precedence operators can be used in predicates (as timing constants) as well as in process timings. For simplicity, we only allow comparison to time points and single identifier in the predicate part and we only introduce continuous precedence operators with precedence for lower time points. Enriching the first-order language of predicates and introducing continuous precedence operators for higher time values are possible extensions of the language.

The semantic rules for our scheduler specification language are given in Figure 10.8. We omitted the semantic rules for operators in common with process specification since they are very similar to those specified in process specification

$$(\textbf{ScA0})\frac{}{(p,\overline{\rho})(0)\sqrt{}} \qquad (\textbf{ScA1})\frac{t' \leq t}{(p,\overline{\rho})(t) \xrightarrow{[(p,false,\overline{\rho})],t'} (p,\overline{\rho})(t-t')}$$

$$(\textbf{Pr0})\frac{P \xrightarrow{M,t} P'}{P \triangleright Q \xrightarrow{M \vee_{neg} enabled(Q),t} P' \triangleright Q}$$

$$(\textbf{Pr1})\frac{Q \xrightarrow{M,t} Q'}{P \triangleright Q \xrightarrow{M,t} P \triangleright Q'} \qquad (\textbf{Pr2})\frac{P\sqrt{} \quad Q\sqrt{}}{P \triangleright Q\sqrt{}}$$

$$(\textbf{NPr0})\frac{P \xrightarrow{M,t} P'}{P \triangleright^n Q \xrightarrow{M \vee_{neg} enabled(Q),t} P'}$$

$$(\textbf{NPr1})\frac{Q \xrightarrow{M,t} Q'}{P \triangleright^n Q \xrightarrow{M,t} Q'} \qquad (\textbf{NPr2})\frac{P\sqrt{} \quad Q\sqrt{}}{P \triangleright^n Q\sqrt{}}$$

$$(\textbf{CPr0})\frac{P(t'') \xrightarrow{M,t'} P'(t'') \quad t'' \in T}{\natural_{t \in T} P(t) \xrightarrow{M \vee_{neg} \exists_t t \in \lfloor T \rfloor_{t''} \wedge enabled(P(t)),t'} \natural_{t \in T} P'(t)}$$

$$(\textbf{CPr1})\frac{t' \in T \quad P(t')\sqrt{}}{\natural_{t \in T} P(t)\sqrt{}}$$

$$(\textbf{NCPr0})\frac{P(t') \xrightarrow{M,t} P' \quad t' \in T}{\natural^n_{t \in T} P \xrightarrow{M \vee_{neg} \exists_t t \in \lfloor T \rfloor_{t'} \wedge enabled(P(t)),t} P'} \qquad (\textbf{NCPr1})\frac{t' \in T \quad P(t')\sqrt{}}{\natural^n_{t \in T} P(t)\sqrt{}}$$

FIGURE 10.8: Semantics of *PARS*, part 2: scheduler specification.

semantics. Rules for the operators common to the process specification language given in Figures 10.3 and 10.4 should be copied here with the following two provisos. First, abstract parallel composition in schedulers has a simpler semantics, since schedulers do not have a deadline operator. Namely, rule **(P′0)** should be replaced with the following rule.

$$\textbf{(P′0)} \; \frac{P \stackrel{M,t}{\to} P'}{\begin{array}{c} P \parallel Q \stackrel{M,t}{\to} P' \parallel Q \\ Q \parallel P \stackrel{M,t}{\to} Q \parallel P' \end{array}}$$

Note that, due to this change, the auxiliary operator $t \gg P$ does not appear in the semantics of schedulers. Second, for simplicity, we did not include action prefixing in the syntax of the language for schedulers. Consequently, rules concerning action transitions need not be copied to the semantics of schedulers. Actions transitions can indeed be useful in modeling interactions within schedulers and among schedulers and processes, but we omitted both for the sake of simpler presentation and keeping the orthogonality between the process and the scheduler language. Note that breaking the orthogonality and merging the two languages opens up the possibility of modeling more complex schedulers which need to communicate with each other and with processes, or need to receive a resource and then distribute it afterward, e.g., servers in a deferable server scheme [7].

The transition relation in the semantics is of the form $\stackrel{M,t}{\to}$, where M is a multiset containing predicates about processes that can receive a certain amount of resources during time t. Elements of multiset M are of the form $(pred, npred, \overline{\rho})$ where *pred* is the positive predicate that the process receiving resources should satisfy, *npred* is the negative predicate that the process should falsify, and $\overline{\rho}$ is the function representing the amount of different resources offered to such a process. In this level, we assume no information about the resource requiring process that the scheduler is to be confronted with. Thus, the resource grant predicates specify the criteria that processes receiving resources should satisfy (being able to match the predicate) and the criteria they should falsify (not being able to perform higher precedence transitions). For example, the positive predicate $\underline{Id}.Id = User$ specifies that the process receiving the resource should have $User$ as its syntactic identifier, and the predicate $\underline{Id}.Dl = t$ specifies that the latest deadline of the receiving process should be t. Once the same predicates appear in the negative side, they mean that no process with identifier $User$ or no process with deadline t is currently able to receive the resource.

Rules **(ScA0)** and **(ScA1)** specify semantics of atomic scheduler actions. Rules **(Pr0)**–**(Pr2)** specify the semantics for precedence operator. In these rules, $M \vee_{neg} pred$ stands for disjunction of negative predicates in all elements of M with predicate *pred*:

$$[(pred_0, npred_0, \overline{\rho})] \vee_{neg} pred \qquad \doteq [(pred_0, npred_0 \vee pred, \overline{\rho})]$$
$$(M + [(pred_0, npred_0, \overline{\rho})]) \vee_{neg} pred \doteq (M \vee_{neg} pred) +$$
$$[(pred_0, npred_0 \vee pred, \overline{\rho})]$$

In the same semantic rules, enabledness of a process term is used as a negative predicate to assure that a lower priority process cannot make a transition when a higher priority one is able to do so. This notion is formally defined as follows:

$$enabled(\delta) \doteq false$$
$$enabled(((pred, \overline{\rho}), t)) \doteq pred$$
$$enabled(P \ ; \ Q) \doteq \begin{cases} enabled(P) \vee enabled(Q) & if \ P\surd \wedge P \rightarrow \\ enabled(P) & if \ \neg(P\surd) \\ enabled(Q) & if \ P\surd \wedge P \nrightarrow \end{cases}$$
$$enabled(P \parallel Q) \doteq enabled(P \parallel\parallel Q) \doteq enabled(P + Q) \doteq$$
$$enabled(P \rhd Q) \doteq enabled(P \rhd^n Q) \doteq enabled(P) \vee enabled(Q)$$
$$enabled(\textstyle\int_{x \in T} P(x)) \doteq$$
$$enabled(\textstyle\oint_{x \in T} P(x)) \doteq enabled(\textstyle\oint_{x \in T}^n P(x)) \doteq \exists_{x \in T} enabled(P(x))$$
$$enabled(\mu X.P(X)) \doteq enabled(P(\mu X.P(X)))$$

In the above definition $P \rightarrow$ stands for the possibility of performing a transition $P \rightarrow P'$ for some P' and $P \nrightarrow$ is the negation of it. Note that using $P \nrightarrow$ and $\neg(P\surd)$ in this definition introduces negative premises to our semantics indirectly. But this is harmless to well-definedness of our semantics since a standard stratification can be found for it. To illustrate the semantics of precedence operator, we give the following example:

Example 10.8
Consider the scheduler specification of Example 10.6. This specification generates a transition system that allows an arbitrary time transition with positive predicate $Id.id = System$. However, according to the rule (**Pr0**), for transitions with positive predicate $Id.id = User$, the predicate of $t \in [0, \infty) \wedge Id.id = System$ is added as a negative predicate as well. Intuitively, this should mean that CPU is provided to a user process if no system process is able to take that transition. Of course, part of this intuition remains to be formalized by the semantics of applying schedulers to processes. $\square$

Rules (**NPr0**)–(**NPr2**) specify the semantics for the non-preemptive precedence operator. The only difference between these rules and their preemptive counterparts, (**Pr0**)–(**Pr2**), is that after one of the two arguments has made a transition, the other argument disappears and thus the argument making the transition can no longer be preempted. Rules (**CPr0**)–(**CPr1**) and (**NCPr0**)–(**NCPr1**) present the semantic rules for preemptive and nonpreemptive continuous precedence operators, respectively.

In rules (**CPr0**) and (**NCPr0**) operator $\lfloor T \rfloor_t$ is defined as follows:

$$\lfloor T \rfloor_t \doteq \{t' | t' \in T \wedge t' < t\}$$

Example 10.9 Specifying scheduling strategies

We specify a few generic scheduling strategies for a single processor to show the usage of the scheduler specification language:

- Nonpreemptive Round-Robin scheduling: Consider a scheduling strategy where a single processor is going to be granted to processes non-preemptively in the increasing order of their identifiers (from 0 to n). The following scheduler specifies this scheduling strategy:

$$Sch_{NP-RR} \doteq (\underline{Id} = 0, \{CPU \mapsto -1\})[0, \infty) \,;$$
$$(\underline{Id} = 1, \{CPU \mapsto -1\})[0, \infty) \,;$$
$$\dots \,; (\underline{Id} = n, \{CPU \mapsto -1\})[0, \infty)$$

- Rate monotonic (RM) scheduling: Consider the following process specification $SysProc$ of several periodic processes composed by the abstract parallel composition operator:

$$SysProc \doteq \|_{i=0}^{n} P_i$$
$$P_i \doteq \mu X.(2i + 1) : (Q) \parallel ((2i) : (\epsilon(t')) \,; X)$$

In the above specification, even identifiers refer to the period of the tasks and odd identifiers refer to the tasks themselves. The following scheduler, specifies the preemptive RM strategy, where processes with shortest period (higher rate), have a priority in receiving CPU time:

$$RMSch(k, t) \doteq (\underline{Id}.id = 2k + 1 \wedge Id_0 = 2k \wedge Id_0.WCET = t,$$
$$\{CPU \mapsto -1\})([0, \infty))$$
$$RMSch \doteq \quad \}_{t \in R^{\geq 0}} RMSch(0, t) + \dots + RMSch(n, t)$$

Identifier $\underline{Id}$ refers to the process receiving the resource (i.e., a task defined by the process Q). Identifier Id_0 is an arbitrary identifier referring to the corresponding period of the task. Hence, $Id_0.WCET$ refers to the period of the task denoted by $\underline{Id}$.

- Earliest deadline first (EDF) scheduling: Consider the previous pattern of processes, then the following expression specifies preemptive EDF scheduling:

$$EDFSch(k, t) \doteq (\underline{Id}.id = 2k + 1 \wedge Id_0 = 2k \wedge Id_0.Dl = t,$$
$$\{CPU \mapsto -1\})([0, \infty))$$
$$EDFSch \doteq \quad \}_{t \in R^{\geq 0}} EDFSch(0, t) + \dots + EDFSch(n, t)$$

$\Box$

Common to the formalism for process specification, one can use the notion of strong bisimilarity to relate schedulers. The definition of strong bisimulation/bisimilarity for schedulers remains the same as Definition 10.1; only items 3 and 4 in this definition should be dropped since schedulers do not have a notion of deadline and WCET.

THEOREM 10.3   ***Congruence of strong bisimilarity for the scheduler language***
   Strong bisimilarity is a congruence with respect to all operators in our scheduler specification language.

PROOF   Again, all our deduction rules for schedulers (including the simplified rule (**P′0**) and those borrowed from Figures 10.3 and 10.4) are in the PANTH format of Ref. [23]. The rules specified for the scheduler are also stratified by the same stratification measure used in the proof of Theorem 10.1. Thus, the set of deduction rules is complete and our notion of strong bisimilarity is a congruence following the metatheorem of Ref. [23].   ▯

Since many of the operators for specifying schedulers are similar to those used for describing processes, they enjoy similar properties. Since schedulers do not have deadline and WCET predicates associated with them, there are additional properties as listed in Theorem 10.4.

THEOREM 10.4   ***Properties of schedulers***
   For arbitrary schedulers P, Q, and R

$$P + P \quad\quad \leftrightarrow\ P$$
$$P + Q \quad\quad \leftrightarrow\ Q + P$$
$$(P + Q) + R \quad\quad \leftrightarrow\ P + (Q + R)$$
$$P + \delta \quad\quad \leftrightarrow\ P$$

$$\delta\ ;\ P \quad\quad \leftrightarrow\ \delta$$
$$\epsilon(0)\ ;\ P \quad\quad \leftrightarrow\ P$$
$$(P + Q)\ ;\ R \quad\quad \leftrightarrow\ (P\ ;\ R) + (Q\ ;\ R)$$
$$(P\ ;\ Q)\ ;\ R \quad\quad \leftrightarrow\ P\ ;\ (Q\ ;\ R)$$

$$\mu X.Sc(X) \quad\quad \leftrightarrow\ Sc(\mu X.Sc(X))$$

$$\textstyle\int_{x \in \varnothing} P(x) \quad\quad \leftrightarrow\ \delta$$
$$\textstyle\int_{x \in T} P(x) \quad\quad \leftrightarrow\ P(t) + \int_{x \in T} P(x) \quad\quad (t \in T)$$
$$\textstyle\int_{x \in T} P(x) \quad\quad \leftrightarrow\ P(x) \quad\quad (x \notin FV(P(x)))$$
$$\textstyle\int_{x \in T}(P(x) + Q(x)) \quad\quad \leftrightarrow\ \int_{x \in T} P(x) + \int_{x \in T} Q(x)$$
$$\textstyle\left(\int_{x \in T} P(x)\right)\ ;\ Q \quad\quad \leftrightarrow\ \int_{x \in T}(P(x)\ ;\ Q) \quad\quad (x \notin FV(Q))$$

$$P \parallel Q \quad\quad \leftrightarrow\ Q \parallel P$$
$$(P \parallel Q) \parallel R \quad\quad \leftrightarrow\ P \parallel (Q \parallel R)$$
$$P \parallel \epsilon(0) \quad\quad \leftrightarrow\ P$$
$$P \parallel \delta \quad\quad \leftrightarrow\ P\ ;\ \delta$$

$$P \mathrel{|||} Q \quad\quad \leftrightarrow\ Q \mathrel{|||} P$$
$$(P \mathrel{|||} Q) \mathrel{|||} R \quad\quad \leftrightarrow\ P \mathrel{|||} (Q \mathrel{|||} R)$$

10.4 Applying Schedulers to Processes

Scheduled systems are processes resulting from applying a number of schedulers to processes. The syntax of scheduled systems is presented in Figure 10.9. In this syntax, P and Sc refer to the syntactic class of processes and schedulers presented in the previous sections, respectively. Term $\langle\!\langle Sys \rangle\!\rangle_{Sc}$ denotes applying scheduler Sc to the system Sys and $\partial_R(Sys)$ is used to close a system specification and prevent it from acquiring resources in R.

The semantics of new operators for scheduled systems is defined in Figure 10.10. In this semantics, the transition relation is the same as the transition relation in the process specification semantics of Section 10.2. Since a process is a system by definition, all semantic rules of that section carry over to the semantics of systems. Moreover, as in schedule specification phase, we reuse the rules of Section 10.2 in a more general sense in order to cover the semantics of sequential, abstract and strict parallel composition, nondeterministic choice of systems, and deadline shift operator.

$$Sys ::= P \mid \langle\!\langle Sys \rangle\!\rangle_{Sc} \mid Sys \; ; \; Sys \mid Sys \parallel Sys \mid Sys \mid\mid\mid Sys \mid Sys + Sys \mid$$

$$\partial_R(Sys) \mid \sigma_t(Sys) \mid \mu X.Sys(X) \mid id : Sys \mid t \gg Sys$$

FIGURE 10.9: Syntax of *PARS*, part 3: syntax of scheduled systems.

$$(\text{Sys0})\frac{P \overset{M,t}{\to} P' \quad Sch \overset{M',t}{\to} Sch'}{\langle\!\langle P \rangle\!\rangle_{Sch} \overset{apply_P(M,M'),t}{\to} \langle\!\langle P' \rangle\!\rangle_{Sch'}}$$

$$(\text{Sys1})\frac{P \overset{a}{\to} P'}{\langle\!\langle P \rangle\!\rangle_{Sch} \overset{a}{\to} \langle\!\langle P' \rangle\!\rangle_{Sch}} \qquad (\text{Sys2})\frac{P\surd}{\langle\!\langle P \rangle\!\rangle_{Sch}\surd}$$

$$(\text{ER0})\frac{Sys \overset{M,t}{\to} Sys' \quad \forall_{(ids,\rho)\in M, r\in R}\rho(r) = 0}{\partial_R(Sys) \overset{M,t}{\to} \partial_R(Sys')}$$

$$(\text{ER1})\frac{Sys \overset{a}{\to} Sys'}{\partial_R(Sys) \overset{a}{\to} \partial_R(Sys')} \qquad (\text{ER2})\frac{Sys\surd}{\partial_R(Sys)\surd}$$

FIGURE 10.10: Semantics of *PARS*, part 3(a): scheduled system specification.

$$\frac{P\dagger_t}{\langle\!\langle P\rangle\!\rangle_{Sch}\dagger_t} \qquad \frac{P\dagger_t}{\partial_R(P)\dagger_t} \qquad \frac{\mho_t(P) \quad \langle\!\langle P\rangle\!\rangle_{Sch}\dagger_{t'}}{\mho_{\min(t,t')}(\langle\!\langle P\rangle\!\rangle_{Sch})} \qquad \frac{\mho_t(P) \quad \partial_R(P)\dagger_{t'}}{\mho_{\min(t,t')}(\partial_R(P))}$$

FIGURE 10.11: Semantics of *PARS*, Part 3(b): Scheduled system specification, deadlines, and WCETs.

To extend the semantics of process specification to the system specification, we need to define deadline and WCET predicates on the newly defined operators. These operator and functions are defined in Figure 10.11.

The application operator $\langle\!\langle P\rangle\!\rangle_{Sch}$ is defined by semantic rules **(Sys0)–(Sys2)** in Figure 10.11. Rules **(ER0)–(ER2)** represent encapsulation of resource usage. In semantic rule **(Sys0)**, the application operator $apply_P : M \times M \to M$ is meant to apply a multiset of resource providing predicates (second parameter) to a multiset of resource requiring tasks (first parameter).

$$apply_P(M, [(pred, npred, \overline{\rho})] + M') \doteq$$

$$apply_P(applyTask_P(M, [(pred, npred, \overline{\rho})], \emptyset), M')$$

$$applyTask_P(\emptyset, [(pred, npred, \overline{\rho})], M) \doteq \emptyset$$

$$applyTask_P([(ids, \rho)], \emptyset, M) \doteq [(ids, \rho)]$$

$$applyTask_P([(ids, \rho)] + M, [(pred, npred, \overline{\rho})], M') \doteq$$

$$\begin{cases} [(ids, max(\overline{0}, \rho + \overline{\rho})]+ & if\ pred(ids, M + M')\wedge \\ applyTask_P(M, [pred, npred, \min(\overline{0}, \overline{\rho} + \rho)], & \neg npred(ids, M + M'))\wedge \\ \qquad M' + [(ids, \rho)]) & \neg engage(P, M, M' + [(ids, \rho)], \\ & (pred, npred, \overline{\rho})) \\ [ids, \rho] + applyTask_P(M, [pred, npred, \overline{\rho} - \rho], & otherwise \\ \qquad M' + [(ids, \rho)]) & \end{cases}$$

The intuition behind this definition is to apply the provided resources to the requirements while satisfying positive predicates and falsifying negative predicates. This is done by taking an arbitrary resource providing predicate, applying it to the resource requiring multisets (by checking its applicability to each task, i.e., pair of identifiers and resource requirements) and proceeding with the rest. In the above definition, $pred(ids, M + M')$ means that there exists a mapping from identifiers of the predicate (containing particularly a mapping from $\underline{Id}$ to a member of ids) that satisfies predicate $pred$. Expressions $min(\overline{0}, \overline{\rho})$ and $max(\overline{0}, \rho)$ are taking point-wise minimum and maximum of $\overline{\rho}(r)$ and $\rho(r)$ with 0, respectively. The predicate *engage* is meant to check that there is no transition from P that can potentially engage with the resources provided by $\overline{\rho}$ and satisfy the negative predicates in the current context. This predicate is defined formally as follows:

$$engage(P, M, M', (pred, npred, \overline{\rho})) \doteq$$
$$\exists_{M'',P',ids',\widetilde{id}',\rho'} P \xrightarrow{M'',t} P' \wedge M' \subseteq M'' \wedge$$
$$(ids', \rho') \in M'' - M' \wedge \rho' \bowtie \overline{\rho} \wedge \widetilde{id}' \in ids' \wedge npred(\widetilde{id}')$$
$$\rho' \bowtie \overline{\rho} \doteq \exists_r \rho'(r) > 0 \wedge \overline{\rho}(r) < 0$$

Note that generally $apply_P$ is not a function and its resulting multiset may depend on the ordering of selecting and applying predicates and tasks. By definition, for all such outcomes, there exists a corresponding transition in the semantics.

THEOREM 10.5   Congruence of strong bisimilarity for the system language
Strong bisimilarity is a congruence with respect to all operators of our scheduled systems language.

PROOF   The deduction rules are in the PANTH format and are stratifiable. Therefore, congruence of strong bisimilarity follows [23].   ☐

To better illustrate the semantics, we give a few examples of system scheduling in the remainder.

Example 10.10   EDF scheduling
Consider the following process specification and three different schedulers:

$$SysProc_{Id} \doteq 1 : (\sigma_1([(CPU \mapsto 1), (Mem \mapsto 50)](1))) \ ||$$
$$2 : (\sigma_2((CPU \mapsto 1)(Mem \mapsto 50)(2)))$$
$$NP - EDF \doteq \mu X.)^n_{t \in R^{\geq 0}}(\underline{Id}.Dl = t)([(CPU \mapsto -2), (Mem \mapsto -100)])(2)$$
$$EDF_1 \doteq \mu X.)_{t \in R^{\geq 0}}(\underline{Id}.Dl = t)[(CPU \mapsto -2), (Mem \mapsto -100)](2)$$
$$EDF_2 \doteq \mu X.)_{t \in R^{\geq 0}} ((\underline{Id}.Dl = t)[(CPU \mapsto -1), (Mem \mapsto -50)](2)) \ |||$$
$$)_{t \in R^{\geq 0}}((\underline{Id}.Dl = t)[(CPU \mapsto -1), (Mem \mapsto -50)](2))$$

It is interesting to observe that according to the semantics of Figure 10.10, in the system $\partial_{Mem}(\langle\!\langle SysProc_{Id}\rangle\!\rangle_{NP-EDF})$ the only possible run follows the following scenario: The scheduler grants both available processors and the whole 100 units of memory for 3 units of time to process with identifier 1 since this is the active process with the least deadline. However, this causes the deadline of the tasks 2 to be shifted for 2 units of time (according to the semantics of parallel composition in Figure 10.4) and thus, process 2 will deadlock, after commitment of process 1.

For the system $\partial_{Mem}(\langle\!\langle SysProc_{Id}\rangle\!\rangle_{EDF_1})$, however, the scenario is different. The scheduler can start providing all available resources to task 1 for one unit of time but after that (since the choice of least deadline remains there) available resources will not be wasted anymore and will be given to process 2. However, the process misses its deadline anyway, since it needs 2 units of time and has a deadline of 1.

In contrast, the system $\partial_{Mem}(\langle\!\langle SysProc_{Id}\rangle\!\rangle_{EDF_2})$ allows for a successful run. In this case, at the first time unit, each of the two processes can receive a CPU and 50 units of memory. This is due to the fact that after providing the required resources of process 1 by one of the schedulers, the other scheduler may assign its resources to process 2 (see definition of operator $apply_P$ and in particular definition of $engage$). It follows from the semantics that after applying one resource offer to process 1 the whole process cannot engage in a resource interaction with a deadline of less than 2 and thus process 2 can receive its required resource.

The above behavior is in-line with the intuition of scheduler specification, as well. Scheduler $NP-EDF$ specifies a nonpreemptive scheduler and thus cannot change its resource grant behavior after making the initial decision. Scheduler EDF_1 suggests that both processes and 100 units of memory should be granted to the process(es) that have the least deadline and thus, disallows other processes with higher deadlines from exploiting the remaining resources. Finally, scheduler EDF_2 specifies that two processes with the two least deadlines may benefit from the provided processor. $\Box$

10.5 Conclusions

In this chapter, we proposed a process algebra with support for specification of resources requirements and provisions. Our contribution to the current real-time or resource-based process algebraic formalisms can be summarized as follows:

1. Defining a dense and asynchronous timed process algebra with resource-consuming processes

2. Providing a (similar) process algebraic language with basic constructs for defining resource providing processes (schedulers with multiple resources)

3. Defining hierarchical application of schedulers to processes and composing scheduled systems

The theory presented in this chapter can be completed/extended in several ways. We have presented some sound axioms for the process and scheduler specification part of *PARS* (which are considerably different from axioms of similar process algebras due to its special properties and constructs such as presence of the abstract parallel composition operator). However, a full axiomatization remains a challenge. As can be seen in this chapter, the three phases of specifications share a major part of the semantics, thus, bringing the three levels of specification closer (for example, allowing for interaction among processes and schedulers or allowing for resource-consuming schedulers) can be beneficial. Such a combination leads to more expressiveness (in that complicated interactions of scheduler and processes can be captured concisely), but is against our design decision to separate the world of schedulers from the world of processes. Furthermore, applying the proposed theory in practice calls for simplification, optimization for implementation, and tooling in the future.

Another interesting extension of our work may be the integration with the algebraic framework for the compositional computation of trade-offs as developed in Ref. [13], which uses the same concepts of requested and granted resources. Such an extension may allow for trade-off analysis in the current algebraic setting.

Acknowledgments

Reinder Brill provided helpful insights on scheduling theory. Useful comments of the anonymous reviewers and the editors are also gratefully acknowledged.

References

[1] L. Aceto and D. Murphy. Timing and causality in process algebra. *Acta Informatica*, 33(4):317–350, 1996.

[2] J. C. M. Baeten and C. A. Middelburg. *Process Algebra with Timing*. EATCS Monographs. Springer-Verlag, Berlin, Germany, 2002.

[3] E. Bondarev, J. Muskens, P. H. N. de With, M. R. V. Chaudron, and J. Lukkien. Predicting real-time properties of component assemblies: A scenario-simulation approach. In *Proceedings of the 20th Euromicro Conference (EUROMICRO'04)*, Rennes, France. IEEE Computer Society, 2004, pp. 40–47.

[4] E. R. V. Bondarev, M. R. V. Chaudron, and P. H. N. de With. CARAT: A toolkit for design and performance analysis of component-based embedded systems. In *Proceedings of the Conference on Design Automation and Test in Europe (DATE'07)*, Nice Acropolis, France. 2007, pp. 1024–1029.

[5] P. Brémond-Grégoire and I. Lee. A process algebra of communicating shared resources with dense time and priorities. *Theoretical Computer Science*, 189(1–2):179–219, 1997.

[6] M. Buchholtz, J. Andersen, and H. H. Loevengreen. Towards a process algebra for shared processors. In *Proceedings of the 2nd Workshop on Models for Timed-Critical Systems (MTCS'01)*, Aalborg, Denmark. *Electronic Notes in Theoretical Computer Science*, 52(3):275–294, 2002.

[7] G. C. Buttazzo. *Hard Real-Time Computing Systems*. Kluwer Academic Publishers, Boston, MA, 2000. Third printing.

[8] F. Corradini, D. D'Ortenzio, and P. Inverardi. On the relationships among four timed process algebras. *Fundamenta Informaticae*, 38(4):377–395, 1999.

[9] F. Corradini, G. Ferrari, and P. Marco. Eager, busy-waiting and lazy actions in timed computation. In *Proceedings of the 4th Workshop on Expressiveness in Concurrency (Express'97)*, Santa Margherita Ligure, Italy. Electronic Notes in Theoretical Computer Science, pp. 133–150, 1997.

[10] M. Daniels. Modelling real-time behavior with an interval time calculus. In *Proceedings of the 2nd International Symposium on Formal Techniques in Real-Time and Fault-Tolerant Systems (FTRTFT'92)*, Nijmegen, the Netherlands. *Lecture Notes in Computer Science*, 571:53–71, 1992.

[11] J. Davies and S. Schneider. A brief history of timed CSP. *Theoretical Computer Science*, 138(2):243–271, Feb. 1995.

[12] A. N. Fredette and R. Cleaveland. RTSL: A language for real-time schedulability analysis. In *Proceedings of the 14th Real-Time Systems Symposium (RTSS'93)*, Raleigh-Durham, NC. IEEE Computer Society Press, 1993, pp. 274–283.

[13] M. C. W. Geilen, T. Basten, B. D. Theelen, and R. H. J. M. Otten. An algebra of Pareto points. *Fundamenta Informaticae*, 78(1):35–74, 2007.

[14] R. van Glabbeek and P. Rittgen. Scheduling algebra. In A. M. Haeberer, (Ed.), *Proceedings of 7th International Conference on Algebraic Methodology And Software Technology (AMAST'99)*, Volume 1548 of *Lecture Notes in Computer Science*, Amazonia, Brazil, pp. 278–292, 1999.

[15] I. Lee, J.-Y. Choi, H. H. Kwak, A. Philippou, and O. Sokolsky. A family of resource-bound real-time process algebras. In *Proceedings of 21st International Conference on Formal Techniques for Networked and Distributed Systems (FORTE'01)*, Cheju Island, Korea, Kluwer Academic Publishers, August 2001, pp. 443–458.

[16] I. Lee, A. Philippou, and O. Sokolsky. Resources in process algebra. *Journal of Logic and Algebraic Programming*, 72:98–122, 2007.

[17] M. R. Mousavi, T. Basten, M. A. Reniers, M. R. V. Chaudron, and G. Russello. Separating functionality, behavior and timing in the design of reactive systems: (GAMMA + coordination) + time. Technical Report 02-09, Department of Computer Science, Eindhoven University of Technology, Eindhoven, the Netherlands, 2002.

[18] M. R. Mousavi, M. A. Reniers, T. Basten, and M. R. V. Chaudron. Separation of concerns in the formal design of real-time shared data-space systems. In *Proceedings of the 3rd International Conference on Application of Concurrency to System Design (ACSD'03)*, Guimarães, Portugal. IEEE Computer Society Press, Los Alamitos, CA, 2003, pp. 71–81.

[19] M. R. Mousavi, M. A. Reniers, T. Basten, and M. R. V. Chaudron. Pars: A process algebra with resources and schedulers. In *Proceedings of the 1st International Workshop on Formal Modeling and Analysis of Timed Systems (FORMATS'03)*, Marseille, France. Revised Papers. Volume 2791 of *Lecture Notes in Computer Science*, Springer, pp. 134–150, 2003.

[20] X. Nicollin and J. Sifakis. The algebra of timed processes ATP: Theory and application. *Information and Computation*, 114(1):131–178, Oct. 1994.

[21] G. D. Plotkin. A structural approach to operational semantics. Technical Report DAIMI FN-19, Computer Science Department, Aarhus University, Aarhus, Denmark, Sept. 1981.

[22] G. D. Plotkin. A structural approach to operational semantics. *Journal of Logic and Algebraic Progamming*, 60:17–139, 2004. Ref.

[23] C. Verhoef. A congruence theorem for structured operational semantics with predicates and negative premises. *Nordic Journal of Computing*, 2(2):274–302, 1995.

Chapter 11

Formal Approach to Derivation of Concurrent Implementations in Software Product Lines

Sergio Yovine, Ismail Assayad, Francois-Xavier Defaut,
Marcelo Zanconi, and Ananda Basu

Contents

11.1   Introduction

High-performance real-time embedded applications, such as HDTV, video streaming, and packet routing, motivate the use of multicore and multiprocessor hardware platforms offering multiple processing units (e.g., VIPER [15], Philips Wasabi/Cake [35], Intel IXP family of network processors [21]). These architectures provide significant price, performance, and flexibility advantages. Besides, such applications are subject to mass customization, as many variations of the same product are delivered to the market with different price, performance, and functionality. The key to mass customization is to capitalize on the commonality and to effectively manage the variation in a software product line [13]. However, in current industrial practices, application requirements and design constraints are spread out and do not easily integrate and propagate through the development process. Moreover, the increasing complexity of applications tends to enlarge the abstraction gap between application description and hardware. Therefore, customization becomes a burdensome and error-prone task. In summary, the complexity of both software and hardware, together with the stringent performance requirements (e.g., timing, power consumption, etc.), makes design, deployment, and customization extremely difficult, leading to costly development cycles which result in products with suboptimal performances.

During the development cycle of applications for multiprocessors, two models of execution should be distinguished. The first one is the abstract model inherent to the specification of the application, which typically corresponds to logically concurrent activities, with data and control dependencies. The second one is the concrete execution model provided by a particular platform (run-time system and hardware architecture). The customization problem consists in exploiting platform capabilities (e.g., multithreading, pipelining, dedicated devices, multiprocessors, etc.) to implement the abstract model, or eventually restricting the latter because of constraints imposed by the concrete model (e.g., synchronous communication, shared memory, single processor, bus contention, etc.). In any case, the programmer must handle both types of execution models during the development cycle.

Therefore, there is a need for design flows for software product lines (1) based on formalisms providing appropriate mechanisms for expressing these models, and (2) supported by tools for formally relating them, in order to produce executable code which (a) is correct with respect to application's logic, and (b) ensures that nonfunctional requirements are met on the concrete execution platform.

In the context of high-performance real-time applications, two questions are particularly important: (1) how to map software's logical concurrency onto hardware's physical parallelism, and (2) how to meet application-level timing requirements with architecture-level resources and constraints. This chapter presents a design flow that provides formal means for coping with concurrency and timing properties from the abstract model all the way down to the concrete realization. Current practices to handle these two issues are summarized below.

11.1.1   Run-Time Libraries and Compiler Directives

A very common practice consists in using a language with no support for concurrency or time (e.g., C), together with specific libraries (e.g., POSIX threads or MPI [19]) or system calls provided by the underlying run-time system, or using compiler directives (e.g., OpenMP [27]).* This approach has several inconveniences. First, there is no way to distinguish between abstract and concrete execution models at program level, and therefore, the reason that motivated the programmer's choice (i.e., application design or platform capability) is not recoverable from program code. This gives rise to a messy development cycle, where application design and system deployment are not handled separately, and application code is customized too early for a specific target, therefore impeding reusability and portability. Second, correctness verification is almost impossible due to system calls (e.g., for threading and resource management [9,31]).

11.1.2   Domain-Specific Programming Languages

Another practice consists in using a language with a (more or less formal) abstract execution model where time and concurrency are syntactic and semantic concepts (e.g., Lustre [20], Ada [11]). It is entirely the role of the compiler to implement the abstract execution model on the target platform. This approach enhances formal analysis. Nevertheless, these languages rely on a fully automatic implementation phase that makes essential customization issues such as targeting, platform exploration, and optimization, very hard to achieve. For instance, a typical industrial practice for exploiting multiprocessor architectures for synchronous programs consists in manually cutting the code into pieces, and adding hand-written wrappers. This practice breaks down formal analysis and suffers from the same inconveniences of the library/directives approach. Although there is ongoing work to solve this problem for specific execution platforms (e.g., Ref. [12]), there is no attempt neither to provide language support nor to develop a general framework.

11.1.3   Modeling Frameworks and Architecture Description Languages

To some extent, some of the above-mentioned problems could be avoided by using domain-specific architecture description languages (ADL) that provide the means to integrate software and hardware models (e.g., Ref. [8]). Still, in all ADL-based

---

* Java provides some mechanisms, but they are typically implemented using platform libraries.

approaches we are aware of, description of the application execution model is tied to a platform-dependent execution model, which consequently is implemented using platform primitives by direct translation of the application code. Model-integrated development [22] also handles requirements composed horizontally at the same level of abstraction. However, it is not well adapted to take care of a primary concern in software product lines, which is the vertical propagation of concurrency and timing requirements across different abstraction layers. Platform-based design [32] is a methodology that supports vertical integration, but it is mainly focused on composing functionality while abstracting away nonfunctional properties. PTOLEMY II [29] is a design framework that supports composition of heterogeneous models of concurrent computation, but it is oriented toward modeling and simulation rather than to application-code synthesis.

11.1.4   Aspect-Oriented Software Development

Aspects could help in bridging the gap between an application's specification and the actual platform-specific implementation. However, to our knowledge, current aspect-oriented approaches require an important programming effort, do not handle timing constraints, and are not specifically focused on code synthesis for different platforms, but are typically used for monitoring and optimization [23].

To overcome the aforementioned problems, we think code-generation tools based on formal languages and models must play the central role of mapping platform independent software into target execution platforms (operating system and hardware), while ensuring at compile time that nonfunctional requirements provided by system's engineers will be met at run-time. Integrating in a design flow, formal analysis, and synthesis techniques for handling nonfunctional constraints and heterogeneous architectures is an innovative way to provide correct-by-construction code. This enables code generation for specific platforms (including software-to-processor mapping and scheduling), and platform-independent functional analysis, to be linked together in the same tool-chain without a semantic gap.

Such a framework will considerably increase the overall quality of industrial systems designed with these tools, guaranteeing the correctness of the resulting solution. This approach enhances the applicability of formal verification and analysis techniques in industrial design flows, leading to a significant reduction in overall system validation time. Nevertheless, building representative models that adequately relate functional and nonfunctional behavior, of both application software and execution platforms, is challenging [34]. Multithreaded software and multicore, multiprocessor architectures bring in additional complexity.

To circumvent this complexity, we propose a design flow consisting of a formal language and its associated compilation chain. The purpose of the language, called FXML [1], is threefold. First, it provides simple and platform-independent constructs to specify the behavior of the application using an abstract execution model. Second, it provides semantic and syntactic support for correctly refining the abstract execution model into the concrete one. Third, the language and the compilation chain are extensible to easily support new concrete execution models, without semantic breakdowns. Besides, the language can be used by the programmer to express program structure, functionality, requirements, and constraints, as well as by the compiler as a representation to be directly manipulated to perform program analyses and program transformations to generate executable code which achieves application requirements and complies to platform constraints.

On one hand, FXML can be regarded as an algebraic language which provides constructs for expressing concurrency and timing constraints, and means for proving whether a term in the algebra is an implementation of another, by term rewriting. On the other, FXML can be seen as a formal coordination language with general-purpose constructs for expressing concurrency (e.g., par, forall), where coordination is thought as managing dependencies between activities. The main difference with other coordination languages and process algebras (see Refs. [6,28] for comprehensive surveys) is that FXML (1) can express rich control and data precedence constraints, and (2) can be gradually extended with more concrete constructs in order to provide synchronization, communication, and scheduling mechanisms for implementing the abstract behavior. Moreover, by design, FXML and its code-generation tool-suite JAHUEL [1], provide an extensible and customizable software production line oriented toward generating code for multiple platforms via domain-specific semantics-preserving syntactic transformations.

11.1.5   Chapter Outline

This chapter presents the components of the design flow shown in Figure 11.1. Gray-colored components constitute the kernel of the design flow. Components in dotted lines on the left are not mandatory, i.e., they may or may not exist in a specific tool based on FXML/JAHUEL. Noncolored elements on the right correspond to platform-dependent components. Sections 11.2 and 11.3 present the gray-colored components. Section 11.2 gives the syntax and semantics of the basic language. FXML can be used as a front-end specification language or obtained from an application's source code in some other language. The role of FXML as formal specification language is illustrated in Section 11.2 with a simple Writer–Reader program and the Smith–Waterman local sequence matching algorithm [16]. Section 11.3 overviews the code-generation approach for FXML implemented in the compilation chain JAHUEL. The Writer–Reader case-study is used to exemplify how C code is

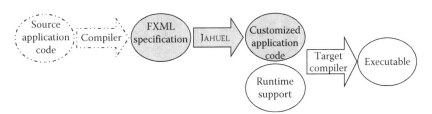

FIGURE 11.1:   Design flow.

generated from FXML for several target run-time platforms, such as *pthreads* and OpenMP [27]. Section 11.4 discusses two applications of FXML in the role of intermediate representation formalism, rather than top-level specification language. The first one presents the integration of FXML and JAHUEL in a C-based compilation tool-suite, where the input language is C extended with pragmas in FXML. The second application consists in using FXML notation to give formal semantics to the programming language StreamIt [36]. As an outcome, JAHUEL can be used to generate code for multiple run-time targets from the FXML-based representation of annotated C or StreamIt programs. Section 11.5 is about another use of JAHUEL as a tool for generating customized code. It explains how to produce component-structured code from FXML, by providing a translation into BIP [4]. This transformation enables, for instance, formal verification via the integrated framework (IF) [10] and execution on sensor networks [5].

11.2 Language: FXML

FXML [1] is a language for expressing concurrency, together with control and data dependencies which can be annotated with properties to restrict parallelism because of timing or precedence constraints.

11.2.1 Abstract Syntax

This section overviews the abstract syntax of FXML used below. The full concrete syntax of FXML *pnodes* used by JAHUEL may be found in Ref. [1] defined as an XML schema.

The *body* of an FXML specification is composed of blocks called *pnodes*. The term *pnode* stands for presentation node. This notion comes from model theory: a *pnode* "presents" an abstract execution.

Basic pnodes:

- `nil` denotes an *empty* set of executions.

- Let X be the set of variables. Variables store values from a set V. An *assignment* α has the form $x_0 = \zeta(x_1, \ldots, x_n)$, where $x_i \in X, i \in [0, n]$, and $\zeta : V^n \to V$ is a computable function. We write α_i for x_i.

  Variables are assumed to be assigned in static single assignment form, that is, there is only one assignment statement for each variable.

- `legacy` B declares a block B of *legacy code* written in, e.g., C, C++, Java, etc.

Conditional pnodes: `if` $\zeta(x_1, \ldots, x_n)$ `then` p `else` q, where p and q are *pnodes*, and $\zeta : V^n \to \mathbb{B}$ is a boolean function.

Sequential composition: `seq` $p_1 \ldots p_n$.

For/while loops: for(i = $init(x_1, \ldots, x_n)$; $test$(i); i = inc(i))⟨per=P⟩ p, and while($test(x_1, \ldots, x_n)$)⟨per=P⟩ p, express iterations: i is the iteration variable, $init : V^n \to \mathbb{N}$ is a computable function that gives the initial value of i, $inc : \mathbb{N} \to \mathbb{N}$ is the increment function, and $test : \mathbb{N} \to \mathbb{B}$ is a boolean function that defines the looping condition. Variable i is assumed not to be modified in p.

The optional declaration per=P states that the loop is *periodic* with period P, that is, there is a loop iteration every P. The execution time of the body of the loop has to be attached to p. Different loop iterations may have different execution times, as long as they are consistent with the loop period P and the execution time interval attached to p (see below).

Parallel composition: par $p_1 \ldots p_n$.

Forall loops: forall(i = $init(x_1, \ldots, x_n)$; $test$(i); i = inc(i)) p where i, $init$, inc and $test$ are as for for-loops specifies parallel executions of p.

Labeling: L: p is a *pnode*.

Dependencies: dep{⟨[a,b]⟩⟨type⟩L$_1$ → L$_2$}p, with $a, b \in \mathbb{N}$, specifies a *dependency* between occurrences of two descendants $L_i : p_i$, $i \in [1, 2]$, of p. The optional declaration *type* annotates the dependency with a type:

- Default type is weak and means that *at least* one occurrence of p_1 must precede *every* occurrence of p_2.

- Type strong means that *every* occurrence of p_1 must precede *at least* an occurrence of p_2.

- Type $(k, f(k))$ means that the $f(k)$th occurrence of p_2 must be preceded by the kth occurrence of p_1.

The optional declaration [a,b] specifies that the timing distance between the corresponding occurrences of p_1 and p_2 falls in the interval $[a, b]$.

Execution times: p[a,b], $a, b \in \mathbb{N}$, means that the execution time of p is in the interval $[a, b]$.

Example 11.1 Writer/Reader
The FXML specification of a simple program where a reader reads and prints out a value written by a writer is as follows:

```
dep [0,15] W -> R
par
  Writer:
    seq
      p = 0 [0,1]
      while(true) per=10
        W: seq {
```

```
            x = p
            p = p + 1
        } [0,1]
  Reader:
    while(true)
      R: seq {
        y = x
        legacy{ printf("%d\n", y); }
      } [0,1]
```

The declaration dep W → R declares the dependency between occurrences of *pnodes* labeled W, in Writer, and R, in Reader. This dependency comes from the fact that variable x must have some value by the time Reader uses it. Since no type is attached to the dependency, it follows that it is of the default type weak. This means that written values may be read zero or more times, but x *must* have been written *at least once* by Writer before it is first read by Reader. The interval [0,15] in dep [0,15] W → R serves for specifying a *freshness constraint*: the value of x cannot be read if the time distance between the write and read operations is greater than 15 time units.

The default dependency behavior can be strengthened with a strong type declaration to require that every written value *must* be read *at least once*. To specify that the value written in the ith iteration of the Writer's loop must be used in the ith iteration of the Reader's loop, the declaration (i,i) has to be added to the dependency dep W → R.

The execution times of p = 0, and of the sequences W and R, are specified to be in the interval [0, 1]. When the execution time of a *pnode* is not given, it means it can take an arbitrary amount of time to execute which is consistent with all timing constraints.

The period declaration attached to the while loop of the Writer states that the body of the loop is executed periodically every 10 time units. Reader, having no period declared, performs continuous iterations. ☐

Example 11.2 Smith–Waterman

The Smith–Waterman [16] local sequence matching algorithm consists of computing the elements of a $N+1$ by $M+1$ matrix A, from two strings S1 and S2 of lengths $N+1$ and $M+1$, respectively.

In FXML, it can be expressed as follows:

```
dep ((i,j), (i+1,j)) LA -> LX
dep ((i,j), (i,j+1)) LA -> LY
dep ((i,j), (i+1,j+1)) LA -> LZ
seq
  forall(j = 0; j <= M; j+1)
    forall(i = 0; i <= N; i+1)
      LI: A[i][j] = 0
```

```
forall(j = 1; j <= M; j+1)
  forall(i = 1; i <= N; i+1)
    seq
      par {
        LX: X = A[i-1][j] + 2
        LY: Y = A[i][j-1] + 2
        LZ: Z = A[i-1][j-1] + (S1[i]==S2[j]?-1:1)
      }
      LA: A[i][j] = MIN(0, X, Y, Z)
```

The dependencies state that the computation of each element (i, j) is a function of its "North" $(i - 1, j)$, "West" $(i, j - 1)$, and "Northwest" $(i - 1, j - 1)$ neighbors A. □

Hereafter, the keyword `seq` will be omitted in the examples.

11.2.2 Semantics

11.2.2.1 Definitions

Before giving semantics to FXML specifications, let us introduce some definitions.

Indexed assignments: An index is a list I of natural numbers and labels. $\langle \ell_1, \ldots \rangle$ denotes the list consisting of elements $\ell_1, \ldots$, and $\circ$ denotes concatenation of lists. An *indexed* assignment is denoted α^I. A set of indexed assignments is denoted $\mathcal{A}$.

Timing: Time is modeled with a timing function $\tau : \mathcal{A} \to \mathbb{R}^+ \times \mathbb{R}^+$. We write, $\tau^b(\alpha^I) = \pi_1(\tau(\alpha^I))$, and $\tau^e(\alpha^I) = \pi_2(\tau(\alpha^I))$, which denote respectively the beginning and ending times of assignment α^I.

τ satisfies $\mathcal{A}$, denoted $\tau \models \mathcal{A}$, iff for each $\alpha^I \in \mathcal{A}$, $\tau^b(\alpha^I) \leq \tau^e(\alpha^I)$.

Dependencies: Let $Out = \{x \in X \mid \exists \alpha^I \in \mathcal{A} : x = \alpha_0\}$ be the set of variables assigned in $\mathcal{A}$.

- The relation $\xrightarrow{d} \subseteq \mathcal{A} \times \mathcal{A}$ models data dependencies: for all $\beta^J \in \mathcal{A}$, for all β_j, if $\beta_j \in Out$, then there exists a *unique* $\alpha^I \in \mathcal{A}$ s.t. $\alpha_0 = \beta_j$ and $\alpha^I \xrightarrow{d} \beta^J$. We write $\alpha^I \xrightarrow{\beta_j} \beta^J$ as a shorthand for $\alpha^I \xrightarrow{d} \beta^J \wedge \alpha_0 = \beta_j$.

$$\tau \models \xrightarrow{d} \text{ iff } \forall \alpha^I, \beta^J \in \mathcal{A} : \alpha^I \xrightarrow{d} \beta^J \implies \tau^e(\alpha^I) \leq \tau^b(\beta^J).$$

- The relation $\xrightarrow{:} \subseteq \mathcal{A} \times \mathcal{A}$ gives an order between indexed assignments in $\mathcal{A}$, thus modeling dependencies derived from the sequential composition.

$$\tau \models \xrightarrow{:} \text{ iff } \forall \alpha^I, \beta^J \in \mathcal{A} : \alpha^I \xrightarrow{:} \beta^J \implies \tau^e(\alpha^I) \leq \tau^b(\beta^J).$$

We define $\longrightarrow = \xrightarrow{d} \cup \xrightarrow{:}$. $\tau \models \longrightarrow$ iff $\tau \models \xrightarrow{d}$ and $\tau \models \xrightarrow{:}$.

Valuations: $\mathcal{A}$ can be seen as a family $\{\mathcal{A}^n\}_{n \in \mathbb{N}}$ of sets of indexed assignments, where $\mathcal{A}^n$ contains only indexed assignments α^I of the form $x_0 = \zeta(x_1, \ldots, x_n)$. Let $\Upsilon = \{v^n\}_{n \in \mathbb{N}}$ be a family of $\mathbb{N}$-indexed functions, with $v^n : \mathcal{A}^n \to V^{n+1}$, $v^n(\alpha^I) = (v_0, v_1, v_2, \ldots, v_n)$, $v_0 = \zeta(v_1, v_2, \ldots, v_n)$. We write $v^n(\alpha^I)_i$ for v_i.

$$\Upsilon \models \xrightarrow{d} \text{ iff } \forall (\alpha^I, \beta^J) \in \mathcal{A}^n \times \mathcal{A}^m \colon \alpha^I \xrightarrow{\beta_j} \implies v^m(\beta^J)_j = v^n(\alpha^I)_0.$$

Executions: An execution e is a tuple $(X, \mathcal{A}, V, \xrightarrow{d}, \xrightarrow{\;:\;}, \tau, \Upsilon)$, such that $\tau \models \mathcal{A}$, $\tau \models \longrightarrow$, $\Upsilon \models \xrightarrow{d}$.

Timing constraints: The starting and ending time of e are, respectively, $\tau^b(e) = \min_{\alpha^I \in \mathcal{A}_e} \tau^b(\alpha^I)$, and $\tau^e(e) = \max_{\alpha^I \in \mathcal{A}_e} \tau^e(\alpha^I)$.

Subexecutions: f is a subexecution of e, denoted $f \subseteq e$, iff $\mathcal{A}_f \subseteq \mathcal{A}_e$, $\tau_f = \tau_e \restriction_{\mathcal{A}_f}$, $v_f^n = v_e^n \restriction_{\mathcal{A}_f^n}$, $\xrightarrow{d}_f = \xrightarrow{d}_e \restriction_{\mathcal{A}_f \times \mathcal{A}_f}$, $\xrightarrow{\;:\;}_f = \xrightarrow{\;:\;}_e \restriction_{\mathcal{A}_f \times \mathcal{A}_f}$, where $\restriction$ is "restricted to."

Partitions: A partition of e, denoted $\&_{i \in I} e_i$, is such that for all $i \in I$, e_i is a nontrivial subexecution of e, and for all $i \neq j$, $\mathcal{A}_{e_i} \cap \mathcal{A}_{e_j} = \emptyset$, and $\bigcup_{i \in I} \mathcal{A}_{e_i} = \mathcal{A}$.

A *sequential* partition of e, denoted $\mathbf{;}_{i \in I} e_i$, is a partition such that $\forall \alpha^I \in \mathcal{A}_{e_i}, \beta^J \in \mathcal{A}_{e_j} : i < j \implies \alpha^I \xrightarrow{\;:\;}_e \beta^J$.

Dependencies: For $e_1, e_2 \subseteq e$, $e_1 \longrightarrow_e e_2$ iff $\forall \alpha^I \in \mathcal{A}_{e_1}, \beta^J \in \mathcal{A}_{e_2} : \alpha^I \longrightarrow_e \beta^J$.

Indexed executions: The indexing of e with K is the execution e^K where $\mathcal{A}_{e^K}$ is defined such that for all $\alpha^I \in \mathcal{A}_e : \alpha^{K \circ I} \in \mathcal{A}_{e^K}$. We write $e \overset{\sim}{=}^K e^K$ to denote that e is the same execution as e^K modulo the indexing with K.

11.2.2.2 Semantic Rules

The semantics of an FXML specification is a set of executions. We use an algebraic definition of the semantics [17]. If p is a *pnode* and e is an execution, $e \models p$ means that e is an execution for p. The semantics of p is $[\![p]\!] = \{e \mid e \models p\}$.

Nil: The semantics of `nil` is the empty execution $(\emptyset, \emptyset, \emptyset, \emptyset, \emptyset, \emptyset, \emptyset)$.

Assignments: $e \models \alpha$ iff $\mathcal{A}_e = \{\alpha^I\}$ for some index I.

Conditional statements: $e \models \text{if } \zeta(x_1, \ldots, x_n) \text{ then } p \text{ else } q$ iff $e = e_1; e_2$ s.t. $e_1 \models \zeta(x_1, \ldots, x_n)$, with $\mathcal{A}_{e_1} = \{\alpha^I\}$, and $e_2 \models p$ if $v_{e_1}^n(\alpha^I) = true$, else $e_2 \models q$.

Sequential composition: $e \models \text{seq } p_1 \ldots p_n$ iff $e = \mathbf{;}_{i \in [1,n]} e_i$, such that $e_i \models p_i$.

Iterations: Let $\mathcal{K} = \{k_j\}_{j \in \mathcal{J}}$ (with $\mathcal{J}$ a finite or infinite interval of $\mathbb{N}$) be the indexed set of the values taken by the iteration variable i. $\mathcal{K}$ is defined by inc, which is increasing, that is, $i < j \implies k_i \leq k_j$, for all $k_i, k_j \in \mathcal{K}$.

- The semantics of `for`-loops is the set of executions defined as follows:
  $e \models$ `for() ` p iff $e = \mathbf{;}_{j \in \mathcal{J}} f_j^{\langle j \rangle}$, where $f_j \models p$, $j \in \mathcal{J}$, and for every $\alpha^I \in \mathcal{A}_{f_j}^m$ (for any $m \in \mathbb{N}$): $\alpha_I^I = $ i $\implies v_{f_j}^m(\alpha^I)_I = k_j$ (that is, the value of the iteration variable i is equal to k_j in f_j).

  If the optional declaration [per=P] is present, e is such that for all $j \in \mathcal{J}$, $[\tau^b(f_j^{\langle j \rangle}), \tau^e(f_j^{\langle j \rangle})] \subseteq [(j-1)P, jP)$.

- For `while`, assignments are indexed using a hidden variable j, whose values are $0, \ldots, N-1$, when the loop stops after N turns:
  $e \models$ `while`$(test(x_1, \ldots, x_n))$ p iff $e = \mathbf{;}_{j \in [0, N-1]}(c_j ; f_j^{\langle j \rangle}) ; c_N$, where $c_j \models test(x_1, \ldots, x_n), j \in [0, N], f_j \models p, j \in [0, N-1]$, and the conditions evaluate to *true* in $c_j, j \in [0, N-1]$, and to *false* in c_N. The semantics of a nonterminating loop is an infinite execution where conditions evaluate to *true* for all c_j.

  If the period declaration is given, the semantics is similar to `for`-loop periods.

Parallel composition: $e \models$ `par` $p_1 \ldots p_n$ iff $e = \&_{i \in [1.n]} e_i$, such that $e_i \models p_i$.

PROPOSITION 11.1
Parallel composition is commutative and associative.

PROPOSITION 11.2
$[\![\texttt{seq } p_1 \ldots p_n]\!] \subseteq [\![\texttt{par } p_1 \ldots p_n]\!]$.

Forall loops: Let $\mathcal{K} = \{k_j\}_{j \in \mathcal{J}}$ be the indexed set of indices defined by inc. $e \models$ `forall() ` p iff $e = \&_{j \in \mathcal{J}} f_j^{\langle j \rangle}$, where $f_j \models p, j \in \mathcal{J}$, and for every $\alpha^I \in \mathcal{A}_{f_j}^m$ (for any $m \in \mathbb{N}$): $\alpha_I^I = $ i $\implies v_{f_j}^m(\alpha^I)_I = k_j$.

PROPOSITION 11.3
$[\![\texttt{for}(i = init(x_1, \ldots, x_n); test(i); i = inc(i))\langle \texttt{per=P} \rangle \ p]\!] \subseteq$
$[\![\texttt{forall}(i = init(x_1, \ldots, x_n); test(i); i = inc(i))\langle \texttt{per=P} \rangle \ p]\!]$.

PROPOSITION 11.4
Let p_i be equal to $p, i \in [1, n]$. $[\![\texttt{par } p_1, \ldots, p_n]\!] \stackrel{\sim \langle i \rangle}{=} [\![\texttt{forall}(i=1; i<=n;]\!]$
$[\![(i+1) \ p]\!]$, where $\stackrel{\sim \langle i \rangle}{=}$ *means equal modulo the indexing given by the iteration variable* i.

Dependencies: $e \models \texttt{dep}\{\langle [\texttt{a}, \texttt{b}] \rangle \langle \texttt{type} \rangle \texttt{L}_1 \rightarrow \texttt{L}_2\}p$, iff $e \models p$, and

- `type` = weak: for every $e_2 \subseteq e$, s.t. $e_2 \models \texttt{L}_2 : p_2$, there exists $e_1 \subseteq e$, s.t. $e_1 \models \texttt{L}_1 : p_1$ and $e_1 \longrightarrow_e e_2$.

- $type$ = strong: e satisfies the condition above and for every $e_1 \subseteq e$, s.t. $e_1 \models L_1 : p_1$, there exists $e_2 \subseteq e$, s.t. $e_2 \models L_2 : p_2$ and $e_1 \longrightarrow_e e_2$.

- $type$ = $(j, f(j))$: for all $j \in \mathcal{J}$, if $e_1^{\langle j \rangle} \subseteq e$ is s.t. $e_1^{\langle j \rangle} \models L_1 : p_1$, and $e_2^{\langle f(j) \rangle} \subseteq e$ is s.t. $e_2^{\langle f(j) \rangle} \models L_2 : p_2$, then $e_1^{\langle j \rangle} \longrightarrow_e e_2^{\langle f(j) \rangle}$.

If the optional timing interval $[a, b]$ is present, e is such that for all $e_i \subseteq e, e_i \models p_i$, $i = 1, 2, e_1 \longrightarrow_e e_2 \implies \tau^b(e_2) - \tau^e(e_1) \in [a, b]$.

PROPOSITION 11.5

Let q be $\mathrm{dep}\{\langle type \rangle L_1 \rightarrow L_2\}p$. We have that
$[\![q[type := (i, i)]]\!] \subseteq [\![q[type := \mathrm{strong}]]\!] \subseteq [\![q[type := \mathrm{weak}]]\!]$.

PROPOSITION 11.6

$\forall [a', b'] \subseteq [a, b]$:
$[\![\mathrm{dep}\{[a', b']L_1 \rightarrow L_2\}p]\!] \subseteq [\![\mathrm{dep}\{[a, b]L_1 \rightarrow L_2\}p]\!] \subseteq [\![\mathrm{dep}\{L_1 \rightarrow L_2\}p]\!]$.

Execution times: $e \models p[a, b]$ iff $e \models p$ and $\tau^e(e) - \tau^b(e) \in [a, b]$.

PROPOSITION 11.7

$\forall [a', b'] \subseteq [a, b]$: $[\![p[a', b']]\!] \subseteq [\![p[a, b]]\!] \subseteq [\![p]\!]$.

Example 11.3 Writer/Reader

Figure 11.2 shows examples of executions of Example 11.1 for different types of dependencies between *pnodes* W and R: (a) weak, (b) strong, and (c) (i, i). Nonlabeled assignments are not shown. The vertical placement of Ws and Rs corresponds to their occurrence in global time, which proceeds from top to bottom. Recall that, by Propoposition 11.5, any execution of type (i, i) is also strong, and any strong is also weak.

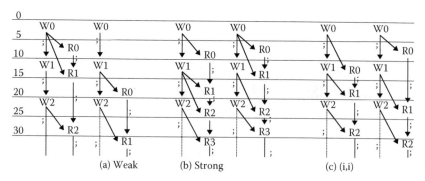

FIGURE 11.2:   Examples of executions of Writer–Reader.

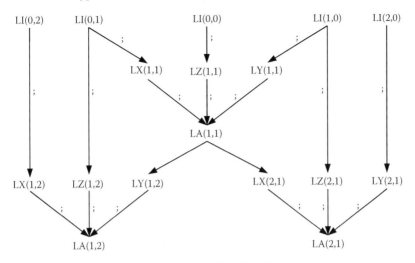

FIGURE 11.3: Examples of executions of Smith–Waterman.

The executions of *pnodes* Writer and Reader are total orders of the form $WO \xrightarrow{\cdot} W1 \cdots$ and $RO \xrightarrow{\cdot} R1 \cdots$, respectively, which are consistent with the timing constraints (Writer's loop period and execution times). Each execution of the composed system contains the union of the executions of *pnodes* Writer and Reader which are consistent with the dependency declaration dep [0,15] $W \rightarrow R$, together with precedences added by it. For instance, in the execution shown in (a)-left, the value written by WO is read by RO and R1. This means that RO and R1 *started* at most 15 time units after WO *terminated*.

However, the occurrence of W1 between RO and R1 does not prevent the value written by WO from being read twice. This execution models a behavior that may occur in a concrete implementation of this program where values are buffered. We see in Section 11.3 how such implementation can be derived from this FXML specification. ⬚

Example 11.4 Smith–Waterman
Figure 11.3 shows a part of the model of the Smith–Waterman program (Example 11.2). ⬚

11.3 Code Generation Chain

11.3.1 Compilation Approach

Compiling an FXML specification consists in transforming it until actual executable code for a specific platform can be generated. Let $\mathcal{L}$ denote a language.

Concretely, $\mathcal{L}$ is given by an XML schema, where each element definition has an associated type.

A *transformation* from $\mathcal{L}$ to $\mathcal{L}'$ is an injective map $\phi : \mathcal{L} \rightarrow \mathcal{L}'$, that is, every element of the XML schema $\mathcal{L}$ is in the set of elements $\mathcal{L}'$. Let $E_{\mathcal{L}}$ be the set of executions of type $\mathcal{L}$, and $F_\phi : E_{\mathcal{L}'} \rightarrow E_{\mathcal{L}}$ be the forgetting function that forgets any information specific to executions of type $\mathcal{L}'$. $\phi : \mathcal{L} \rightarrow \mathcal{L}'$ satisfies that for all executions $e' \models_{\mathcal{L}'} \phi(p)$ it follows that $F_\phi(e') \models_{\mathcal{L}} p$.

The compilation process is a sequence of transformations $\mathcal{L}_0 \mapsto^* \mathcal{L}_0 \mapsto \mathcal{L}_1 \mapsto^* \dots \mathcal{L}_n$, where $\mathcal{L}_0$ is basic FXML. $\mathcal{L}_i \mapsto^* \mathcal{L}_i$ is a sequence of transformations from $\mathcal{L}_i$ to $\mathcal{L}_i$, resulting in a sequence of programs $p_i^1 \dots p_i^n$, such that $[\![p_i^{k+1}]\!] \subseteq [\![p_i^k]\!]$. Examples of transformations from $\mathcal{L}_0$ to $\mathcal{L}_0$ are replacing weak dependencies by strong or (i, i) ones, or replacing par and forall by seq and for, respectively.

Transformations from $\mathcal{L}_0$ to $\mathcal{L}_0$: Let us define the following transformations:

- $\phi_;$ s.t. $\phi_;(p) = $ seq $p_1 \dots p_n$, if $p = $ par $p_1 \dots p_n$, else $\phi_;(p) = p$.

- ϕ_{for} s.t. $\phi_{\text{for}}(p) = $ for$(\dots)q$, if $p = $ forall$(\dots)q$, else $\phi_{\text{for}}(p) = p$.

- ϕ_{strong} s.t. $\phi_{\text{strong}}(p) = p[^{\text{weak}}/_{\text{strong}}]$, if $p = $ dep$\{$weak $L_1 \rightarrow L_2\}$ q, else $\phi_{\text{strong}}(p) = p$.

- $\phi_{(i,i)}$ s.t. $\phi_{(i,i)}(p) = p[^{\text{type}}/_{(i,i)}]$, with type $=$ weak or type $=$ strong, if $p = $ dep$\{$type $L_1 \rightarrow L_2\}$ q, else $\phi_{(i,i)}(p) = p$.

PROPOSITION 11.8

For all pnode p, $[\![\phi_*(p)]\!] \subseteq [\![p]\!]$, with $\phi_* \in \{\phi_;, \phi_{\text{for}}, \phi_{\text{strong}}, \phi_{(i,i)}\}$.

Transformations of the form $\mathcal{L}_i \mapsto \mathcal{L}_{i+1}$ *add* information not expressible in $\mathcal{L}_i$. An example consists in inserting communication and synchronization mechanisms (e.g., semaphores, queues, etc.) to ensure dependencies are met.

11.3.2 Tool: JAHUEL

A sequence of transformations *defines* the steps to be carried out to perform a specific *customization* of the product decided by the designer. The goal is to have a tool which (1) provides the appropriate transformations, and (2) *automatically* performs a specified sequence of them. Moreover, the tool must be extensible, in the sense that it should be possible to *add* new transformations to it.

For this purpose, we have developed JAHUEL, an FXML-based code-generation chain, constructed to be easily extended to cope with new execution models by extending the basic FXML XML-schema, and by adding transformations.

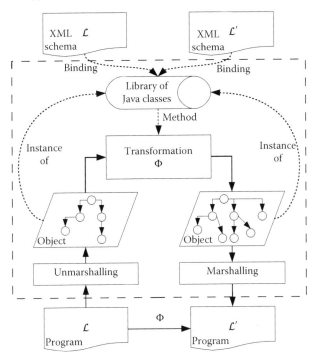

FIGURE 11.4: Schematic architecture and flow of the implementation of a transformation in JAHUEL.

JAHUEL is implemented in Java, using the Java Architecture for XML Binding (JAXB) API * to manipulate XML documents. FXML and its extensions are defined by XML schemas. Using JAXB, each language is bound to a Java class which provides the appropriate data representation and manipulation methods. Transformations are implemented on top of these Java classes.

The architecture and flow of the implementation of a transformation in JAHUEL is shown in Figure 11.4. The flow of a transformation $\phi : \mathcal{L} \to \mathcal{L}'$ is as follows. The input specification in $\mathcal{L}$ is given in the lower left as an XML file according to the $\mathcal{L}$ schema (upper left). The XML input file is unmarshalled to obtain its internal representation as a Java object, to which the method implementing the transformation is applied. The result is an object which is then marshalled into the XML output file according to $\mathcal{L}'$ schema, which can be used by a subsequent transformation. This strategy ensures traceability of implementation choices. The ultimate code generation phase for the target platform is done via a stylesheet (not shown). A configuration of JAHUEL consists in applying a sequence of transformations. This is done through a configuration file (not shown).

Currently, JAHUEL provides some general transformations which can be customized for different execution and simulation platforms. We have instantiated them to

---

* http://java.sun.com/developer/technicalArticles/WebServices/jaxb/.

generate code for Java, C with *pthreads*, SystemC, and P-Ware [3]. The compilation chain must be instantiated with the sequence of transformations to be applied.

11.3.3   Examples of Code Generation

We illustrate here the use of JAHUEL in Examples 11.1 and 11.2.

11.3.3.1   Generic Transformations

In order to generate executable code, one customization decision that needs to be made is to determine the active *components* of the system, which will become processes, threads, etc., depending on the target programming language and execution platform. For instance, in Example 11.1, it is natural to consider *pnodes* Writer and Reader as components.

Components: Let ϕ_c such that $\phi_c(p, \mathtt{L}) = \mathtt{L:component}\ q$, if $p = \mathtt{L}:q$, else $\phi_c(p, \mathtt{L}) = p$. We define $[\![\mathtt{component}\ q]\!] = [\![q]\!]$. Trivially, $F_{\phi_c}([\![\phi_c(p)]\!]) = [\![p]\!]$.

Let w and r be *pnodes* Writer and Reader, respectively. Then, ϕ_c allows transforming w and r as follows:

$$\phi_c(w, \mathtt{Writer}) = \mathtt{Writer:component}\ w$$
$$\phi_c(r, \mathtt{Reader}) = \mathtt{Reader:component}\ r$$

Besides, most synchronization mechanisms have the same kind of behavior: a component implementing a *pnode* will *wait for* some condition to hold before executing a piece of code involved in a dependency, and it will *signal* the other activities concerned by the dependency that something has happened after executing it.

Synchronization: Let ϕ_{wn} such that $\phi_{wn}(\mathtt{L}:q) = \mathtt{seq}\{\mathtt{waitfor}\ \mathtt{L}:q\ \mathtt{signal}\}$ if q is a descendant of some p with $\mathtt{dep}\{\mathtt{W} \rightarrow \mathtt{L}\}p$ or $\mathtt{dep}\{\mathtt{L} \rightarrow \mathtt{R}\}p$, else $\phi_{wn}(p) = p$. We define $[\![\phi_{wn}(p)]\!] = [\![p]\!]$. Trivially, $F_{\phi_{wn}}([\![\phi_{wn}(p)]\!]) = [\![p]\!]$.

Let us first consider the Writer–Reader specification without timing constraints. We will take care of timing constraints later. The transformed specification obtained by applying ϕ_c and ϕ_{wn}, is as follows:

```
dep W -> R
par
  Writer: component
    p = 0
    while(true)
      waitfor
      W: { x = p
           p = p + 1 }
      signal
```

```
Reader: component
  while(true)
    waitfor
    R: { y = x
         legacy{ printf("%d\n", y); } }
    signal
```

These generic transformations have no effect on the semantics, but only annotate the specification with useful information for easing further transformations.

11.3.3.2 Threads with Lock, Unlock, Wait, and Notify Primitives

JAHUEL provides a transformation of an FXML specification into a C program where concurrency is implemented using the *pthreads* library, expressed as ϕ_{luwn}:

PROPOSITION 11.9

Let ϕ_{luwn} be the transformation that translates waitfor *and* signal *into* lock, unlock, wait, *and* notify. *For all p, $F_{\phi_{luwn}}(\llbracket \phi_{luwn}(p) \rrbracket) \subseteq \llbracket p \rrbracket$.*

Roughly speaking, it works as follows. Suppose now we would like to generate code for an execution platform providing *threads*, *mutexes*, and *condition variables*, such as the *pthreads* library. The generated code will consist of two threads, sharing variable x. Concurrent accesses to x must be ensured to be mutually exclusive, and for a weak dependency, x must be written at least once by Writer before Reader could read it. In order to do this, basic FXML is extended with the appropriate constructs to handle these notions, independent of the actual API provided by the run-time. The transformed specification looks as follows:

```
dep W -> R
par
  Writer: thread
    p = 0
    while(true)
      mcx.lock
      W: { x = p
           p = p + 1 }
      mcx.notify(1)
  Reader: thread
    while(true)
      mcx.wait(1)
      R: { y = x
           legacy{ printf("%d\n", y); } }
      mcx.unlock
```

The statement thread specifies that component Writer will later become a *thread*. The translation of this statement into actual C code with *pthreads* requires a rather involved transformation which is beyond the scope of this chapter.

mcx is a structure composed of a *mutex* mx and a *condition variable* cx to protect accesses to the shared variable x. In *pnode* Writer, waitfor is implemented by mcx.lock, since a weak dependency does not require Writer to wait, but the implementation of FXML variable x as a shared C variable imposes mutual exclusion. mcx.lock can be directly translated into the corresponding *pthreads* operation, e.g., pthread_mutex_lock(&(mcx.mx)).

The code generated for the notification is mcx.notify(1), which consists in setting the value of a flag attached to the condition variable to 1. The implementation of this statement in *pthreads* looks like

```
/* The mutex has already been acquired */
mcx.b=1;
pthread_cond_signal(&(mcx.cx));
pthread_mutex_unlock(&(mcx.mx));
```

In Reader, the waitfor statement is translated into mcx.wait(1), which consists in waiting for the condition variable to be equal to 1. It can be implemented in *pthreads* as follows:

```
pthread_mutex_lock(&(mcx.mx));
while(mcx.b==0) pthread_cond_wait(&(mcx.cx), &(mcx.mx));
```

The signal statement is translated into mcx.unlock, since no notification is required, and implemented as pthread_mutex_unlock(&(mcx.mx)).

11.3.3.3 Threads Communicating through Buffers

JAHUEL provides another transformation that allows implementing FXML variables as buffers. This leads to another extension of FXML, expressed as ϕ_{buf}:

PROPOSITION 11.10

Let ϕ_{buf} be the transformation that translates waitfor *and* signal *using operations on buffers. For all p, $F_{\phi_{buf}}(\llbracket \phi_{buf}(p) \rrbracket) \subseteq \llbracket p \rrbracket$.*

The transformed Writer–Reader specification is as follows:

```
dep W -> R
par
  Writer: thread
    p = 0
    while(true)
      W: { xbuf.put(p)
           p = p + 1 }
  Reader: thread
    while(true)
      R: { y = xbuf.get()
           legacy{ printf("%d\n", y); } }
```

The FXML variable x is implemented by a shared buffer xbuf. Writing and reading x become xbuf.put(e) and xbuf.get(), respectively.

The actual implementation (e.g., array, queue, socket, etc.) and size are to be determined later, by a subsequent transformation. The abstract behavior depends on the type of the dependency. This is captured in the specification by attaching the buffer to the dep declaration:

- For a weak dependency, the buffer is only requested to produce values in a way consistent with the order of writes and reads, that is, the value returned by the $(i + 1)$th call to get() must not have been put before the value returned the ith time.

- For strong, get() is required to deliver *all* written values. This imposes a fairness constraint, which can be realized, for instance, by implementing get() so as to return the value inserted right after the one delivered in the previous call, if it exists, otherwise the last returned one.

- (i, i) case can be implemented with a blocking FIFO buffer.

11.3.3.4   Translation into OpenMP

The definition of FXML has been actually inspired by OpenMP [27], and therefore, there is a natural translation into it. The basic idea consists in encapsulating sequential *pnodes* in sections, and compiling par and forall into *parallel* sections and for-loops, respectively. However, since intra-forall dependencies are not allowed, it is necessary to perform the adequate transformations beforehand to rule them out.

Example 11.5 Smith–Waterman
The dependencies in the Smith–Waterman program characterize the well-known wavefront-like scheduling where the matrix elements on a diagonal are computed in parallel, using elements on matrix diagonals previously computed: all elements (i, j) such that $i + j = d$, for all $d \in [2, M + N]$, can be simultaneously computed, as they only depend on elements (i', j'), with $i' + j' = d' < d$ (Figure 11.5). This behavior can be expressed in FXML *without* intra-forall dependencies:

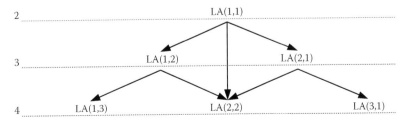

FIGURE 11.5:   Smith–Waterman.

```
for(d = 2; d <= M+N; d+1)
  forall(k = 1; cond(k, d, M, N); k+1) {
    i = indexi(k, d, M, N);
    j = indexj(k, d, M, N);
    LX: X = A[i-1][j] + 2;
    LY: Y = A[i][j-1] + 2;
    LZ: Z = A[i-1][j-1] + (S1[i]==S2[j]?-1:1;
    LA: A[i][j] = MIN(0, X, Y, Z);
  }
```

where `cond()`, `indexi()`, and `indexj()` are appropriately defined functions. The resulting C+OpenMP code looks as follows:

```
#pragma omp section
for(int d = 2; d <= M+N; d=d+1)
  #pragma omp parallel for
  for(k = 1; cond(k, d, M, N); k+1) {
    i = indexi(k, d, M, N);
    j = indexj(k, d, M, N);
    LX: X = A[i-1][j] + 2;
    LY: Y = A[i][j-1] + 2;
    LZ: Z = A[i-1][j-1] + (S1[i]==S2[j]?-1:1;
    LA: A[i][j] = MIN(0, X, Y, Z);
  }
```

Indeed, this and other code transformations issued from research on loop parallelization [14] could be specified in terms of FXML transformations.   ⬚

11.3.3.5   General Code-Generation Flow

So far, we have left aside several important issues, such as, whether the FXML semantics is ensured by the target platform, how we cope with target language limitations, and how timing constraints are handled by JAHUEL.

Concerning semantics, our approach relies on the existence of an *abstract formal model* of the target concrete execution platform, onto which basic FXML can be translated, by performing successive transformations whose correction is proved formally inside the FXML semantic world. The underlying assumption is that the concrete platform is indeed an implementation of the model, and that this relationship can be proved by some other means such as theorem proving or model checking.

Every transformation from a language $\mathcal{L}$ to $\mathcal{L}'$ has to deal with the question of how $\mathcal{L}$ statements are implemented in terms of $\mathcal{L}'$. If this cannot be done in general, then the transformation is typically defined only for a translatable subset of $\mathcal{L}$. Like any other compiler, JAHUEL performs static sanity checks on the input specification and rejects those which do not conform to the restrictions imposed by the transformation.

Nevertheless, our approach handles semantics, language limitations, and non-functional constraints (which are sometimes not directly supported by the target

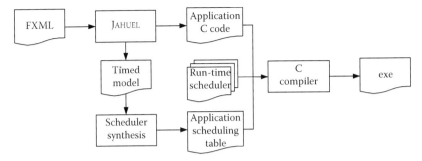

FIGURE 11.6:   General code-generation flow.

execution platform) homogeneously, following the general code-generation flow shown in Figure 11.6. The basic approach has been first proposed in Ref. [25], and implemented for Java in Ref. [24], in the context of a Java-to-native code-generation tool-chain.

The idea is as follows. JAHUEL generates two outputs, namely the application code and a timed model of it, both generated from the FXML specification. The former, rather than calling platform primitives directly, calls a generic primitive `jahuel_call()`, implemented on top of the target platform, which is responsible for ensuring the correct behavior. For a *thread*-based implementation, the pseudo-code of `jahuel_call()` looks like this:

```
void jahuel_call(th_id tid, state curr, call_op cop, params p)
{
    system_lock(scheduler_mutex);
    jahuel_scheduler(tid, curr, cop, p);
    system_unlock(scheduler_mutex);
}
```

where `system_lock()` and `system_unlock()` are the corresponding platform lock and unlock functions, `scheduler_mutex` is a platform mutex used to insure mutual exclusion access to the scheduler by concurrent application threads, and `jahuel_scheduler()` is the platform-dependent function that performs the appropriate scheduling in order to preserve the semantics. This includes, in particular, ensuring that timing constraints are met. For this, the application provides some reflective information, such as its *thread id* (`tid`) and its *current state* (`curr`), as well as the primitive to be called (`cop`) together with its parameters (`p`).

The timed model is fed into a *scheduler synthesis* algorithm, based on the *controller synthesis* approach presented in Ref. [26], which generates an *application scheduling table*, if a scheduler exists. This table determines for each application's state with which the scheduler is called, which action has to be taken to preserve the semantics. Typically, such action consists in choosing a thread to execute and updating its state according to the scheduling table. The schematic view of the run-time is shown in Figure 11.7. If a scheduling table cannot be computed, the algorithm returns useful information which can be used by the designer to understand the reasons for this and to modify its design accordingly.

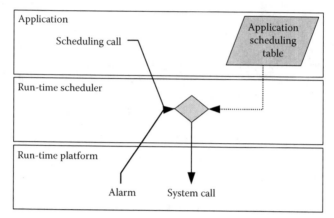

FIGURE 11.7:   Schematic architecture and flow of the run-time.

In this setting, the generated code for the Writer–Reader looks as follows:

```
dep [0,15] W -> R
par
  Writer: thread
    state(wr0)
    jahuel_execute(Writer, wr0, {p = 0}, [0,1])
    jahuel_setclock(WPER)
    while(true)
      state(wr1)
      jahuel_call(Writer, wr1, lock)
      state(wr2)
      W: jahuel_execute(Writer, wr2,
             { x = p; p = p + 1; }, [0,1])
      jahuel_setclock(CLK)
      state(wr3)
      jahuel_call(Writer, wr3, notify)
      state(wr4)
      jahuel_waitforperiod(Writer, wr4, WPER, 10)
  Reader: thread
    while(true)
      state(rd0)
      jahuel_call(Reader, rd0, wait, CLK, [0,15])
      state(rd1)
      R: jahuel_execute(Reader, rd0,
           { y = x; legacy{ printf("%d\n", y); } }, [0,1])
      state(rd2)
      jahuel_call(Reader, rd2, notify)
```

Roughly speaking, calls with arguments `lock` and `notify` will behave like for *pthreads*. The call `jahuel_setclock(CLK)` sets clock CLK to 0. Thus, CLK counts the time elapsed since the last occurrence of W. `jahuel_call(Reader,` rd0, wait, CLK, [0,15]) will block `Reader` if the value of CLK is not in the interval [0,15]. In this case, `Reader` will be awakened the next time `Writer` executes jahuel_setclock(CLK). The call jahuel_waitforperiod() makes `Writer` wait until the value of clock WPER reaches its loop's period (10), and resets WPER to 0 when it returns, to start counting another period. If execution times are satisfied, the scheduler synthesis algorithm ensures that `Writer` never misses its period.

The function jahuel_execute() executes a block of code, updates the thread state, and checks whether specified execution times are respected. In principle, execution times should be checked to hold using worst- and best-case execution-time analysis techniques (e.g., Ref. [30]). In this case, run-time monitoring of execution times could be disabled.

In this example, the scheduling table generated by the synthesis algorithm depends on the type of the dependency, either weak, strong, or (i,i), as well as on the timing constraints. The reader is referred to Refs. [24,25] for more detailed information about this technique.

11.4   From Code to FXML

FXML can also be used as an *intermediate form* to represent program behavior. Besides providing formal semantics, translating a language into FXML enables performing program transformations and compiling programs to different target platforms, as explained in Section 11.3, and also doing, for instance, performance-driven design-space exploration [3], application-oriented scheduler synthesis [25], etc.

In this section, we study two examples of this approach. The first one consists in extending C with FXML-driven pragmas. The second one is about giving formal semantics to StreamIt [36].

11.4.1   C with Pragmas

A common embedded-software programming practice in industry consists in using pragmas to annotate the program with extra-functional information about the behavior of the program and the target execution platform (run-time system and hardware). Such annotations are used to produce optimized code by the industrial C-compiler FlexCC2 [7] developed by STMicroelectronics. This approach has two major drawbacks. First, pragmas typically do not have formally defined meanings. Second, they are compiler dependent. Using FXML allows one to overcome these two issues.

To illustrate the idea, let us use again the Writer–Reader application. The following writer-reader.c C program is given in the actual syntax used by FlexCC2:

```
int x = 0;                              {
                                            p = 0;
#pragma code_block                      }
#pragma dependency
main.writer.write -> (x)                #pragma code_block
  main.reader.read [0,15] us            #pragma execution_time [0,1] us
#pragma parallel writer reader          void write()
main()                                  {
{                                           x = p++;
  writer(); /* /&/ */ reader();         }
}
                                        #pragma code_block
#pragma code_block                      void reader()
void writer()                           {
{                                           while (1) { read(); }
  int p;                                }
  write_init();
                                        #pragma code_block
  #pragma period 10 us                  #pragma execution_time [0,1] us
  while (1) { write(); }                void read()
}                                       {
                                            int y = x;
#pragma code_block                          printf("%d\n", y);
#pragma execution_time [0,1] us         }
void write_init()
```

FlexCC2 analyzes the program (pragmas and C code) and extracts a description of it in the concrete XML syntax of FXML. The FXML specification obtained is similar to the one used in Section 11.2. The main difference is that assignments and calls to C functions are considered to be legacy code:

```
dep [0,15]
main.writer.write -> main.reader.read
main:
  legacy{ #include writer-reader.h }
  par
    main.writer:
      legacy{ int p; write_init(); } [0,1]
      while(true) per=10
        main.writer.write: legacy{ write(); } [0,1]
    main.reader:
      while(true)
        main.reader.read: legacy{ read(); } [0,1]
```

Then, JAHUEL can be used to generate code for different execution platforms as explained in Section 11.3. An industrial application of the tool-chain composed of FlexCC2 and JAHUEL is presented in Ref. [2].

11.4.2   StreamIt to FXML

StreamIt [36]* is a language designed for programming streaming applications. StreamIt semantics is defined by its compiler infrastructure which provides native compilation for the MIT Raw machine [18] and code-generation into a C++ run-time library for execution on general-purpose processors.

Here, we provide a translation of a subset of StreamIt into FXML.

11.4.2.1   StreamIt Syntax

StreamIt is built around the notion of *stream*. A stream is an ordered (unbounded) sequence of data. Streams are implicit, that is, they do not have a name and can only be accessed through specific built-in functions. A StreamIt process S is defined as follows.

Filters: The basic StreamIt process is the filter. Filters are (endless) loops with (at most) one input stream and (at most) one output stream. The syntax is as follows: **filter init** C_{init} **work pop** k_1 **peek** k_2 **push** k_3 C_{work}. C is a *block* of sequential code. A filter executes the **init** block C_{init} once and the **work** block C_{work} at every iteration.

Filters manipulate the input stream via the function pop() that returns and removes the first element of the stream, and the function peek(i) that returns the ith element. Filters manipulate the output stream via the function push(data) that appends data to the output stream.

The number of inserted, peeked, and deleted elements at each iteration, that is the pop, peek, and push *rates*, are specified by the **pop**, **peek**, and **push** declarations. The peek rate specifies the *maximum* index that is allowed to be peeked in any iteration. It is required to be greater than or equal to the pop rate.

Pipelines: Processes can be grouped in a *pipeline* connected through input/output streams: **pipeline** S_1 ... S_n. Connections are made sequentially following the declaration order.

Example 11.6 Writer/Reader
   In StreamIt, the writer–reader application of Example 11.1 is as follows:

```
void -> void pipeline PC() {
  add Writer();
  add Reader();
}
void -> int filter Writer() {
  init int p = 0;
  work push 1 {
```

---

* http://www.cag.csail.mit.edu/streamit/index.shtml.

```
      push(p);
      p++;
   }
}
int -> void filter Reader() {
   work pop 1 {
      print(pop());
   }
}
```

Writer and Reader are implemented as *filters* connected in a *pipeline*. A **filter** repeatedly executes the **work** function: Writer pushes a value of p at a time, expressed by a *push* rate of 1, while Reader pops one element at each iteration, expressed by a *pop* rate of 1, and in the same order. Clearly, this StreamIt program behaves like the FXML one in Example 11.1, with a dependency of type (i, i).   □

Example 11.7 Writer/Reader with peek
   Now consider the following StreamIt program where Reader peeks two values from the input stream and sums them up:

```
void -> void pipeline PC2() {
   add Writer();
   add Reader();
}
void -> int filter Writer() {
   int p = 0;
   work push 1 {
      push(p);
      p++;
   }
}
int -> void filter Reader() {
   work pop 1 peek 2 {
      print(peek(0) + peek(1));
      pop();
   }
}
```

In this case, the translation to FXML is not as simple as before. A compositional and systematic way of doing it consists in adding a buffer in-between *:

```
var int soutP
var int sinC[2]
dep (0,0) push -> init
dep (i+1,i) push -> get
```

---

* To enhance readability, we explicitly declare variables through the var statement.

```
dep (i,i) put -> peek
par
  Writer:
    var int p = 0
    while(true)
      push: soutP = p
      p++
  Buffer:
    init: sinC[1] = soutP
    while(true)
      put: {
        shift: sinC[0] = sinC[1]
        get: sinC[1] = soutP
      }
  Reader:
    var int x[]
    while(true)
      peek: for(k=0; k<2; k++) x[k] = sinC[k]
      print(x[0]+x[1])
```

This example provides the basis for a systematic translation of StreamIt
into FXML. ▯

StreamIt provides other constructs, such as **splitjoin**, which allows an input stream
to be split in several copies to be handled by multiple filters simultaneously, and
feedbackloop, which allows making the output stream available as input stream. For
simplicity, we do not consider these operators here since their translation into FXML
is more involved, but can be done following the same ideas.

11.4.2.2 StreamIt Semantics in FXML

Let S be a syntactically correct StreamIt program. We assume that for every **work**
construct in a filter F pop rate $popr(F)$ and peek rate $peekr(F)$:

- Pops are grouped at the end in a loop of the form:

$$\text{for}\,(k = 0;\ k < popr(F);\ k\text{++})\,\textbf{pop}().$$

We denote this block $\textbf{pop}(popr(F))$.

- Peeks appear in a loop of the form:

$$\text{for}\,(k = 0;\ k < peekr(F);\ k\text{++})\,\{x[k] = \textbf{peek}(k)\},$$

where $x[]$ is a local array of dimension $peekr(F)$. We denote this block
$\textbf{peek}(peekr(F))$.

The translation from StreamIt to FXML is as follows.

Filter: A filter F is a sequential *pnode*, with two associated variables s_F^{in} and s_F^{out}:

$$\Gamma(\textbf{filter init } C_{init} \textbf{ work } rates\ C_{work}) \overset{\Delta}{=} \texttt{seq}\ \Gamma(\textbf{init } C_{init})$$
$$\Gamma(\textbf{work } rates\ C_{work})$$

Work: The translation of **work** functions is independent of the push, peek, and pop rates:

$$\Gamma(\textbf{work } rates\ C) \overset{\Delta}{=} \texttt{while(true)}\{\Gamma(C)\}.$$

Push: Pushing a value in the output stream of a filter F is translated into storing the value in the variable s_F^{out}:

$$\Gamma(\textbf{push}(f(\dots))) \overset{\Delta}{=} \texttt{push}_F : s_F^{out} = \Gamma(f(\dots)).$$

Pop: Popping values is done in the corresponding buffer. Then

$$\Gamma(\textbf{pop}(n)) \overset{\Delta}{=} \texttt{nil}.$$

Peek: $\Gamma(\textbf{peek}(n)) \overset{\Delta}{=} \texttt{peek}_F : \texttt{forall}\,(k = 0;\ k < n;\ k\texttt{++})\,\{\,x[k] = s_F^{in}[k]\,\}.$

Pipeline: The *pnode* of a pipeline consists in composing its processes with a `par` and connecting them with intermediate buffers: $\Gamma(\textbf{pipeline } S_1 \dots S_n)$ is the *pnode*

$$\texttt{dep } D_{1,\dots,n} \texttt{ par } \Gamma(S_1)\ B_{S_1,S_2}, \dots, B_{S_{n-1},S_n}\ \Gamma(S_n)$$

with

$$B_{S_i,S_{i+1}} \overset{\Delta}{=} B(s_{S_i}^{out}, s_{S_{i+1}}^{in}, peekr(S_{i+1}), popr(S_{i+1}))$$

where `B(bin, bout, peekr, popr)` is the *pnode*.

```
B:
  for(k = 0; k < peekr - popr; k++)
    init: bout[k + popr] = bin
  while(true)
    put: {
      forall(k = 0; k < peekr - popr; k++)
        shift: bout[k] = bout[k + popr]
      for(k = 0; k < popr; k++)
        get: bout[k + peekr - popr] = bin
    }
```

such that `bin` is instantiated with the output variable $s_{S_i}^{out}$ of the process S_i pushing values into the stream, `bout[]` is instantiated with the corresponding vector $s_{S_{i+1}}^{in}[]$ to store the pushed values, and `peekr` and `popr` are the peek $peekr(S_{i+1})$ and pop $popr(S_{i+1})$ rates of S_{i+1}, respectively.

B starts by initializing the vector `bout[]` with `peekr-popr` elements, from index `popr`. Afterward, it keeps shifting the contents of the vector left forever, which corresponds to popping `popr` items and inserting `popr` new ones.

$D_{1,...,n} \triangleq \bigcup_{1 \le i \le n-1} D_{i,i+1}$ with $D_{i,i+1}$ the set of dependencies from $\Gamma(S_i)$ to $B_{i,i+1}$ and from $B_{i,i+1}$ to $\Gamma(S_{i+1})$, defined as follows:

- The first k pushed values, $k \in [0, peekr(S_{i+1}) - popr(S_{i+1}))$ serve to initialize the buffer, that is, $init^{\langle k \rangle}$ in $B_{i,i+1}$ depends on $push^{\langle k \rangle}$ in $\Gamma(S_i)$:

$$\{(k, k) \mid k \in [0, peekr(S_i) - popr(S_i)]\} \, push_{S_i} \rightarrow init_{B_{i,i+1}}$$

- Every occurrence $peek^{\langle j \rangle}$ in $\Gamma(S_{i+1})$ is required to be preceded by $put^{\langle j \rangle}$ in $B_{i,i+1}$ in order to ensure that $peekr(S_{i+1})$ values have been pushed:

$$(j, j) \, put_{B_{i,i+1}} \rightarrow peek_{S_{i+1}}$$

- Every occurrence $get^{\langle j,k \rangle}$ in $B_{i,i+1}$, $k \in [0, peekr(S_{i+1}) - popr(S_{i+1}))$, gets the value pushed by $\Gamma(S_i)$ in the assignment $push^{\langle h \rangle}$, with index $h = peekr(S_{i+1}) - popr(S_{i+1}) + j \cdot popr(S_{i+1})$:

$$\{(h, (j, k)) \mid h = peekr(S_{i+1}) - popr(S_{i+1}) + j \cdot popr(S_{i+1})$$
$$\wedge k \in [0, peekr(S_{i+1}) - popr(S_{i+1}))\} \, push_{S_i} \rightarrow get_{B_{i,i+1}}$$

The translation of StreamIt into FXML gives a formal semantics to StreamIt and enables verification and scheduler synthesis. Besides, it enables the use of JAHUEL to generate code for target platforms other than those supported by the StreamIt compiler infrastructure.

11.5 FXML to BIP

For embedded software product lines, where new functionalities and services are continuously developed, the main challenge is to provide design frameworks capable of supporting software componentization to ease integration and evolution. BIP (Behavior, Interaction, Priority) [4] has been designed to overcome the difficulties of state-of-the-art component-based approaches [33]. BIP provides a language and a theory for incremental composition of heterogeneous components, ensuring correctness-by-construction for essential system properties such as mutual exclusion, deadlock-freedom, and progress. Besides, it enables verification through model-checking via the IF tool-suite [10].

Nevertheless, many high-performance embedded applications, such as video compression (e.g., MPEG-4), are not programmed following a component-based approach, but most likely a data-flow one. These applications are better described using languages such as StreamIt and FXML. Here, we provide an automated method for generating componentized implementations in BIP of data-driven applications specified in FXML. We illustrate the concept with an industrial MPEG-4 video encoder [2].

11.5.1   BIP Language

BIP is formally defined in Ref. [33]. It supports a methodology for building components from atomic ones, using connectors, to specify interaction patterns between ports of atomic components, and priority relations, to select amongst possible interactions. Here, we review the basic concepts through illustrative examples.

Figure 11.8 shows an atomic component with two ports *in*, *out*, local variables *x*, *y*, and control states *empty*, *full*. Ports are action names used for synchronization with other components. Control states denote locations at which the components await for synchronization. Variables are used to store local data. Transitions model atomic computation steps. In general, a transition is a tuple of the form (s_1, p, g_p, f_p, s_2), representing a step from control state s_1 to s_2. It can be executed if the guard g_p is true and some interaction including port p is offered. Its execution is an atomic sequence of two microsteps: (1) an interaction including p, which involves synchronization between components with possible exchange of data, followed by (2) an internal computation specified by the function f_p. In the example, component *Reactive* can take the transition labeled *in* at *empty* if $0 < x$. When an interaction through *in* takes place, the variable *x* is eventually modified and a new value for *y* is computed. From control state *full*, the transition labeled *out* can occur. The omission of guard and function for this transition means that the guard is true and the internal computation microstep is empty.

A compound component is a component consisting of atomic or compound subcomponents. An example of a compound component named *System* is shown in Figure 11.9. It is the connection of three instances of *Reactive*.

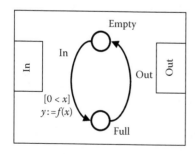

```
component Reactive
    port in, out
    data int x, y
    behavior
        state empty
            on in provided 0 < x
            do y:=f(x) to full
        state full
            on out to empty
    end
end
```

FIGURE 11.8:   Atomic component.

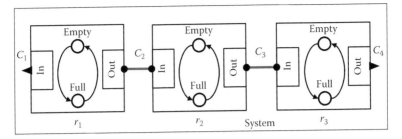

```
component System
    contains Reactive r₁, r₂, r₃
    connector C₁ = r₁.in
    complete = r₁.in
    connector C₂ = r₁.out|r₂.in
    behavior
        on r₁.out|r₂.in do r₂.x := r₁.y
    end
    connector C₃ = r₂.out|r₃.in
    behavior
        on r₂.out|r₃.in do r₃.x := r₂.y
    end
    connector C₄ = r₃.out
    complete = r3.out
    priority P₁ r₁.in < r₂.out|r₃.in
    priority P₂ r₁.in < r₃.out
    priority P₃ r₁.out|r₂.in < r₃.out
end
```

FIGURE 11.9: Compound component.

Components are connected through *connectors*, which are sets of ports that contain at most one port from each atomic component. An *interaction* is any nonempty subset of a connector. In *System* there are four connectors: C_1, consisting of port $r_1.in$ alone, C_2 consisting of ports $r_1.out$ and $r_2.in$, and so forth. There are two types of interactions, namely complete and incomplete. An interaction of a connector is feasible if it is complete or if it is maximal. We denote graphically an incomplete interaction by a bullet and a complete one by a triangle. For instance: C_1 is complete, meaning that *System* can engage in an interaction containing port *in* if r_1 can; C_2 is maximal, meaning that components r_1 and r_2 must synchronize on $r_1.out$ and $r_2.in$, that is, neither one can proceed alone on transitions labeled *out* and *in*, respectively. Connectors may have behavior specified as for transitions, by a set of guarded commands associated with feasible interactions. For instance, whenever the interaction $r_2.out|r_3.in$ takes place, $r_2.x$ receives the value of $r_1.y$. In general, guards and statements are C expressions and statements, respectively.

Priorities are used to choose amongst simultaneously enabled interactions. They are a set of rules, each consisting of an ordered pair of interactions associated with a condition. When the condition holds and both interactions are enabled, only the higher-priority one is possible. Conditions can be omitted for static priorities. The rules are extended for composition of interactions, e.g., $b_1 < b_2$ means that any interaction of the form $b_2|\alpha$ has higher priority than all interactions of the form $b_1|\alpha$, for all interactions α. In our example, $r_1.in < r_2.out|r_3.in$ means that *System* will not take any transition where $r_1.in$ is involved, whenever the synchronization between $r_2.out$ and $r_3.in$ is enabled. Indeed, the priorities specified in *System* enforce a causal order of execution as follows: once there is an *in* through C_1, data are processed and propagated sequentially through subcomponents r_1, r_2, and r_3, finally producing an *out* through C_4 before a new *in* occurs through C_1. This is achieved by a priority order which is the inverse of the causal order.

11.5.2   Translation Scheme

To illustrate the idea of the FXML-to-BIP translation scheme, let us start with the writer-reader FXML specification of Example 11.1, without timing constraints. *Pnodes* Writer and Reader become BIP components:

```
component Writer
  port out
  data int x, p
  behavior initial do p = 0; to S
    state S
      on out do x = p; p = p + 1;  to S
  end
end
```

```
component Reader
  port in
  data int y
  behavior initial to S
    state S
      on in do y = x; {# printf("%d\n", y); #} to S
  end
end
```

Communication between the two is done through a component Buffer, which is added for several reasons: (1) to encapsulate x, because BIP does not allow shared variables; (2) to realize the synchronization protocol ensuring the dependency $W \rightarrow R$, in order to comply with FXML semantics; and (3) to implement the buffering scheme. In BIP, legacy code is written between "{#" and "#}."
The composed system in BIP is

```
component System
  contains Buffer B
  contains Writer P
  contains Reader C
```

```
    connector C1 = P.out | B.put
        behavior do B.x = P.x; end
    complete P.out | B.put
    connector C2 = C.in | B.get
        behavior do C.x = B.x; end
    complete C.in | B.get
end
```

Connectors C1 and C2 implement the data transfer. The behavior of Buffer depends on the dependency type and the storage policy. For a weak dependency with single-storage (Figure 11.10), Buffer is

```
component Buffer
  port put, get
  data int x, b
  behavior initial do b=0; to S
    state S
        on put do b=1; to S
        on get provided (b==1) to S
    end
end
```

For a strong dependency with single-storage (Figure 11.11), Buffer uses variable r to notify whether the latest written value of x has been read, and therefore whether a new put can be accepted, and w, to notify whether x has been written (at least once), to condition interactions on get. The BIP model of Buffer (Figure 11.11) is as follows:

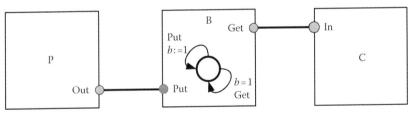

FIGURE 11.10: BIP model for weak dependency.

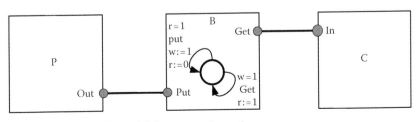

FIGURE 11.11: BIP model for *strong* dependency.

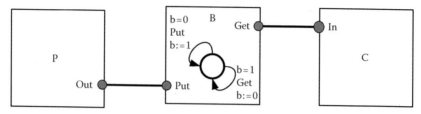

FIGURE 11.12: BIP model for (i, i) dependency.

```
component Buffer
  port put, get
  data int x, b
  behavior initial do r=1; w=0; to S
    state S
      on put provided (r==1) do w=1; r=0; to S
      on get provided (w==1) do r=1; to S
  end
end
```

For a (i, i) dependency with single-storage, `Buffer` (Figure 11.12):

```
component Buffer
  port put, get
  data int x, b
  behavior initial do b=0; to S
    state S
      on put provided (b==0) do b=1; to S
      on get provided (b==1) do b=0; to S
  end
end
```

The writer–reader example provides a basis for a general translation scheme.

- Consider the case of multiple dependencies incoming into an assignment ℓ : $y = f(x_1, \ldots, x_n)$ of a *pnode* C from assignments $\ell_i : x_i = \ldots$ in *pnodes* $P_i, i \in [1, n]$. The FXML semantics is the conjunction of constraints imposed by the dependencies. In BIP, this can be modeled by setting up buffer components B_i, one for each dependency $\ell_i \to \ell, i \in [1, n]$, whose role is to realize the corresponding control policy depending on the dependency type, as well as to implement the desired buffering policy (if any). Without loss of generality, we assume B_i is a single storage buffer, with a local variable $B_i.x, i \in [1, n]$. Other buffering policies only require changing the behavior of connectors. The conjunctive semantics is ensured in BIP by the maximal interaction $in|get_1|\ldots|get_n$, where buffered values $B_i.x$ are copied to C-local variables $C.x_i$ (Figure 11.13).

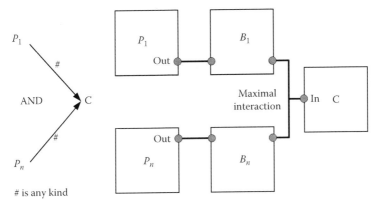

FIGURE 11.13:   Multiple incoming dependencies.

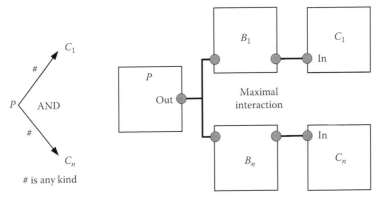

FIGURE 11.14:   Multiple outgoing dependencies.

- Other paradigmatic case consists of multiple dependencies $\ell \rightarrow \ell_i$ outgoing from a (writer-like) *pnode* P, executing the assignment $\ell : x = \ldots$, to many (consumer-like) *pnodes* C_i, computing $y_i = f_i(x), i \in [1, n]$. The translation is similar to the previous case where a buffer B_i is used for each dependency $\ell \rightarrow \ell_i, i \in [1, n]$. The conjunctive semantics is ensured by the maximal interaction $out|put_1|\ldots|put_n$, whose behavior is to set $B_i.x = P.x$ for all $i \in [1, n]$ (Figure 11.14).

11.5.3   Case Study: MPEG-4 Encoder

In this section we apply the translation scheme presented previously to a MPEG-4 video encoder. For lack of space, we only present here a significant part of the FXML and BIP models. The full FXML specification is given in Ref. [2]. This model describes all the existing concurrency in the compression algorithm at the macroblock level. Such concurrency does not appear in the simplified MPEG-4 block diagram shown in Figure 11.15.

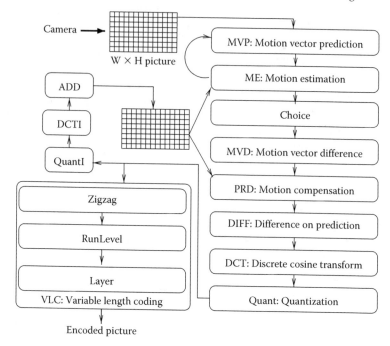

FIGURE 11.15: MPEG block diagram view.

The specification is composed of `forall` nodes, C-code blocks of the corresponding MPEG-4 computations, and dependencies of the MPEG-4 phases. The basic data structure is a matrix of $W \times H$ macroblocks. FXML specification uses x and y as the iteration variables, i.e., $x \in [0, W)$ is the row, and $y \in [0, H)$ the column, of the macroblock. A *pnode* specifying the behavior of an MPEG-4 computation step s (e.g., MVP, ME, etc.) has the following structure:

```
forall(x = 0; x < W; x + 1)
  forall(y = 0; y < H; y + 1)
    legacy{ M_s[x][y] = F_s( ... ); }
```

where `M_s` is the *output* matrix of step s, and `F_s` is the computation applied at step s. `F_s` depends on a matrix computed in a preceding step, e.g., `M_MVP[x][y]` depends on `M_ME[x][y]`, `M_ME[x-1,y]`, `M_ME[x,y-1]`, and `M_ME[x-1,y-1]`.

For readability, we use a tree-like representation, instead of a textual pseudocode, where *pnodes* are labeled with numbers in brackets (Figure 11.16). We note $(1,(x, y))$ the computation corresponding to the execution of the ME (motion estimation) phase on the frame macroblock at position (x, y). The arrows indicate dependencies between these computations. There are three types of dependencies : (1) data dependencies resulting from the MPEG-4 standard specification (e.g., in Figure 11.16, $(1,(x, y)) \rightarrow (3,(x, y))$ is a data dependency expressing that the ME

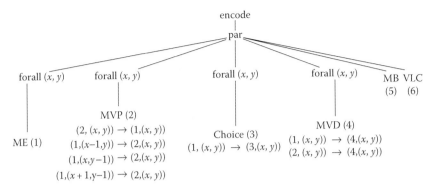

FIGURE 11.16:   FXML specification of encode.

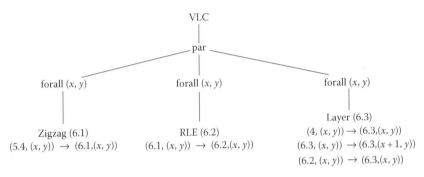

FIGURE 11.17:   FXML specification of VLC.

phase on macroblock (x, y) must finish before starting the Choice phase on the same macroblock), (2) functional dependencies necessary for the correct functioning of the application (e.g., there is a functional dependency from macroblock (x, y) to macroblock $(x + 1, y)$ in the specification of variable length coding (VLC) (Figure 11.17) because generated headers and blocks are sequentially written in the output bitstream), and (3) dependencies resulting from implementation decisions (e.g., using input and output buffers with one-frame capacity) of encoding frames one after another.

The overall specification is 7650 lines of FXML, including 7500 lines of C code corresponding to encoding computations. Indeed, the FXML specification can be obtained from the sequential C code annotated with special purpose pragmas [1] using FlexCC2 [7].

Before applying the translation scheme to obtain a BIP model, we perform several FXML-to-FXML transformations. The main problem to face is to determine the granularity of the componentization. Here, we take each MPEG-4 computation step to be implemented inside a single atomic component C_s. To achieve this, the parallelism inside each step is eliminated, by making each `forall` statement to become a for. To code the nested for in BIP, we need to add two complete ports tau and stop, together with two singleton connectors, to model internal component transitions. Therefore, the component C_s is as follows:

```
component C_s()
  port in, out
  port complete tau, stop
  data int x, y
  behavior initial x = 0; y = 0; to WAIT
    state WAIT
      on in provided ((x < W) && (y < H)) to WORK
      on tau provided ((x < W) && (y == H))
        do x = x + 1; y = 0; to WAIT
      on stop provided (x == W) do x = 0; to WAIT
    state WORK
      on out do {# M_s[x][y] = F_s(...); #}; y = y + 1; to WAIT
  end
  connector conn_tau = tau
    behavior
    end
  connector conn_stop = stop
    behavior
    end
end
```

C_s has a local matrix variable M_s[][] to store the computed value at each iteration. For each component C_s, there is a single output buffer B_s, which encapsulates the output matrix of macroblocks. This is because there is only one outgoing dependency out of each component. The transfer of each macroblock M_s[x][y] from the (writer) component C_s to the buffer B_s is done in the connector connecting ports C_s.out and B_s.put. From B_s to the (reader) component C_s', corresponding to the step s' following s, the transfer occurs in the connector connecting ports B_s.get and C_s'.in. Figure 11.18 depicts the schematic view of the BIP components for VLC, where cylinders represent buffers.

The connector T_4_62_63 defining the interaction between B4, B62, and C63, is as follows:

```
connector T_4_62_63 = B4.get,B62.get,C63.in
  behavior
    do C63.M_4[x][y] = B4.M[x][y], C63.M_62[x][y] = B62.M[x][y]
  end
```

The other connectors are obtained similarly.

The resulting BIP model consists of 37 components, 21 of which are buffers, and 32 connectors. The BIP source code has 8150 lines, including 650 lines of pure BIP and the 7500 lines of C code of the original sequential encoder.

It is worth noticing that the BIP model characterizes the set of all possible schedulings of the computations. This nondeterministic behavior can be constrained by

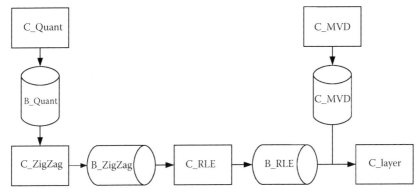

FIGURE 11.18:   Variable length coding.

adding priorities, eventually resulting in a sequential execution of the encoder. To start with, we have considered single-storage buffers, so producers are prevented from rewriting values until consumer(s) have read them. Clearly, by changing the size of buffers and their policy, we can obtain a higher degree of parallelism. Moreover, this change does not affect the FXML specification, but only the last phase of the code generation chain. The latter just implies regenerating the behavior of the buffer component in BIP, but not its ports and connectors, thus achieving modular code generation.

We have also compared the BIP model obtained from the FXML description with a BIP model of the encoder written by hand. The latter consists of 15% fewer lines of BIP (548 instead of 650), almost 50% less components (11 MPEG components and 9 buffers), and 33 connectors. The larger number of components is mainly due to the fact that the hand-written model encapsulates VLC and QuantI/DCI in single components. Besides, most dependencies are between two successive components and hence all connectors are of arity two (creating a chain of dependencies). In the FXML specification, dependencies allow more parallelism and relate more than two components, resulting in more than one incoming buffers and n-ary connectors in some cases (with $n > 2$) (see Figure 11.18, for instance).

11.6   Conclusions

FXML is a formal language for specifying concurrent real-time applications. It has a simple abstract execution model based on the notion of assignments and dependencies. It can be incrementally extended with information related to refinements of the abstract model into more concrete ones. FXML can be used as a modeling language by itself, that is, FXML specifications can be directly written by designers, or as semantic framework for other languages, such as StreamIt, where FXML specifications are obtained by a compiler.

An FXML-compiler is a sequence of transformations going from a language (or model) to another (more concrete one). Based on this idea, we have developed the compilation chain JAHUEL which implements several translation phases that can be easily customized for different platforms. Hence, JAHUEL provides tool support for handling concurrency and timing constraints in software product lines. In particular, we have shown how to generate code for several execution platforms, such as *pthreads* and OpenMP. JAHUEL provides an FXML-to-BIP transformation which enables formal verification via the IF.

The FXML and JAHUEL are grounded on well-established notions from process algebras, program analysis and transformation, refinement, and scheduler synthesis. The main contribution of this work is to have shown that these techniques could be put together into a preindustrial, extensible, and customizable code-generation chain for software product lines, without semantic breakdowns from an abstract specification all the way down to executable code.

Ongoing work includes applying FXML and JAHUEL in other industrial applications, strengthening the integration into an end-to-end industrial design flow, and generating code for other platforms. Special effort is being put on generating code for the simulation infrastructure P-Ware [3]. The main motivation for this is early prototyping, verification, and testing of embedded applications on simulated hardware platforms, since automated generation of both executable and simulation code from the same formal model ensures simulation results are trustworthy.

Acknowledgments

We thank the anonymous reviewers for their valuable comments. We are indebted to Valérie Bertin, Philippe Gerner, Christos Kloukinas, and Olivier Quévreux for their contribution to the early stages of the definition and development of FXML and JAHUEL, and to Joseph Sifakis who motivated the FXML-to-BIP transformation. The work reported in this chapter has been partially supported by projects MEDEA+ NEVA, Minalogic SCEPTRE, Crolles-II ANACONDA, STIC-AmSud TAPIOCA, IST SPEEDS, and RNTL OpenEmbeDD. This chapter has been written while Sergio Yovine was a visiting professor at DC/FCEyN/UBA, and UADE, Argentina, partially funded by BID-ANPCyT, project PICTO-CRUP 31352 and MINCyT "César Milstein" grant.

References

[1] I. Assayad, V. Bertin, F. Defaut, Ph. Gerner, O. Quévreux, and S. Yovine. JAHUEL: A formal framework for software synthesis. In *7th International Conference on Formal Engineering Methods, (ICFEM 2005)*, Manchester, U. K., November 1–4, 2005. *Lecture Notes in Computer Science* 3785, Springer 204–218.

[2] I. Assayad, Ph. Gerner, S. Yovine, and V. Bertin. Modelling, analysis, implementation of an on-line video encoder. In *1st International Conference on Distributed Frameworks for Multimedia Applications*, (*DFMA 2005*), Besançon, France, February 6–9, 2005, IEEE Computer Society, 2005, pp. 295–302.

[3] I. Assayad and S. Yovine. P-Ware: A precise, scalable component-based simulation tool for embedded multiprocessor industrial applications. In *10th Euromicro Conference on Digital System Design: Architectures, Methods and Tools*, (*DSD 2007*), Lübeck, Germany, August 29–31, 2007 IEEE Computer Society, 2007, pp. 181–188.

[4] A. Basu, M. Bozga, and J. Sifakis. Modeling heterogeneous real-time components in BIP. In *4th IEEE International Conference on Software Engineering and Formal Methods*, Pune, India, September 11–15, 2006 (*SEFM 2006*), IEEE Computer Society, 2006, pp. 3–12.

[5] A. Basu, L. Mounier, M. Poulhiès, J. Pulou, and J. Sifakis. Using BIP for modeling, verification of networked systems: A case study on TinyOS-based networks. In *6th IEEE International Symposium on Network Computing and Applications* (*NCA 2007*), IEEE Computer Society, 2007, pp. 257–260.

[6] J.A. Bergstra, A. Ponse, and S.A. Smolka. *Handbook of Process Algebra*. Elsevier, Amsterdam, the Netherlands, 2001.

[7] V. Bertin, J.M. Daveau, P. Guillaume, T. Lepley, D. Pilat, C. Richard, M. Santana, and T. Thery. FlexCC2: An optimizing retargetable C compiler for DSP processors. In *2nd International Conference on Embedded Software*, (*EMSOFT 2002*). Grenoble, France, October 7–9, 2002, *Lecture Notes in Computer Science*, 2491: 382–398.

[8] P. Binns and S. Vestal. Formalizing software architectures for embedded systems. In *1st International Workshop on Embedded Software*, (*EMSOFT 2001*). Tahoe City, CA, October 8–10, 2001, *Lecture Notes in Computer Science*, 2211: 451–468.

[9] H. Boehm. Threads cannot be implemented as a library. In *ACM SIGPLAN 2005 Conference on Programming Language Design and Implementation*, (*PLDI 2005*), Chicago, IL, June 12–15, 2005 ACM, 2005, pp. 261–268.

[10] M. Bozga, S. Graf, Il. Ober, Iul. Ober, and J. Sifakis. The IF Toolset. In *Formal Methods for the Design of Real-Time Systems*, (*SFM 2004*), Bertinoro, Italy, September 13–18, 2004. *Lecture Notes in Computer Science*, 3185: 237–267.

[11] A. Burns and A. Wellings. *Concurrency in Ada*. Cambridge University Press, New York, 1998.

[12] P. Caspi, A. Curic, A. Maignan, C. Sofronis, S. Tripakis, and P. Niebert. From Simulink to SCADE/Lustre to TTA: A layered approach for distributed embedded applications. In *Conference on Languages, Compilers, and Tools for Embedded Systems*, (*LCTES 2003*), San Diego, CA, June 11–13, 2003 ACM, 2003, pp. 153–162.

[13] P. C. Clements and L. Northrop. *Software Product Lines: Practices, Patterns.* Addison-Wesley, Boston, MA, 2001.

[14] A. Darte, Y. Robert, and F. Vivien. *Scheduling, Automatic Parallelization.* Birkhäuser, Boston, MA, 2000.

[15] S. Dutta, R. Jensen, and A. Rieckmann. Viper: A multiprocessor SOC for advanced set-top box, digital TV systems. *IEEE Design, Test of Computers*, 18(5):21–31, Sept./Oct. 2001.

[16] K. Ebcioglu, V. Sarkar, T. El-Ghazawi, and J. Urbanic. An experiment in measuring the productivity of three parallel programming languages. In *3rd Workshop on Productivity and Performance in High-End Computing*, (*PPHEC 2006*), Austin, February 12, 2006, IEEE Computer Society, 2006, pp. 30–36.

[17] J. Goguen and G. Malcolm. *Algebraic Semantics of Imperative Programs.* MIT Press, Cambridge, MA, 1996.

[18] M. I. Gordon, W. Thies, and S. P. Amarasinghe. Exploiting coarse-grained task, data, and pipeline parallelism in stream programs. In *12th International Conference on Architectural Support for Programming Languages and Operating Systems*, (*ASPLOS 2006*), San Jose, CA, October 21–25, 2006 ACM, 2006, pp. 151–162.

[19] W. Groppa, E. Lusk, and A. Skjellum. *Using MPI*. Scientific, Engineering Computation, 2nd ed. MIT Press, Cambridge, MA, November 1999.

[20] N. Halbwachs, P. Caspi, P. Raymond, and D. Pilaud. The synchronous dataflow programming language Lustre. *Proceedings of the IEEE*, 79(9), Sept. 1991.

[21] D. Meng, R. Gunturi, and M. Castelino. Intel® IXP2800 Network Processor IP Forwarding Benchmark Full Disclosure Report for OC192-POS. Intel Corporation, 2003.

[22] G. Karsai, J. Sztipanovits, A. Ledeczi, and T. Bapty. Model-integrated development of embedded software. *Proceedings of the IEEE*, 91(1), 2003.

[23] M. Kersten. Comparison of the leading AOP tools. In *4th International Conference on Aspect-Oriented Software Development*, (*AOSD 2005*), Chicago, IL, March 14–18, 2005 Industry track. Invited talk. ACM, 2005.

[24] Ch. Kloukinas, Ch. Nakhli, and S. Yovine. A methodology and tool support for generating scheduled native code for real-time java applications. In *3rd International Conference on Embedded Software*, (*EMSOFT 2003*). Philadelphia, PA, October 13–15, 2003 *Lecture Notes in Computer Science*, 2855: 274–289.

[25] Ch. Kloukinas and S. Yovine. Synthesis of safe, QoS extendible, application specific schedulers for heterogeneous real-time systems, In *15th Euromicro Conference on Real-Time Systems*, (*ECRTS 2003*), Porto, Portugal, July 2–4, 2003, IEEE Computer Society, 2003, pp. 287–294.

[26] O. Maler, A. Pnueli, and J. Sifakis. On the synthesis of discrete controllers for timed systems. In *12th Annual Symposium on Theoretical Aspects of Computer Science*, (*STACS 1995*), Munich, Germany, March 2–4, 1995. *Lecture Notes in Computer Science*, 900 Springer 1995, pp. 229–242.

[27] B. Chapman, G. Jost, and R. van der Pas. Using Open MP: Portable Shared Memory Parallel Programming. The MIT Press, 2007.

[28] G. A. Papadopoulos and F. Arbab. Coordination models and languages. *Advances in Computers*, 46, 1998.

[29] J. Davis, R. Galicia, M. Goel, C. Hylands, E. A. Lee, J. Liu, X. Liu, L. Muliadi, S. Neuendorffer, J. Reekie, N. Smyth, J. Tsay, and Y. Xiong, Ptolemy II - Heterogeneous Concurrent Modeling and Design in Java, Technical Memorandum UCB/ERL No M99/40, University of California, Berkeley, CA, July 19, 1999.

[30] P. Puschner and A. Burns. Guest editorial: A review of worst-case execution-time analysis. *Real-Time Systems* 18(2/3):115-128, 2000.

[31] M. C. Rinard. Analysis of multithreaded programs. In *8th International Symposium Static Analysis*, (*SAS 2001*), Paris, France, July 16–18, 2001. *Lecture Notes in Computer Science*, 2126:1–19.

[32] A. Sangiovanni-Vincentelli. Defining platform-based design. *EEDesign*, February 5, 2002. Available at http://www.gigascale.org/pubs/141.html.

[33] J. Sifakis. A framework for component-based construction. In *3rd IEEE International Conference on Software Engineering and Formal Methods*, (*SEFM 2005*), Koblenz, Germany, September 7–9, 2005, IEEE Computer Society, 2005, pp. 293–300.

[34] J. Sifakis, S. Tripakis, and S. Yovine. Building models of real-time systems from application software. *Proceedings of the IEEE*, 91(1):100–111, January 2003.

[35] P. Stravers. Homogeneous multiprocessing for the masses. In *2nd Workshop on Embedded Systems for Real-Time Multimedia*, (*ESTImedia 2004*), Stockholm, Sweden, September 6–7, 2004, IEEE Computer Society, 2004, p. 3.

[36] W. Thies, M. Karczmarek, and S. Amarasinghe. StreamIt: A language for streaming applications. In *11th International Conference on Compiler Construction*, (*CC 2002*), Grenoble, France, April 8–12, 2002. *Lecture Notes in Computer Science*, 2304:179–196.

Index

A

ABP, *see* Alternating bit protocol
Abstract data types (ADTs), 164, 290
Abstract specification, 304–306
ACSR, *see* Algebra of communicating
    shared resources
Action systems formalism
    derivation method
        exchange expressions rules, 85
        merge-into-loop rule, 84
        sequence-to-loop, 83–84
    parallel programming method
        action systems, 81
        partitioned action systems,
          81–82
ADL, *see* Architecture description
    languages
Agent tracker application
    information retrieval, 272
    initial and final runtime states,
      275–276
    specification and implementation
      process, 273
    tracker registration and
      delivery, 274
Algebra of communicating shared
    resources (ACSR), 332
Alternating bit protocol (ABP)
    components, 114
    K and L channels, 114–116
    LTS structure, 118
    S and R channels, 117–118
Ambient calculus, 238, 279
Architecture description languages
    (ADL), 167, 361–362

Aspect-oriented software development,
    362–363
Associative memory, *see* Content
    addressable memory
Asynchronous π-calculus
    bisimulation scheme, 136
    labeled transition system, 134
    structural congruence axioms, 135
    transition rules, 134–135
    vs. standard π-calculus, 136

B

Base model, SOMAM
    congruence rules, 206
    initial and final states, 205–206
    reduction rules
        halt provider execution, 209
        launching new service provider,
          206–207
        method invocation, 207–208
        program thread termination, 209
        service-oriented process
          mobility, 208–209
    syntactic assumptions, 204–205
    syntax
        heap reference, 203
        identifiers, 202
        network and program, 204
        syntactic categories, 202–203
BIP language, 388–390
    atomic component, 388
    compound component, 389
    MPEG-4 encoder, 393–397
    multiple dependencies, 392–393
    translation scheme, 393–395

403